LIFE SCIENCE WORK-TEXT

Revised Edition

Milton S. Lesser

Former Assistant Principal, Supervisor,
General and Biological Sciences
Abraham Lincoln High School, New York City

AMSCO SCHOOL PUBLICATIONS, INC.
315 Hudson Street / New York, N.Y. 10013

The author wishes to acknowledge the helpful contributions of the following consultants in the preparation of this book:

Judy Dembrow
Teacher, Biology and Physical Science
New Rochelle High School, New Rochelle, N.Y.

Dr. Harold Friend
Teacher, Biology and General Science
Adrien Block Intermediate School 25, New York City

Kenneth Parente
Teacher, Life Science
Willoughby Junior High School 162, New York City

Lane Schwartz
Former Principal (Retired)
Glen Cove High School, Glen Cove, New York
Former Assistant Principal, Science
John Jay High School, New York City, New York

Mike Weinick
Assistant Principal for Science
George C. Tilyou Junior High School 303, New York City

Science, Technology and Society features written by Carl Proujan and Lane Schwartz.

Science, Technology and Society feature photos: p. 15, courtesy of People for the Ethical Treatment of Animals; p. 150, neg. #129205, painting by C. R. Knight, courtesy of Department of Library Services, American Museum of Natural History; pp. 47, 175, 213, 233, 299, and 315, UPI/Bettmann; pp. 67 and 269; Reuters/Bettmann; pp. 95 and 121, Bettmann.

Please visit our Web site at: ***www.amscopub.com***

When ordering this book, please specify:
either **R 777 W** *or* **Life Science Work-Text**

ISBN 0-87720-191-9

Copyright © 2004, 1993 by Amsco School Publications, Inc.

No part of this book may be reproduced in any form without written permission from the publisher.

1 2 3 4 5 6 7 8 9 10 08 07 06 05 04 03

Printed in the United States of America

Contents

The World of Science

LABORATORY INVESTIGATION

USING A COMPOUND MICROSCOPE

A. One of the most important tools of a scientist is the microscope. Work with a partner to learn how to use a *compound microscope*.

Refer to Fig. 1-1 to identify the parts of a microscope. Locate the *arm* of the microscope. The arm supports the *stage* and the *body tube*. (The body tube lets light pass through the lenses.) When you carry a microscope, grasp it by the arm and the base. When you use a microscope, place it down with the arm facing toward you.

B. Identify the two *objectives* (lenses), which are attached below the *nosepiece*.

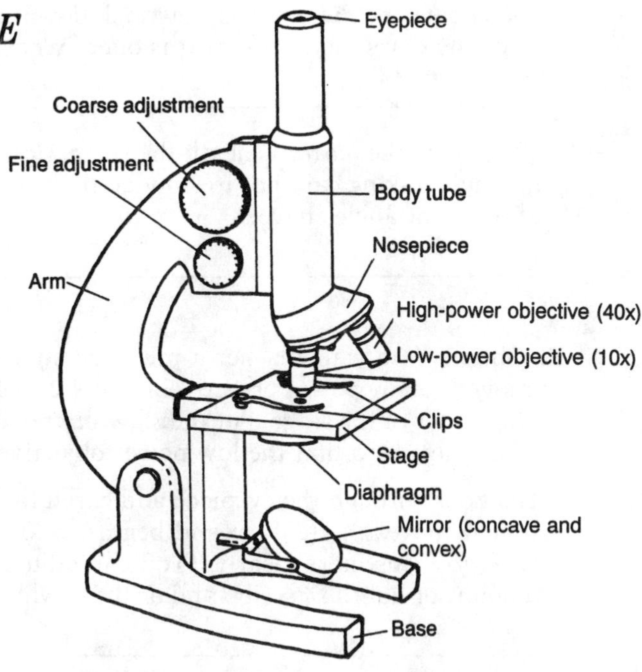

Fig. 1-1. The compound microscope.

1. Which objective is for low power (10×) magnification?

---

2. What do you think the label "10×" means?

---

3. Which objective is for high power?

C. Locate the *coarse adjustment* and *fine adjustment* knobs. Turn the nosepiece so that the low power objective clicks in place below the body tube.

4. Carefully turn the coarse adjustment knob back and forth. What happens?

____

5. Carefully turn the fine adjustment knob back and forth. What happens?

____

6. How is the movement of each knob different from the other one?

____

7. What do you think the purpose of each of these knobs is?

____

D. Use a piece of lens tissue to wipe the objective lenses and the *eyepiece*. The eyepiece holds the microscope's top lens.

8. Locate the *diaphragm*. Now look down into the eyepiece, through the hole in the stage. Turn the diaphragm so that it is open. What does the diaphragm control?

____

9. Locate the *mirror* beneath the stage. Holding the mirror at opposite edges, tilt it to adjust its angle, taking care not to smudge the surface with your fingers. What effect does tilting it at different angles have?

____

E. Obtain from your teacher a prepared slide of the letter *e*. Use the coarse adjustment knob to move the low power objective about 1/2 inch above the stage. Place the slide over the hole, in the center of the stage. Put the *stage clips* over the ends of the slide to hold it in place. Adjust the knob again so that the low power objective is just over, but not touching, the slide.

10. Look through the eyepiece and carefully turn the coarse adjustment knob toward you so that it moves the lens up and begins to focus it on the slide. Now use the fine adjustment knob to focus more clearly. (You can adjust the diaphragm and the mirror to get the right amount of light.) Describe and/or draw what you see.

____

11. *a.* Move the slide slightly to the left. What appears to happen as you look through the eyepiece?

____

b. Move the slide slightly to the right. What happens?

____

c. Move the slide toward you. What happens?

____

d. Move the slide away from you. What happens?

____

12. Now, making sure there is enough space between the lens and the slide, carefully turn the nosepiece so that the high power objective is in place over the slide. (Look sideways at the slide—not through the eyepiece—as you do this, to avoid hitting it.) **Note:** Use only the fine adjustment knob to focus when viewing through the high power objective. Focus carefully while looking through the eyepiece. Describe and/or draw what you see.

____

13. Explain why you think the microscope is such an important tool for scientists.

____

The Study of Science

EARLY SCIENTIFIC THOUGHT

For thousands of years, people have tried to explain and understand the world around them. Many myths and legends were told that attempted to answer such questions as: what causes people to get sick; why do volcanoes erupt; where does the wind come from; and how did life originate. People would observe events around them and try to make some connections between the possible causes and effects.

The explanations that early scientists came up with were not always correct, even if their ideas seemed to be supported by observations. For example, hundreds of years ago, people knew that maggots developed on rotting meat, so they assumed that the maggots came directly from the meat. In fact, the maggots came from tiny eggs that flies had laid on the rotting meat. It was only through careful observation that the truth was discovered. (You will read about the experiments that proved this fact in Chapter 9.)

TABLE 1-1. SOME SUBBRANCHES OF SCIENCE

Branch	*Subbranch*	*Areas of Study*
Biology	Genetics Botany Zoology Paleontology Microbiology	Heredity Plants Animals Fossils Microscopic life forms
Chemistry	Organic chemistry Polymer chemistry Nuclear chemistry Biochemistry	Carbon compounds Very large molecules The atom's nucleus Chemistry of living things
Physics	Astrophysics Nuclear physics Mechanics Optics Acoustics	Stars and objects in space Atomic structure and power Forces Light Sound
Earth Science	Seismology Meteorology Oceanography Geology Astronomy	Earthquakes Weather Oceans Earth's surface Stars and galaxies

Another example of how myth and science have gotten confused is found in the discovery of dinosaur bones. Because there were already many stories about the existence of dragons, the first dinosaur bones uncovered were assumed to belong to long-dead dragons. Of course, scientific studies later proved this to be wrong—although dinosaurs were probably as hard to believe in as dragons! Over the years, careful observations and logical reasoning have led to the accumulation of knowledge about our world. Such information is known as *scientific knowledge.*

WHAT SCIENCE IS

Science is commonly defined as the body of knowledge about our universe and the process by which that knowledge has been acquired. This body of knowledge has been divided by scientists into several major branches and subbranches. The major branches of science are:

1. Biology—the study of living things.
2. Chemistry—the study of the structure and properties of matter.
3. Physics—the study of the interactions of matter and energy.
4. Earth Science—the study of the earth and its place in the universe.

Often, the different branches overlap, leading to new branches of science, such as biochemistry—the chemistry of living things. In addition, there are many subbranches of scientific study (see Table 1-1 for a partial list).

HOW A SCIENTIST WORKS

One basic trait of all scientists is curiosity. In order to discover the truth about our world, scientific researchers first must be curious enough to wonder how and why things are as they are. In addition, scientists must have the imagination to think of possible answers to the questions they ask about the universe.

However, imaginative explanations are not enough to solve a scientific problem. Scientists must examine whatever evidence exists that could explain a relationship that they are investigating. And scientists have to be ready to change their explanations when the evidence shows that they are probably wrong.

Scientists do not work alone. They share their ideas and discoveries. The great body of scientific knowledge that we have today is the result of the continued and combined research efforts of many scientists.

It may take years of research by many men and women to actually prove a **theory**, which is a scientific explanation of facts, that was first put forth by an earlier scientist. And the ideas and discoveries of scientists working separately may, when studied together, lead to an understanding of important laws and principles. For example, *Gregor Mendel's* (1822–84) discoveries about heredity, *Charles Darwin's* (1809–82) theory of evolution, and more recent research in genetics all fit together to give us a pretty clear picture of how life forms change and pass along their traits.

THE SCIENTIFIC METHOD

How do scientists go about their work of examining evidence and solving problems? They conduct research and follow an organized approach to problem-solving. This approach is referred to as the **scientific method.** In this method, some general steps are followed in a systematic way to solve problems. These steps are: state the problem; collect information; form a hypothesis; test your hypothesis; record your observations; check your results; draw your conclusions; and communicate your results.

State the Problem. Before a scientist begins an investigation, he or she states the exact problem being researched. By putting the problem in the form of a question, scientists can focus clearly on what they want to find out. For example, if you want to determine whether or not plant food containing nitrates causes geraniums to increase their rate of growth, you might state the problem as: Does nitrate-containing plant food increase the growth rate in geraniums?

Collect Information About the Problem. It is important to be aware of any information that relates to the problem you are studying. You would refer to books and scientific journals, for instance, to learn about plant growth in general, the effects of nitrates on plant growth, and characteristics of growth in geraniums in particular. This information may help you research your problem more accurately, and avoid repeating work already done by other scientists.

Form a Hypothesis. A **hypothesis** is an idea or possible answer to a problem that can be tested. When you form a hypothesis, you suggest a solution to the problem. Sometimes a hypothesis is simply a guess based on the information you currently have. For example, in the case of the geraniums, you might hypothesize that those geraniums which are fed nitrate-containing plant food will grow faster and larger than those geraniums which are not fed nitrates.

Test Your Hypothesis. You must test your hypothesis to see whether it is true. To do that, you must design and perform an experiment. A hypothesis proposes one *factor*, called the **variable,** that may be the solution to the problem being investigated. The variable in your plant experiment would be the nitrate-containing plant food. To be sure that the only factor, or variable, being tested is the one proposed in the hypothesis, a scientist performs a **controlled experiment.**

In a controlled experiment, two groups are tested; they are identical except for the one variable, also known as the **experimental factor.** The subject group that is exposed to the variable being tested is called the **experimental group.** The other group, which is kept under identical conditions but is not exposed to the variable, is called the **control group.**

In your investigation, the experimental group would receive the plant food; the control group would receive no plant food. All other conditions, such as the amount of light, soil, and water, would remain the same for both groups. In this way, you can be fairly certain that any change in the growth rate of the plants in the experimental group is due to the variable alone.

Record Your Observations. When conducting an experiment, a scientist must carefully make and record observations. This record of scientific observations, or **data,** is important for understanding the experimental results. Data may include diagrams and charts, as well as notes about the investigation.

There also should be a record of the equipment used to carry out the experiment, and how long the procedure lasted. Such information is necessary in case the researcher, or another scientist, wants to redo the experiment to confirm the results.

The record for the above experiment should include the following: the problem; the hypothesis; the materials being used (for example, ten potted geranium plants, water, plant food); the step-by-step procedure used to conduct the experiment; and observations made as the experiment proceeded (for example, the differing rates of plant growth in the control and experimental groups).

Check Your Results. Scientists check their results to see if they support the hypothesis. They repeat the experiment to be absolutely sure the results are valid, or correct. If the results do not support the hypothesis, they will reject that hypothesis, form a new hypothesis, and conduct other experiments.

Draw Your Conclusions. A scientist draws his or her conclusion based on the data and results of the experiment. For example, in your experiment, the results may support the hypothesis, and you may conclude that a certain amount of plant food causes a specific increase in the growth rate of geraniums. Your conclusion from this experiment may have importance for other scientific problems being researched.

Communicate Your Results. When scientists are sure that they have tested their hypothesis thoroughly, and that their results are valid, they communicate their results to other scientists. Ways to do this include publishing written reports in scientific magazines and giving lectures at science conferences. Other scientists can then repeat the experiment, if they want to confirm the results. Or, they can use some of the information to better understand a problem they may be researching.

Thus, data obtained by solving one problem often inspire new research ideas. For example, after concluding that nitrate-containing plant food promotes growth in geraniums, you may want to find out if it also promotes growth in other types of plants. Refer to Fig. 1-2 to review the steps of the scientific method.

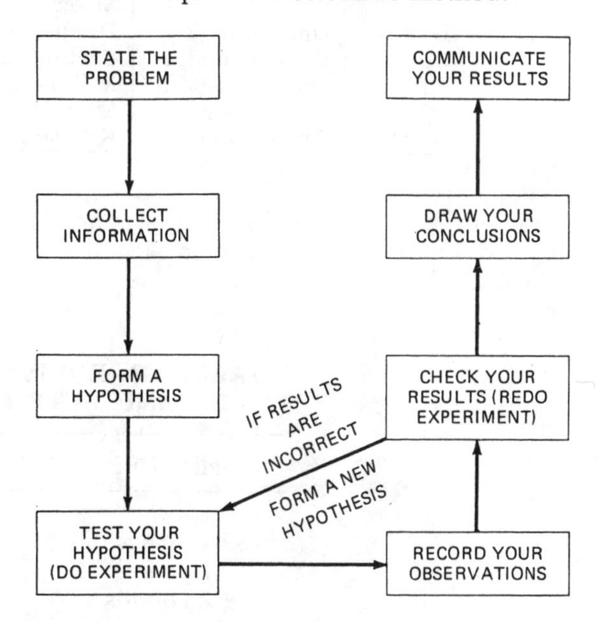

Fig. 1-2. The scientific method.

UNITS OF MEASUREMENT

When conducting experiments, scientists often have to measure the length, mass, volume, and temperature of various objects. Instruments such as rulers, balances, graduated cylinders, and thermometers are used for making these measurements. However, measurements are not always exact because the researcher's ability to observe measurements is rarely perfect. The instruments used are seldom perfect either. Therefore, all measurements are said to be somewhat *uncertain*, that is, not perfectly precise. However, scientists try to measure as precisely as possible, and they need to use a measurement system that can be understood by others, too.

The English System. The system of measurement that you probably use most often is called the **English system.** The English system uses the foot as the unit of length, the pound as the unit of weight, and the quart as the unit of volume. There is one major problem with this system—the large units cannot be divided easily into smaller units. For example, in units of length, 12 inches equals 1 foot; 3 feet equals 1 yard; and 5280 feet equals 1 mile. As a result, you have to do calculations to determine, for example, how many inches are in a mile. (5280 feet/mile multiplied by 12 inches/foot = 63,360 inches/mile.)

The Metric System. A simpler system of measurement, called the **metric system,** is used by scientists. This system, first officially used in France in 1837, is the one used by most nations today. The metric system uses the meter as the unit of length, the kilogram as the unit of mass, and the liter as the unit of volume. This system is easier to use than the English system, because any of the units can be converted into larger or smaller units of the same kind by dividing or multiplying by 10 (or multiples of 10). Because this system is based on multiples of 10, it is also called the *decimal system* (*deci* meaning ten).

For example, an object that has a mass of 5000 grams can also be expressed as 5 kilograms, since a kilogram equals 1000 grams. Likewise, a length of 10,000 meters can be written more compactly as 10 kilometers, since a kilometer equals 1000 meters (*kilo* meaning thousand); and 100 centimeters can be written as 1 meter (*centi* = hundredth). Refer to Table 1-2 to study the prefixes and units of the metric system.

TABLE 1-2. PREFIXES OF THE METRIC SYSTEM

Prefix	*Meaning*	*Unit (length)*	*Decimal Notation*
		Meter	1 meter = 1.0 meter
Deci	One-tenth	Decimeter	1 decimeter = 0.1 meter
Centi	One-hundredth	Centimeter	1 centimeter = 0.01 meter
Milli	One-thousandth	Millimeter	1 millimeter = 0.001 meter
Micro	One-millionth	Micrometer	1 micrometer = 0.000001 meter
Kilo	One thousand	Kilometer	1 kilometer = 1000 meters

TABLE 1-3. EQUIVALENTS: ENGLISH AND METRIC SYSTEMS

English Unit	*Metric Equivalent*
1 inch	2.54 centimeters
39.4 inches	1 meter
1 pound	454 grams
2.2 pounds	1 kilogram
1.06 quarts	1 liter

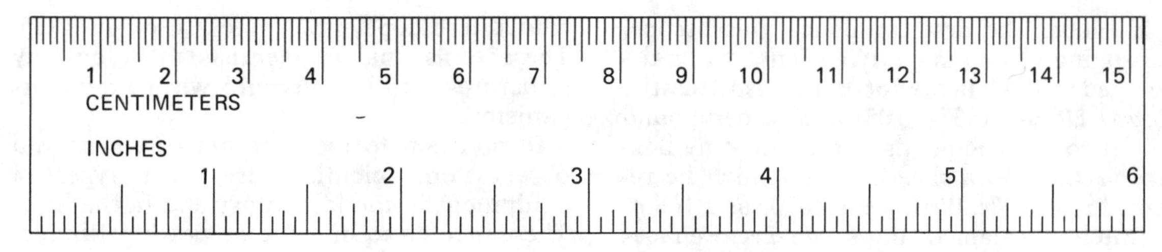

Fig. 1-3. Converting inches to centimeters.

Although scientists prefer to use the metric system, they sometimes find it necessary to use the English system as well. See Table 1-3 to learn some common equivalents in the two systems. (See Fig. 1-3 to compare inches and centimeters.) Both systems of measurement use the second as the unit of time.

To measure temperature, most scientists use the *Celsius* temperature scale, which is a convenient system. For example, on the Celsius scale, the freezing point of water is 0°C, and the boiling point of water is 100°C. You are probably more familiar with the *Fahrenheit* temperature scale, in which the freezing point of water is 32°F and the boiling point is 212°F. (See Fig. 1-4.)

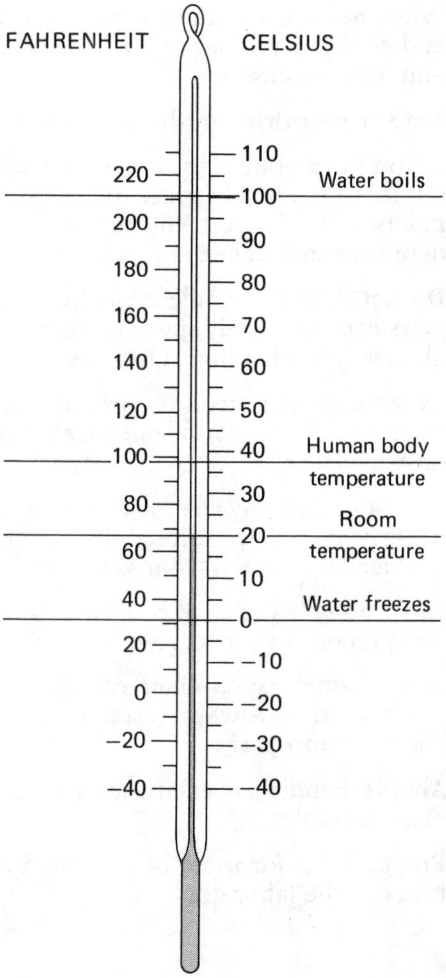

Fig. 1-4. Fahrenheit and Celsius temperature scales.

THE TOOLS OF SCIENTISTS

The various instruments used for making measurements, such as the rulers, thermometers, and balances mentioned above, are all tools of scientists. In your laboratory investigation, you learned how to use one of the most important tools—the **compound microscope.** The first simple microscope, consisting of a single magnifying lens, was developed in the

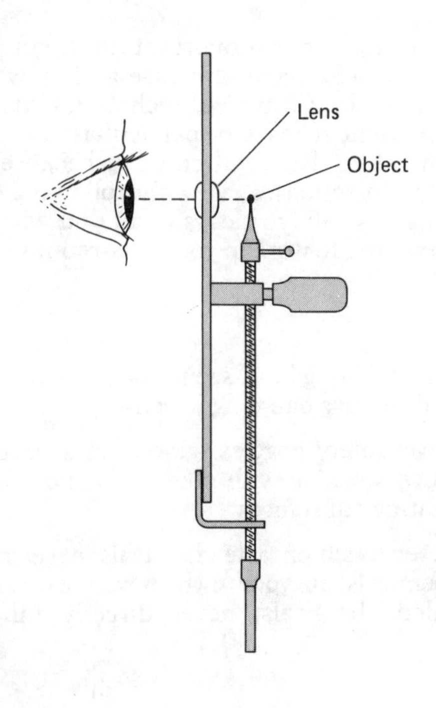

Fig. 1-5. One of Leeuwenhoek's single-lens microscopes.

mid-1400s. The first compound microscope, consisting of two magnifying lenses, was developed in 1590. In the 1660s, English scientist *Robert Hooke* (1635–1703) used a compound microscope when he discovered the tiny box-like sections in a slice of cork, which he referred to as *cells*. During the 1670s and 1680s, a Dutchman named *Anton van Leeuwenhoek* (1632–1723) produced very fine single lenses that provided clearer and larger images (see Fig. 1-5, page 7). As a result, Leeuwenhoek was able to observe and draw tiny one-celled organisms, or **microorganisms,** thus opening the way for the study of microscopic life forms.

Nowadays, scientists work with very complicated types of microscopes, such as the **electron microscope,** first widely used in 1935. These tools enable scientists to study tiny structures within cells and within microorganisms.

In addition to tools for measurement and observation, scientists use other types of equipment in the laboratory and in the field. These include equipment for viewing, filming, and recording events and organisms; various vessels for holding and transferring materials, such as flasks, beakers, and test tubes; and equipment for heating substances during an experiment, such as the Bunsen burner. See Fig. 1-6, which shows a variety of laboratory tools.

LABORATORY SAFETY AND SKILLS

Your skill in handling scientific equipment and your knowledge of safety procedures go hand in hand in a laboratory setting. You must know not only how to use a tool correctly, but how to use it safely. For example, when heating a substance in a test tube, always point the open end of the tube away from yourself and others. In case of an accident, you should know what to do right away, and be prepared to take action. For example, if a chemical is spilled, you should clean it up immediately and then dispose of the wastes in the proper place.

In addition, it is important to clean all equipment before and after use and to work with *sterile*, or germ-free, techniques at all times. A scientist has to be particularly careful when handling bacterial cultures because of the risk of infections. Review the following list of laboratory safety rules, so that you will be well-prepared to work in your classroom's laboratory:

1. Tie back long hair; secure baggy clothing; and remove dangling jewelry.
2. Wear safety goggles, gloves, and aprons when working with chemicals and when heating substances.
3. Never touch or taste chemicals; never mix chemicals on your own; never use unlabaled chemicals; never directly inhale chemical vapors. (Always clean up and dispose of spilled chemicals.)
4. When heating substances, point the open end of the container away from yourself and other students.
5. Never eat or drink in the laboratory.
6. If you burn your skin or spill a chemical on it, immediately rinse the area with plenty of cold water, and report the accident to your teacher.
7. Do not directly handle heated or broken glassware. (Use only special heat-resistant glassware for heating substances.)
8. Observe proper clean-up procedures and store all equipment and chemicals in their proper places.
9. Do not handle any chemical or equipment until you receive directions from the teacher or person in charge to do so.
10. Report any instance of broken or unusual equipment to your teacher.
11. Never pour excess chemicals back into stock bottles. Always dispose of excess chemicals properly.
12. Always handle all chemicals and equipment carefully.
13. Proper behavior *must* be exhibited at all times in the laboratory.

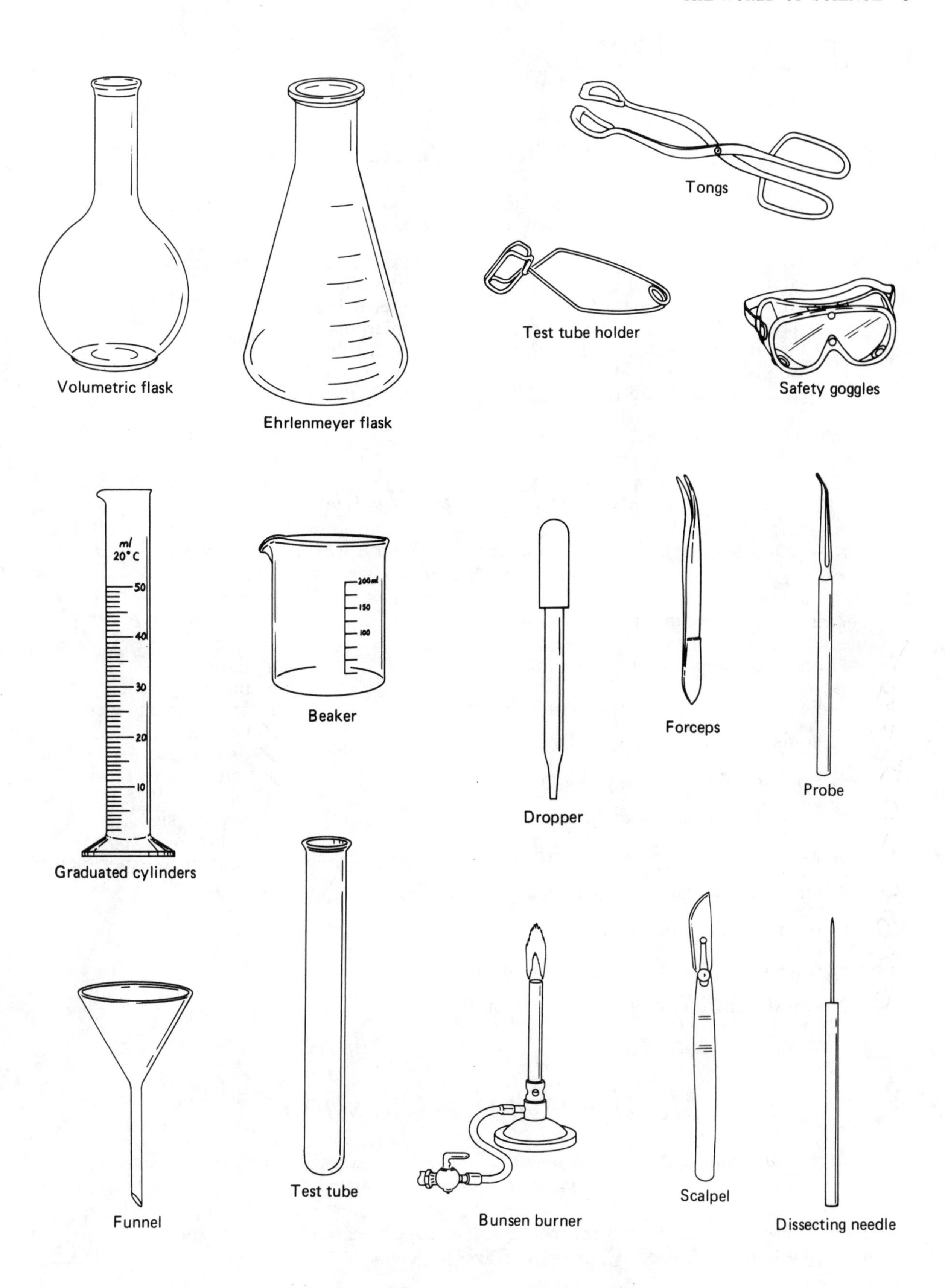

Fig. 1-6. Fifteen tools used by scientists.

CHAPTER REVIEW

Science Terms

The following list contains all of the boldfaced words found in this chapter and the page on which each appears.

compound microscope (p. 7)
control group (p. 5)
controlled experiment (p. 5)
data (p. 5)
electron microscope (p. 8)
English system (p. 6)
experimental factor (p. 5)
experimental group (p. 5)
hypothesis (p. 4)
metric system (p. 6)
microorganisms (p. 8)
scientific method (p. 4)
theory (p. 4)
variable (p. 5)

Matching Questions

On the blank line, write the letter of the item in column B which is most closely related to the item in column A.

Column A

____ 1. organized approach to problem-solving
____ 2. factor that differs from other conditions
____ 3. holds the microscope's top lens
____ 4. suggested answer to a problem
____ 5. body of information about our world
____ 6. what scientist does to test an idea
____ 7. temperature scale on which water boils at 100°
____ 8. record of scientific observations
____ 9. measurement system based on units of 10
____ 10. adjusts amount of light on microscope stage

Column B

a. data
b. metric system
c. scientific knowledge
d. Celsius
e. diaphragm
f. scientific method
g. variable
h. hypothesis
i. Fahrenheit
j. controlled experiment
k. eyepiece

Multiple-Choice Questions

On the blank line, write the letter preceding the word or expression that best completes the statement or answers the question.

1. The two objective lenses on a compound microscope are attached below the
a. eyepiece *b.* nosepiece *c.* arm *d.* stage 1____

2. The coarse adjustment and fine adjustment knobs of a microscope are located on the
a. stage *b.* diaphragm *c.* arm *d.* body tube 2____

3. Scientific knowledge about the world is based on
a. ancient myths and legends
b. imaginative explanations
c. careful observations and reasoning
d. educated guesses only
3 ____

4. The scientific study of the structure and properties of matter is known as
a. biology *b.* chemistry *c.* physics *d.* earth science
4 ____

5. An organized approach to problem-solving is known as
a. a theory *b.* a hypothesis *c.* an observation *d.* scientific method
5 ____

6. The first step of the scientific method is to
a. collect information
b. state the problem
c. form a hypothesis
d. record observations
6 ____

7. A suggested answer to a problem that can be tested is a
a. theory *b.* hypothesis *c.* factor *d.* variable
7 ____

8. A scientist performs an experiment in order to
a. test a hypothesis
b. form a hypothesis
c. collect information
d. state the problem
8 ____

9. In an experiment, the subject group that is exposed to the variable being tested is called the
a. experimental factor
b. control group
c. experimental group
d. controlled experiment
9 ____

10. The record of experimental observations may include all the following items *except* for
a. diagrams and charts
b. written notes about the experiment
c. list of equipment used in the experiment
d. conclusions about the results
10 ____

11. The system of measurement that uses the foot as the unit of length and the pound as the unit of weight is called the
a. metric system
b. decimal system
c. English system
d. scientific system
11 ____

12. The system most often used by scientists to measure length, volume, and mass is called the
a. metric system
b. English system
c. Fahrenheit scale
d. Celsius scale
12 ____

13. In the metric system of measurement, a kilometer is equal to
a. 10 meters
b. 100 meters
c. 1000 meters
d. 1/1000th meter
13 ____

14. On the Celsius scale of measurement, the freezing point of water is
a. 0° *b.* 10° *c.* 32° *d.* 100°
14 ____

15. The first simple microscopes were developed during
a. prehistoric times
b. the 1400s
c. the 1600s
d. the 1900s
15 ____

16. The man who first observed tiny organisms through the fine lenses he produced was
a. Robert Hooke
b. Charles Darwin
c. Gregor Mendel
d. Anton van Leeuwenhoek
16 ____

17. The very complicated tool now used by scientists to study microorganisms and the insides of cells is the
a. single-lens microscope
b. compound microscope
c. electron microscope
d. compound telescope
17 ____

18. When heating a substance in a test tube, always point the open end of the tube
  a. toward yourself, so that you can view the substance
  b. toward other students, so they can observe the process
  c. directly upward, so the contents can evaporate to the ceiling
  d. away from yourself and other students, for your protection — 18 d

Modified True-False Questions

In some of the following statements, the italicized term makes the statement incorrect. For each incorrect statement, write the term that must be substituted for the italicized term to make the statement correct. For each correct statement, write the word "true."

1. The part of a microscope onto which a slide is placed is called the *lens*. — 1 ____________
2. Careful observations about the world have led to the accumulation of scientific *knowledge*. — 2 ____________
3. The study of living things is known as *earth science*. — 3 ____________
4. Scientific explanations that are proved true based on years of research are known as *theories*. — 4 ____________
5. When researchers follow a series of steps to solve a problem, they are using the scientific *hypothesis*. — 5 ____________
6. The variable that differs from all other conditions in an experiment is called the experimental *group*. — 6 ____________
7. The system of measurement based on units of 10 is known as the decimal or *English* system. — 7 ____________
8. The basic unit of length used in the decimal system is the *liter*. — 8 ____________
9. Robert Hooke discovered the tiny structures he called cells by observing them through a *simple* microscope. — 9 ____________
10. The tiny one-celled creatures that were first observed through fine lenses in the 1600s are called *microorganisms*. — 10 ____________

Testing Your Knowledge

1. What is the difference between *forming* a hypothesis and *testing* a hypothesis?

____

____

2. Explain the difference between the two terms in each of the following pairs:
  a. coarse adjustment knob and fine adjustment knob

____

b. theory and hypothesis

theory-something thats been around for years and people know about
hypothesis-is an educated guess

c. control group and experimental group

control group-stays the same
experimental group-stays changing

d. metric system and English system

metric system

e. Celsius and Fahrenheit

f. simple microscope and compound microscope

3. Complete the table of measurement units.

Measurement	*Metric Units*	*English Units*
Volume		
Length		
Mass/Weight		

4. Why is it important for a scientist to *state the problem* before beginning a research project?

5. Why is it important to have all conditions the same in an experiment, except for the one *experimental factor*?

6. Explain why it is important for scientists to clean all laboratory equipment before and after use and, in particular, to be sure that their hands, tools, and lab area are as free of germs as possible.

Animal Experimentation: Should It Continue?

Upwards of 70 million animals (mostly vertebrates) are used annually in the United States in various kinds of scientific experiments. The uses of animal experimentation vary widely but the major outcome of animal experimentation is (1) treatment for cancer, Alzheimer's disease, AIDS, heart disease; (2) antibiotics and vaccines; (3) regeneration of damaged nerves and brain cells; (4) surgeries; (5) heart transplant research; and (6) to test for the possible dangers of all sorts of products to human beings.

Unfortunately, many of the animals used for these tests die, are killed, or suffer pain. Most are mammals such as mice, rats, cats, dogs, and primates (monkeys and chimpanzees). Many people question whether experiments that are harmful to animals even need to be performed.

Some people say that animals should never be used for these purposes. Other people say that animals should be used only for certain tests, like medical experiments. Still others believe that there should be no restrictions on such uses. And there are a lot of opinions in-between.

Before you can begin to form your own opinion, you need to know about some kinds of experiments that are performed on animals. Here are some examples of medical experiments:

Chimpanzees have been used to try to find ways to treat hepatitis—a liver disease—in people. A vaccine for malaria, a tropical disease that kills from one to two million children a year, may soon be available thanks to experiments with mice. Dogs have been used to test a new device to treat diabetes, a serious human disease. And experiments with rats have yielded clues to the transplantation of organs that won't be rejected by the people who get them.

In each of these experiments, some animals suffered pain, died, or were killed. Healthy dogs were given diabetes by removing a vital organ from their bodies. Not all the dogs survived. Infant chimps—rare animals in the wild—were taken from their mothers, who often were killed because they wouldn't let go of their young. Many other chimpanzees have suffered and died in captivity. To test the malaria vaccine, scientists exposed a group of mice to malaria. But to tell whether or not the vaccine worked, some mice in the group were not given the vaccine. These mice became sick. The rats who received organ transplants had to have their own organs removed first.

No one wants to harm animals. But are there other ways to obtain the knowledge and results described above without experimenting on animals? Yes, say people concerned with "animal rights." Some of these people recommend the "3 Rs"—replacements, reductions, and refinements.

Replacements involve substitutions for animals. For example, cells grown in glass dishes can be used to explore ways of keeping viruses from infecting cells. Reduction involves using fewer animals in a particular experiment or test. And refinements include changing the way experiments are performed on animals so that any discomfort to them is minimized.

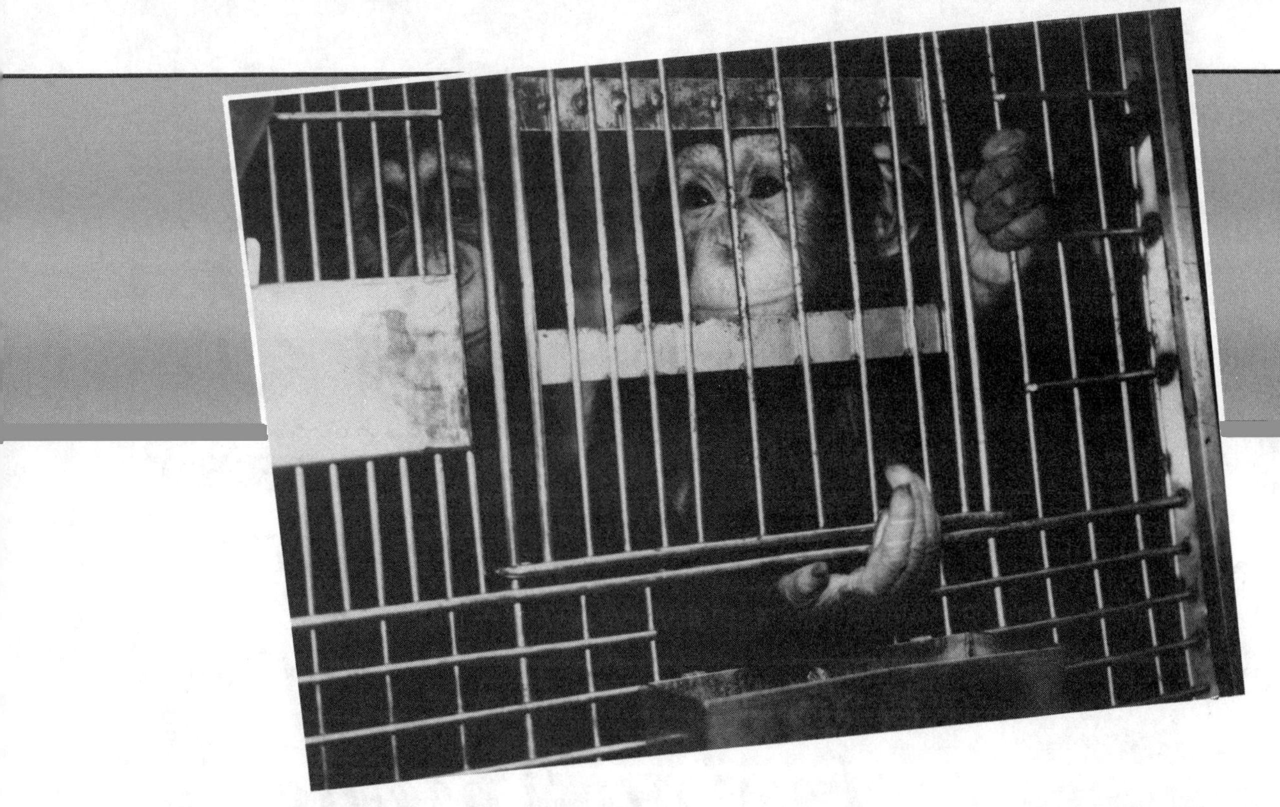

Few scientists would argue with the 3 Rs and most try to achieve them. Yet many scientists contend that in certain cases there simply are no substitutes for animals. You can't vaccinate a cell, they say, since the whole body is involved in combating an infectious disease. Nor can you just study cells to uncover the mysteries of certain mental conditions.

What kinds of experiments, if any, should be performed on animals? Should animals be used for cosmetic and household products testing? How might we otherwise gain knowledge vital to the health of people? What do you think?

1. *In the opening paragraph, three purposes of animal experimentation are numbered and described. This article focuses on*

a. purpose 1.
b. purpose 2.
c. purpose 3.
d. purposes 2 and 3.

2. *Mice have been used to*

a. study hepatitis.
b. develop a vaccine for malaria.
c. treat diabetes.
d. transplant organs.

3. *In an experiment, which of the following would be considered a replacement?*

a. Using dogs instead of chimpanzees.
b. Using 10 chimps rather than 20 chimps.
c. Using cells instead of chimps.
d. Using sick chimps instead of healthy ones.

4. *Explain why most experimental animals used are mammals.*

____

____

____

5. *Two tests on animals that are highly controversial are LD50 tests and Draize tests. LD50 (lethal dose 50) tests are used to determine the dosage of a drug, medicine, or chemical that will kill 50% of the animals that receive it. The Draize test, which is performed on rabbits, is used to determine the possible effects of specific cosmetics on a person's eyes. Do library research to find out more about these tests. Choose one and write a short essay explaining your view, either in favor of or against the test.*

2

The Activities and Chemistry of Living Things

LABORATORY INVESTIGATION

TESTING FOR DIGESTION OF STARCH

A. Add a pinch of starch to a beaker containing 75 ml of water. Heat the mixture gently over a Bunsen burner for a few minutes but do not boil it (see Fig. 2-1). Allow the mixture to cool and then divide it equally between two test tubes. Label the tubes A and B. (See Fig. 2-2.)

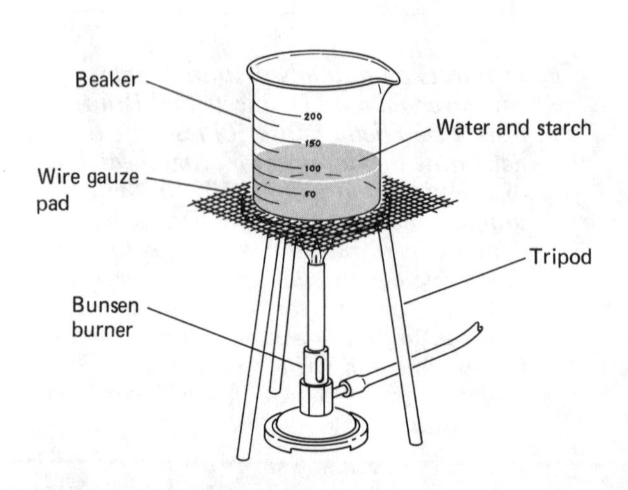

Fig. 2-1. Setup for heating beaker of starch and water.

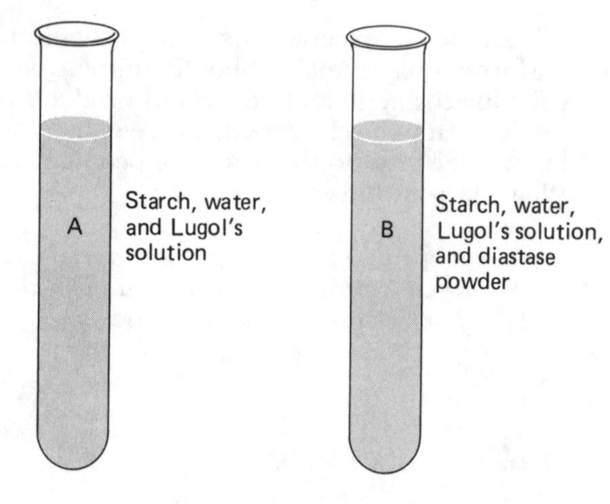

Fig. 2-2. Test tubes A and B for starch experiment.

B. Add one drop of Lugol's (dilute iodine) solution to each of the test tubes. Lugol's solution tests for the presence of starch; the solution turns blue-black when starch is present. (**Caution:** Be careful when working with iodine. It is poisonous if swallowed, and can stain your skin and clothing.)

1. What is the color of the contents of each test tube after adding the drop of Lugol's solution?

____

C. Add a pinch of diastase powder (a plant extract that digests starches) to test tube B. Cover the opening of each tube with your thumbs and gently shake the tubes for about one minute each. Then warm both tubes by holding them in your hands for about 10 minutes. (**Caution:** Wash your hands thoroughly after this investigation.)

2. What is the color of the contents of each tube after 10 minutes?

____

3. Describe the difference that finally appears between tubes A and B.

____

4. What is the purpose of using test tube A?

____

5. Explain what may have happened to the starch in test tube B.

____

Living Things

In your daily life, you encounter three types of matter: (1) living matter; (2) matter that was once part of a living thing; and (3) matter that was never alive nor part of any living thing. **Matter** is anything that takes up space and has mass.

A living thing, such as a dog, ameba, or rosebush, is called an **organism.** The matter that makes up an organism is called *organic matter.* Some examples of organic matter are wool, honey, cotton, sugar, and wood. Matter that was never alive nor part of any organism is

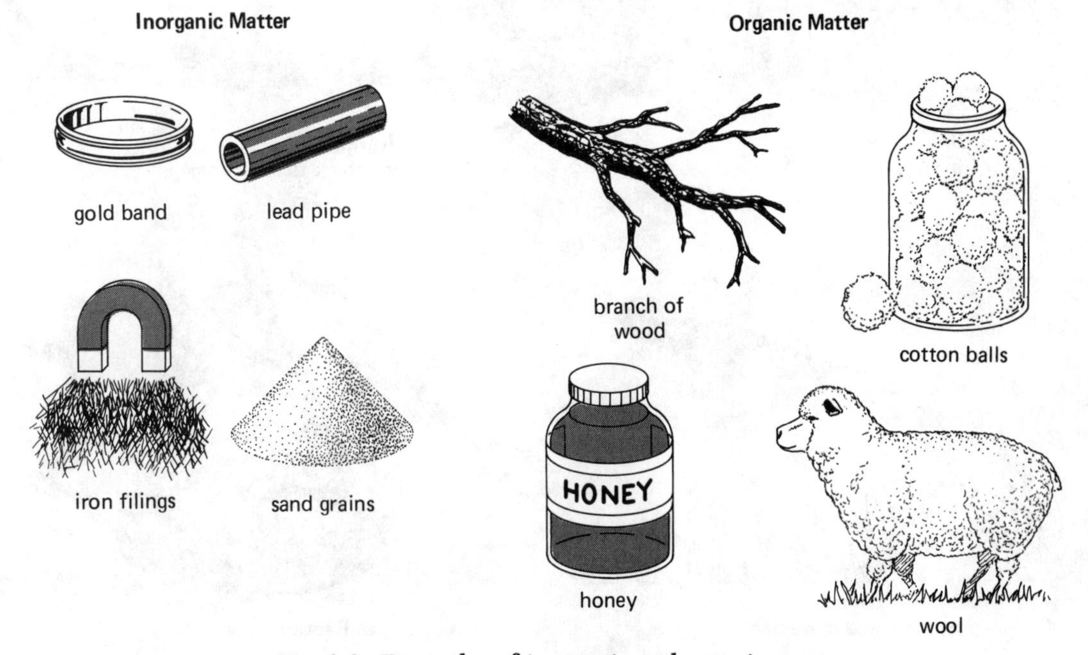

Fig. 2-3. Examples of inorganic and organic matter.

called *inorganic matter.* Some examples of inorganic matter are sand, iron, gold, and lead. (See Fig. 2-3.)

When an organism is compared with any inorganic object, an important difference can be noted. Organisms carry out certain activities that inorganic objects do not. These activities, which enable the organisms to stay alive, are called **life activities,** or *life processes.*

THE LIFE ACTIVITIES

To carry out their life activities, plants, animals, and other living things require energy. Some of the life activities are involved with producing energy. Other life activities use up this energy. If energy production in a living thing were to cease, all other life activities would stop and the organism would die. In addition to needing energy, living things require a proper temperature range in their environment in order to carry out their life activities.

There are several life activities that are basic to all living things. All organisms carry out the processes of **nutrition** (which includes ingestion, digestion, absorption, circulation, and assimilation), growth and self-repair, respiration, excretion, movement, and response. Organisms are also capable of the process of reproduction. (See Fig. 2-4, which shows some life activities of a bird.)

Ingestion (Food-getting). All organisms require food for growth and repair, for the production of energy, and for the proper functioning of the body. The part of nutrition known as **ingestion,** or *food-getting,* is the taking in of food by organisms. Most animals, and many one-celled organisms, actively seek and take in their food. Fungi, such as mushrooms, also have to get their food from other living matter in the environment. Plants and algae, however, do not have to seek or take in food, since they make their own food by using carbon dioxide and energy from the sun. In addition, all organisms need water to survive.

Digestion. In general, food that has been taken in by an organism, or made by a plant and stored, is in a form that the organism's body cannot use. **Digestion** is the breaking down of food into simpler products that the body can use. In your laboratory investigation, you saw that starch (a complex com-

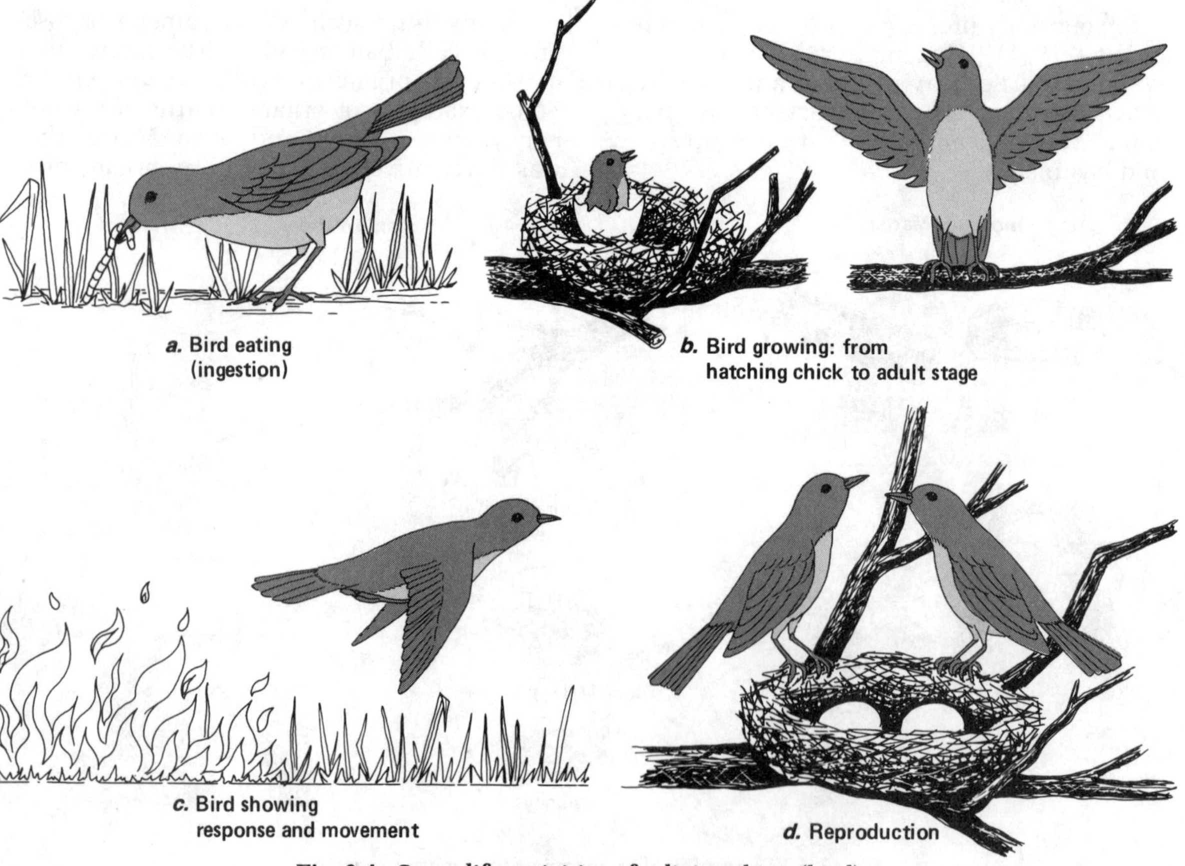
a. Bird eating (ingestion)

b. Bird growing: from hatching chick to adult stage

c. Bird showing response and movement

d. Reproduction

Fig. 2-4. Some life activities of a living thing (bird).

pound) was broken down, or digested, by *diastase* into sugar (a simple compound). Although diastase is not alive, it is the product of living things, and it is essential for the life of many plants. *Salivary amylase* is another nonliving substance produced by living things. Found in the *saliva* (or spit) of many animals, salivary amylase is important for their survival because it, too, aids in the process of digestion.

Absorption. The passage of simple substances into the internal parts of a plant or animal is called **absorption.** In plants, absorption occurs when water and minerals from the soil enter the roots. In the higher animals, absorption occurs in the small intestine when digested foods pass into the bloodstream.

Circulation. The movement of digested food and other materials throughout the body of an organism is called **circulation.** In the higher plants, a system of tubes transports these materials. In the higher animals, the bloodstream transports them to the cells of the body.

Assimilation. The changing of digested food into new living material is called **assimilation.** This chemical process takes place in the body's cells.

Growth and Self-Repair. The increase in size of an organism is called *growth*. The process of assimilation provides the new materials for growth. In addition, the process of *development* occurs as an organism grows from its early (juvenile) stages into its mature (adult) form. The rebuilding of worn-out or injured body tissues is called *self-repair*. Assimilation also provides the new materials for self-repair.

Respiration. Most organisms take in oxygen from the air or oxygen that is dissolved in water. In the chemical process of **oxidation,** the oxygen reacts with digested foods. As a result of this reaction, energy is released. In some organisms, energy is released by the reaction of digested foods with elements other than oxygen. **Respiration** is the release of energy as a result of the chemical reactions involving food. The energy released enables all other life activities to continue.

Excretion. As a result of respiration and other life activities, the body of an organism produces wastes. Among such wastes are carbon dioxide, excess water, and urea (in urine). These wastes must be eliminated from the body. **Excretion** is the process of ridding the body of wastes that result from the other life activities. By contrast, the process by which an organism's body produces useful substances (like saliva) is called **secretion.**

Movement. All living things show some *movement*. Most animals and one-celled animallike organisms move freely from place to place by the process called **locomotion.** Most plants and fungi cannot locomote, but they do move very slowly as a result of growth. Although some organisms do not appear to move, they often have internal structures that do move. Movement helps an organism find food, shelter, and mates, and avoid danger.

Response. All living things react to changing conditions, or **stimuli,** in their environment. Among such conditions are heat and light. Organisms also react to other living and nonliving things around them. The ability of an organism to detect changes in its surroundings is called *sensitivity*. The reaction to such changes is called a **response.** This ability to respond helps an organism survive. For example, an animal responds negatively to a fire by fleeing from it. A plant responds positively to light by turning its leaves toward the sun. (This type of response by plants is called a *tropism*.) (See Fig. 2-5.)

Fig. 2-5. Plant tropism—a positive response to light.

Reproduction. The process by which an organism produces offspring like itself is called **reproduction.** Organisms only reproduce living things of their own kind. Reproduction is the only life activity that is not necessary to keep an individual organism alive.

All organisms have a *life span*—they are born, they grow and develop, and they die. However, many organisms reproduce and start a new generation of their kind. Thus, although reproduction is not necessary for the survival of an individual organism, it is necessary for the continuation of life of each kind of organism from one generation to the next.

THE CHEMISTRY OF LIFE: ELEMENTS, COMPOUNDS, AND MIXTURES

The substances that make up living things can be broadly classified as elements, compounds, and mixtures. In order to understand how an organism functions, it is necessary to know the properties of these three types of substances. There are some important differences between elements and compounds, and between compounds and mixtures.

Elements. An **element** is a substance that cannot be broken down into any other substance by ordinary means. For example, oxygen is an element; processes such as heating or dissolving in water cannot break down oxygen into any substance other than oxygen. There are 92 naturally occurring elements and, at present, an additional 19 elements have been made in laboratories.

Atoms. The smallest unit of an element that still has the properties of the element is called an **atom.** An individual atom is too small to be seen. However, many millions of identical atoms may join together to form a visible substance or object. Atoms of the same element are alike, but are different from the atoms of every other element.

Compounds. Many elements tend to unite chemically with one another. When two or more elements unite, they do so in a definite proportion by mass of each element. For example, in the formation of water, or H_2O, no matter how much hydrogen is used, eight times as much oxygen is always needed. In other words, two grams of hydrogen will unite completely only with 16 grams of oxygen. Thus, a **compound** is a substance composed of two or more elements united chemically in a definite proportion by mass.

The properties of a compound are different from those of the elements that make up the compound. For example, sodium is a highly reactive metal, and chlorine is a poisonous gas. But when united, they form sodium chloride—commonly known as table salt—a compound essential for many life activities. (See Fig. 2-6.)

Molecules. The smallest unit of a compound that can exist by itself and still have all the properties of the compound is called a **molecule.** Molecules, like atoms, are too small to be seen. A molecule of water is called a compound because it is composed of two different elements—hydrogen and oxygen—united chemically. Sometimes, two atoms of the same element, such as oxygen, join to form a molecule. Thus, two atoms of oxygen form one molecule of oxygen gas, millions of which are breathed in by organisms every minute.

Mixtures. When two or more substances (elements or compounds) are mixed together and the substances do not unite chemically, the resulting material is called a **mixture.** Unlike a compound, the substances in a mixture can be mixed in any proportion, and the substances will always retain their original properties.

Mixtures may consist of solids, liquids, or gases. An example of a mixture is table salt (the compound sodium chloride) dissolved in water. The result is a saltwater solution—like seawater—that can be changed back to salt and plain water by simple physical means. A *solution* is a type of mixture in which one substance, the *solute* (usually a gas or a solid), is dissolved in another substance, the *solvent* (usually a liquid).

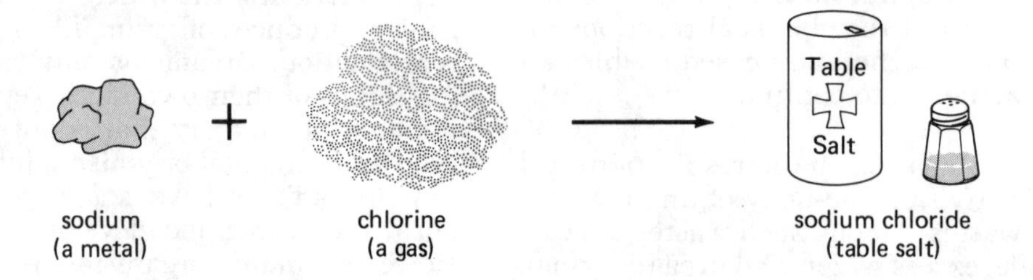

Fig. 2-6. The compound sodium chloride—table salt.

ORGANIC AND INORGANIC COMPOUNDS

Organic Compounds. Compounds that contain carbon atoms (except for carbon dioxide, carbon monoxide, and the carbonates) are called **organic compounds.** Organic compounds are important because they are made by living things. Examples of organic compounds are sugar, fat, and protein.

Inorganic Compounds. Compounds composed of the atoms of other elements (including carbon dioxide, carbon monoxide, and the carbonates) are called **inorganic compounds.** Inorganic compounds are not made by living things. Other examples are table salt (sodium chloride) and water.

ELEMENTS AND COMPOUNDS IN ORGANISMS

There are few special elements or compounds that distinguish living matter from nonliving matter. In fact, many of the elements present in living matter are the same as those found in nonliving matter. Yet living things are made up almost entirely of only four elements—carbon, oxygen, hydrogen, and nitrogen. The most abundant compound in living things is water, which is made up of hydrogen and oxygen (H_2O).

Many of the organic and inorganic compounds present in living matter can be made in a chemical laboratory. Natural organic compounds, such as blood hemoglobin, however, are more complex than the organic compounds, like nylon, made in laboratories.

Interestingly, evidence from laboratory experiments indicates that life may have formed from early natural organic compounds. It is thought that energy from lightning, sunlight, and other sources broke apart simple compounds and recombined them to form complex organic compounds.

The following tables list the most common elements and compounds found in living matter, and their percentages by mass (Table 2-1 and Table 2-2).

TABLE 2-1. THE MOST COMMON ELEMENTS IN LIVING MATTER

Element	*Symbol*	*Percent by Mass*
Oxygen	O	65.0
Carbon	C	18.5
Hydrogen	H	9.4
Nitrogen	N	3.1
Calcium	Ca	1.5
Phosphorus	P	1.0
Potassium	K	0.4
Sulfur	S	0.3
Magnesium	Mg	0.1
Sodium	Na	0.1
Chlorine	Cl	0.1
Iron	Fe	trace
Iodine	I	trace

TABLE 2-2. SOME COMPOUNDS IN LIVING MATTER

Compound	*Percent by Mass*
Water	80
Proteins	12
Carbohydrates, Nucleic acids, Vitamins, and Other compounds	3
Fats	3
Mineral salts	2

CHEMICAL REACTIONS

All life processes depend on chemical reactions. A few of the important types of chemical reactions that occur in both living and nonliving things are listed below.

Combination. When two or more elements unite and form a compound, the reaction is called *combination.* For example, the combination of carbon and oxygen produces the compound carbon dioxide (CO_2).

Oxidation. The union of an element or compound with oxygen is called *oxidation.* This type of combination reaction releases energy. Examples of oxidation are the combinations of carbon with oxygen and of hydrogen with oxygen.

Decomposition. When a compound is broken down into its elements, the reaction is called *decomposition.* For example, a molecule of water (H_2O) can be broken down, or decomposed, into hydrogen gas and oxygen gas.

Catalytic Reactions. Chemical reactions take place only when heat or some other form of energy is supplied. For example, ordinary hydrogen peroxide (H_2O_2) can be decomposed slowly by heat or light into water and oxygen. This reaction can be made to go rapidly, however, by adding a pinch of the compound manganese dioxide. Substances like manganese dioxide that speed up reactions by lowering the energy required are called **catalysts** (or *catalytic agents*). A catalyst is not changed in a reaction, nor is it part of the chemical reaction. Upon completion of a reaction, a scientist can recover all of the catalyst. Manganese dioxide is an example of an *inorganic catalyst.* An inorganic catalyst is a compound that does NOT contain the element carbon but still alters the speed of the chemical reaction.

ENZYMES

The life activities of organisms depend on many very complicated chemical reactions. These reactions are able to occur at relatively low (body) temperatures because catalysts are present. The catalysts that are present in living things are special proteins called **enzymes** (or *organic catalysts*). An organic catalyst is a compound that CONTAINS the element carbon and still alters the speed of the chemical reaction.

Enzymes take part in every life process. For example, in your laboratory investigation, you learned that the plant extract diastase digests starches; diastase is an enzyme. Likewise, salivary amylase is an enzyme that helps digest starches in animals.

Many other enzymes decompose proteins into simpler compounds. In respiration, certain enzymes take part in the oxidation reactions that cause the release of energy. The number of enzymes has not been determined, but it is known that each enzyme aids a different reaction in an organism.

Although enzymes are organic compounds, they are not alive. Many reactions that are catalyzed by enzymes in the body can also be carried out in test tubes (as you did in the diastase experiment). However, the ability to manufacture and use enzymes is a property of living things and distinguishes life from nonlife.

CHAPTER REVIEW

Science Terms

The following list contains all of the boldfaced scientific terms found in this chapter and the page on which each appears.

Matching Questions

On the blank line, write the letter of the item in column B which is most closely related to the item in column A.

Column A	Column B
____ 1. transport of materials in the body	*a.* digestion
____ 2. a reaction to stimuli	*b.* atom
____ 3. substance that cannot be broken down	*c.* excretion
____ 4. ridding of wastes by the body	*d.* circulation
____ 5. breaking down of food in the body	*e.* response
____ 6. smallest unit of an element	*f.* ingestion
____ 7. rebuilding of worn-out body tissues	*g.* element
____ 8. smallest unit of a compound	*h.* enzyme
____ 9. process of food-getting	*i.* self-repair
____ 10. catalyst that speeds reactions	*j.* molecule
	k. respiration

Multiple-Choice Questions

On the blank line, write the letter preceding the word or expression that best completes the statement or answers the question.

1. An example of an organism is
a. wood *b.* a maple tree *c.* a lettuce leaf *d.* steel 1 ____

2. Unlike a living thing, an automobile *cannot*
a. move *b.* use energy *c.* reproduce *d.* get rid of wastes 2 ____

3. Which group contains examples of *only* inorganic matter?
a. water, sugar, salt, pepper *c.* milk, rubber, iron, copper
b. onion, water, celery, milk *d.* steel, sand, copper, water 3 ____

4. When you bite into an apple, you are carrying out the life activity called
a. absorption *b.* circulation *c.* ingestion *d.* excretion 4 ____

5. Assimilation is the change of
a. complex compounds to simple compounds
b. oxygen to carbon dioxide
c. wastes to useful substances
d. digested food into living material 5 ____

6. Digested foods are carried throughout the body in the process called
a. circulation *b.* assimilation *c.* digestion *d.* absorption 6 ____

7. Starch is changed into sugar in the life activity called
a. ingestion *b.* digestion *c.* secretion *d.* excretion 7 ____

8. Material for an organism's growth results from the life activities of
a. ingestion, digestion, absorption, assimilation
b. reproduction, respiration, excretion, digestion
c. assimilation, response, excretion, movement
d. movement, assimilation, circulation, reproduction
8 A

9. Which group contains examples of *only* organic matter?
a. hair, glass, steel, paper
b. paper, sand, salt, wool
c. hair, starch, sugar, wood
d. steel, salt, rubber, wool
9 C

10. Energy for performing the life activities is produced during the process of
a. excretion *b.* respiration *c.* digestion *d.* circulation
10 D

11. The one life process that is *not* necessary for an individual's survival is
a. respiration *b.* reproduction *c.* circulation *d.* absorption
11 B

12. A substance that cannot be broken down by ordinary means is
a. a compound *b.* a molecule *c.* an element *d.* a mixture
12 C

13. The smallest unit of an element that still has the properties of that element is
a. an atom *b.* a molecule *c.* a compound *d.* an enzyme
13 B

14. A substance composed of two or more elements that have been united chemically is called a
a. mixture *b.* catalyst *c.* molecule *d.* compound
14 A

15. Two or more substances together that are *not* united chemically make up
a. a molecule *b.* a compound *c.* a mixture *d.* an enzyme
15 B

16. The element that is present in *all* organic compounds is
a. hydrogen
b. carbon
c. sodium
d. calcium
16 B

17. Living things are almost entirely made up of the four elements carbon, oxygen, hydrogen, and
a. potassium
b. calcium
c. nitrogen
d. sodium
17 B

18. When two or more elements unite to form a compound, the reaction is called
a. combination *b.* oxidation *c.* decomposition *d.* reduction
18 A

19. When an element or compound unites with oxygen and releases energy, the reaction is called
a. decomposition *b.* oxidation *c.* combination *d.* secretion
19 B

20. An enzyme is a type of
a. organic catalyst
b. inorganic catalyst
c. element
d. starch
20 A

Modified True-False Questions

In some of the following statements, the italicized term makes the statement incorrect. For each incorrect statement, write the term that must be substituted for the italicized term to make the statement correct. For each correct statement, write the word "true."

1. In order to stay alive, all living things require *energy*. 1 True

2. The life activity that enables living things to continue from generation to generation is ~~*self-repair*~~. reproduction 2

3. In the body of a bird, food is changed into a form that the bird's body can use by the process of *circulation*. 3 true

4. The passage of digested food into the bloodstream is called *absorption*. 4 ____________

5. A cat climbs a tree when it sees a dog. The cat's behavior is an example of the life activity called *assimilation*. 5 ____________

6. Ingestion, digestion, and absorption are all part of an organism's process of *excretion*. 6 ____________

7. Wastes formed by the body are gotten rid of by the process of *secretion*. 7 ____________

8. Two or more elements united in a definite proportion by mass make up a *mixture*. 8 ____________

9. A type of mixture in which one substance is dissolved in another substance is called a *solution*. 9 ____________

10. Compounds containing carbon that are formed by living things are called *inorganic* compounds. 10 ____________

11. The most abundant compound in living things is *protein*. 11 ____________

12. During respiration, energy is released as a result of the reaction called *oxidation*. 12 ____________

13. A solution of table salt and water makes up a *compound*. 13 ____________

14. Substances that speed up chemical reactions, but are not part of the reactions, are called *elements*. 14 ____________

15. When hydrogen and oxygen in water molecules are separated, the reaction is known as *decomposition*. 15 ____________

Testing Your Knowledge

1. Explain why a snail is considered to be a living thing whereas an electric iron is not.

____

____

2. Explain the difference between the two terms in each of the following pairs:

a. circulation and respiration

____

____

b. secretion and excretion

____

____

c. absorption and assimilation

d. growth and self-repair

e. organic matter and inorganic matter

f. compound and mixture

g. organic compound and inorganic compound

h. organic catalyst and inorganic catalyst

i. combination and decomposition

3. Of what value is the process of circulation to an animal such as a cow?

4. Of what value is the process of circulation to the leaves of a maple tree?

5. Explain why the process of respiration is important to your muscles.

6. Many chemical reactions require a lot of heat, or other form of energy, to help them take place. Explain the role of catalysts in reactions, and why enzymes in particular are very important catalysts.

7. List five examples of elements.

8. A student pours some sand into a beaker of water.
a. Is this an example of a mixture or a compound?

b. Explain why.

c. Can the two substances be easily separated out again?

d. Explain how this might be done.

9. Explain the importance of water to living organisms.

10. Why are organic compounds usually larger molecules than inorganic compounds?

3

Living Things Under the Microscope: Cells

LABORATORY INVESTIGATION

MICROSCOPIC STUDY OF LETTUCE SKIN

A. With a slightly twisting motion, tear a lettuce leaf. Snip off a 1/2-cm square of the very thin, delicate, transparent tissue that appears on one of the torn edges. This tissue is the skin (or *epidermis*) of the leaf. Mount the square in a drop of water on a microscope slide. Place a cover slip over the drop and examine the specimen with the unaided eye.

1. Describe the color and general appearance of the specimen of lettuce skin.

____

____

B. Focus the specimen under the low power of the microscope.

2. Describe the color and general appearance of the specimen now.

____

____

Each brick-shaped structure is a *cell.* Each large round structure is composed of two cells that control the size of a *pore* (opening) that leads into the internal tissue of the leaf. Each of these cells is called a *guard cell.*

3. Describe the color and shape of a guard cell.

____

____

C. Mount another piece of lettuce skin in a drop of Lugol's solution. Cover the specimen and examine it under low power. Study a brick-shaped cell. (**Caution:** Wash your hands after working with iodine.)

4. What effect does the Lugol's have on the appearance of the cell?

____

____

D. Examine a few cells under high power. Sketch a few cells that are close together. Put in the details of at least one cell.

Compare your sketch with Fig. 3-1 below Label your sketch to correspond with the figure.

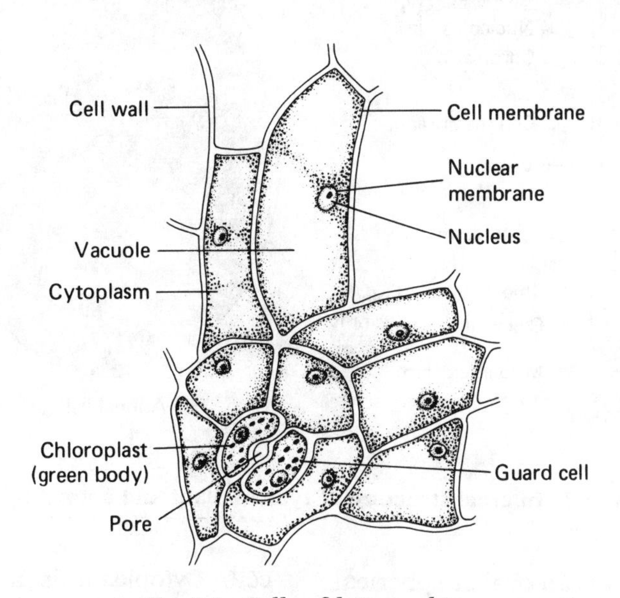

Fig. 3-1. Cells of lettuce skin.

Cells

By using the microscope, we are able to learn about the structure of living things. Your study of lettuce skin shows that the skin is constructed of small units called **cells.** If you were to observe specimens from another part of this plant or of any other plant, you would find that they have a similar structure.

Specimens of animal tissue show that they, too, are constructed of cells. In many-celled organisms, different types of cells aid in performing all of the life activities.

The most important characteristics of cells are described in the *cell theory,* which states that all living things are made up of cells; that the cell is the smallest unit of a living thing that carries out the basic life activities; and that all cells come from other living cells.

The bodies of organisms are composed of cells in much the same way that buildings are composed of bricks and that matter is composed of atoms. Larger organisms have more cells than smaller organisms. Some organisms, such as the ameba, are microscopic in size and consist of only one cell. Other organisms, like humans, are relatively large and consist of millions and millions of cells.

STRUCTURE OF CELLS

Although cells vary greatly in size and shape, their internal structure is basically the same (Fig. 3-2). Every cell—except for bacteria—has three main parts: *nucleus, cytoplasm,* and *cell membrane.* These structures are alive. **Protoplasm** is the name given to all the living material in a cell.

Plant cells have a structure, the *cell wall,* that is not found in animal cells. Because the cell wall is not composed of living material, it is not considered part of the protoplasm.

3. *Nucleolus*—a round body containing proteins and the *nucleic acid* **RNA.** The nucleolus sends chemical (RNA) "messages" out into the cytoplasm. These messages instruct the cell to make proteins to carry out growth, digestion, excretion, and other life activities. Some nuclei contain several nucleoli.

Cytoplasm. All of the material in the cell that surrounds the nucleus is called the **cytoplasm.** Under the direction of the genes, the cytoplasm carries out most of the activities in the

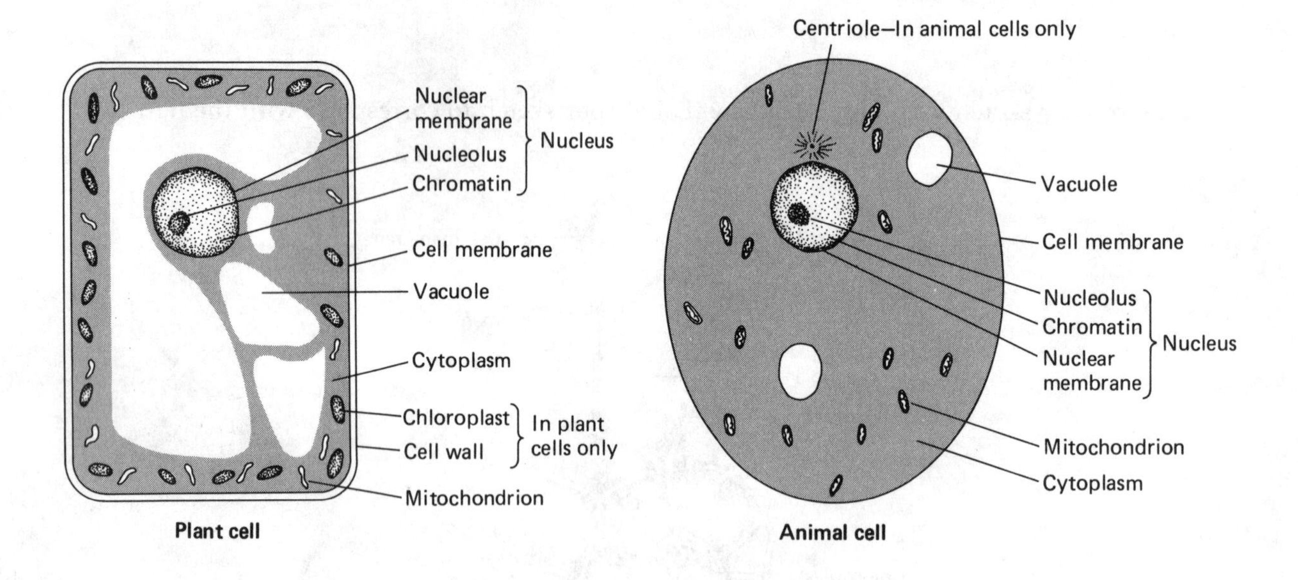

Fig. 3-2. Internal structure of typical plant and animal cells.

Nucleus. The **nucleus** is an oval or spherical body that, in animal cells, is often located near the middle of a cell. If a nucleus is removed from a cell, the cell dies. The nucleus regulates all life activities of the cell and takes part in cell division and reproduction. As Fig. 3-2 shows, the nucleus consists of a *nuclear membrane, chromatin,* and a *nucleolus.*

1. *Nuclear membrane*—the outermost part of the nucleus. Materials that pass between the nucleus and the cytoplasm go through the nuclear membrane.
2. *Chromatin*—composed of granules that are part of very fine coiled threads. Before cell division and reproduction, these threads shorten and thicken into dense rods called **chromosomes.** Chromosomes are composed of genes and proteins. The **genes** control all activities in the cell and determine the heredity of an organism.

cell. Cytoplasm is a thick colloid (a type of mixture) in which tiny bodies, called **organelles,** can be seen. Living cytoplasm shows a constant streaming motion, which carries materials throughout the cell.

Cytoplasm contains the following organelles:

1. *Vacuoles*—liquid-filled sacs that store materials the cell has taken in or produced. The vacuoles of animal cells are generally smaller than those of plant cells.
2. *Chloroplasts*—small bodies found in the cells of plants and algae. Chloroplasts contain the green pigment (coloring matter) called *chlorophyll,* which is essential for photosynthesis.
3. *Mitochondria*—oblong-shaped structures scattered in the cytoplasm. A mitochondrion contains enzymes that act in cell oxidation of food molecules and produce en-

ergy. Accordingly, mitochondria are often called the "powerhouses" of the cell.

4. *Ribosomes*—very tiny, round granules made of RNA. These are best seen with the tremendous magnification of the electron microscope. Ribosomes take part in making enzymes and other proteins inside the cells.
5. *Centrioles*—a pair of small organelles that lie near the nucleus. Most animal cells have centrioles; most plant cells do not. Where centrioles are present, they take part in cell division. Where they are absent, cell division goes on normally without them.
6. *Endoplasmic Reticulum*—network of tube-like membranes along which the ribosomes are located.
7. *Golgi Bodies*—areas for secretion, packaging, and storage of chemicals in the cell.

Cell Membrane. The **cell membrane** is the outermost *living* structure of all cells. The cell membrane encloses the cytoplasm and controls the passage of materials into and out of the cell.

Cell Wall. All plant cells, and most algae cells, have a rigid **cell wall** around the cell membrane. All animal cells lack cell walls. The cell wall of plants and most algae is composed mainly of *cellulose,* a nonliving compound produced by the cytoplasm. Fungi have a cell wall composed of *chitin,* also a hard substance. Since it is made of rigid material, the cell wall protects the inner, living parts of the cell and helps maintain its shape.

EXCHANGE OF MATERIALS IN CELLS

Diffusion. The process by which molecules of two or more substances spread and mix together is called **diffusion.** The molecules spread from a region where they are more concentrated (greater in number) to a region where they are less concentrated (fewer in number), until they are evenly distributed.

For example, when a drop of red ink is placed in a glass of water, the ink molecules spread from the drop, the region of higher concentration, through the water, the region of lesser concentration. In time, the molecules of the ink and water become so well mixed that the solution becomes evenly colored. Likewise, some substances within a cell spread to all parts of the cytoplasm by diffusion.

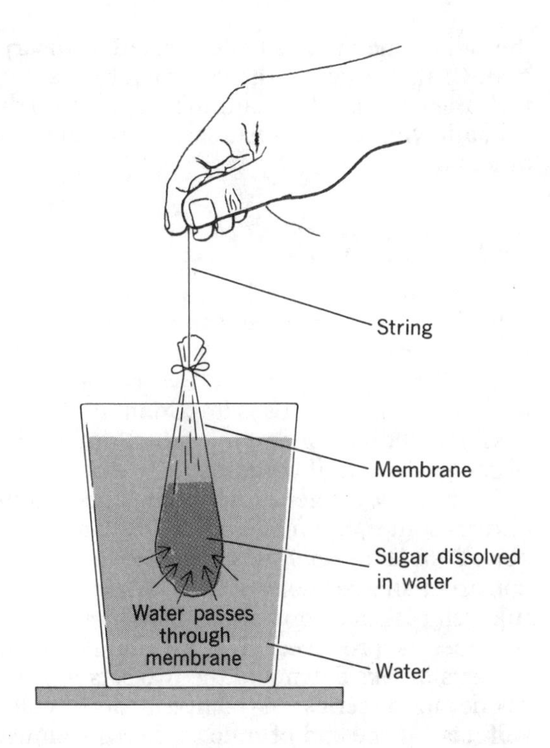

Fig. 3-3. Osmosis.

Osmosis. The diffusion of water through a membrane is called **osmosis.** Many membranes have tiny pores. These pores act like the openings in a sieve and permit only particles smaller than themselves to pass through. Water molecules are usually small enough to pass through most membranes. The water molecules move from the region where they are more concentrated to the region where they are less concentrated.

For example, if a solution of sugar is poured into a bag made of a cellophane membrane and the bag is suspended in a glass of water, water from the glass enters the bag of sugar solution (Fig. 3-3). The sugar molecules are too large to pass through the membrane into the water.

Osmosis can take place through membranes that are either living or nonliving. Although water can enter a living cell by osmosis through the cell membrane, water does not ordinarily pass out of a cell by this process. When water passes into a cell, the cell does not have to use energy. However, a cell has to use up some of its energy to get rid of excess water.

Selective Absorption. While water may enter a cell by osmosis through the cell membrane, dissolved substances such as oxygen and glucose cannot. These substances may enter the cell by diffusion. Other substances may enter the cell by the process called **selective absorption.** In this process, the cell membrane controls the passage of certain substances into the cell, allowing some substances in and keeping others out.

CELL DIVISION

All organisms begin life as a single cell. As cells divide (or reproduce) and enlarge, the organism develops and grows. In **cell division,** one cell becomes two, the two become four, and so on. The nucleus is important in the process of cell division and reproduction because of the genes that it contains.

Chemically, the genes, which are areas on the chromosomes, are made up of **DNA,** a nucleic acid. DNA is the only substance known that can construct an exact copy of itself. When the DNA in a cell makes a copy of itself, an identical set of genes is produced. The membrane of the nucleus breaks down and the two sets of genes divide in a series of complex steps called **mitosis.** At the end of mitosis, two new nuclei form. This process is followed by a division of the cytoplasm. After the cytoplasm has divided, two **daughter cells** are formed—each with a complete set of identical genes.

Steps in Mitosis. By the time a cell begins to divide, its genes have already been copied. The chromatin threads (see Fig. 3-4*a*), which contain the copied genes, change into dense rods—the chromosomes (see Fig. 3-4*b*). Each chromosome is composed of an original set of genes and a copied set of genes, so every chromosome is actually doubled.

While the chromosomes are forming within the nucleus, the nucleolus disappears. Fibers form on either side of the nucleus. The nuclear membrane disappears (see Fig. 3-4*c*). Some fibers from the ends of the cell (at the centrioles) become attached to each doubled chromosome (see Fig. 3-4*d*). The doubled chromosomes line up in the center of the cell and are then pulled apart by the contacting fibers. This action separates each original part of a chromosome from its copy (see Fig. 3-4*e*). A new cell membrane forms between these two sets of chromosomes, isolating one set of chromosomes in each of the two newly formed daughter cells. In a short time, the chromosomes thin

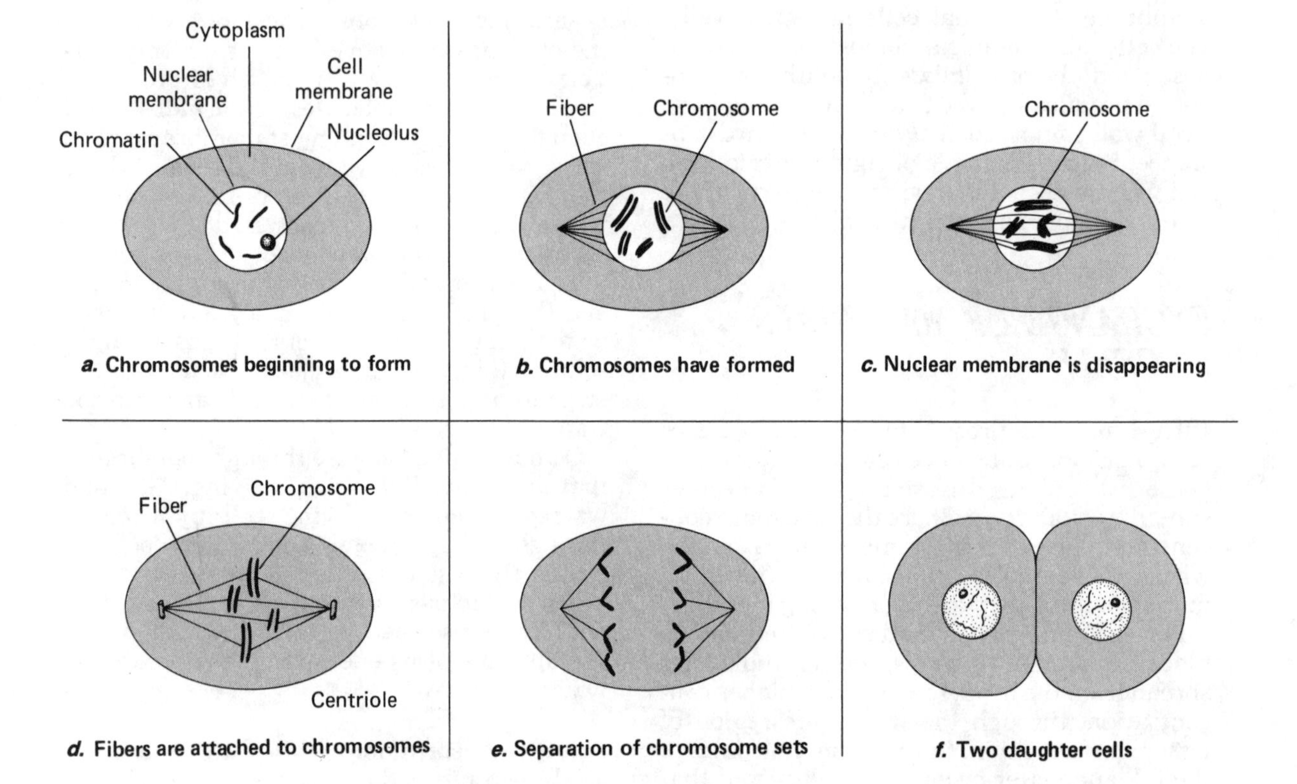

Fig. 3-4. Steps in mitosis.

out again into strands of chromatin, the nucleolus and nuclear membrane reappear, and two complete cells are produced (see Fig. 3-4*f*).

After the genes in each set of chromosomes are copied again, the cells may divide once more.

The number and the kind of chromosomes found in each kind of organism are definite. For example, an onion cell always has sixteen chromosomes, whereas a human cell has forty-six. Although some very different organisms may have the same number of chromosomes, the genes within their chromosomes are not the same.

CHAPTER REVIEW

Science Terms

The following list contains all of the boldfaced scientific terms found in this chapter and the page on which each appears.

cell division (p. 32)
cell membrane (p. 31)
cell wall (p. 31)
cells (p. 29)
chromosomes (p. 30)
cytoplasm (p. 30)
daughter cells (p. 32)
diffusion (p. 31)
DNA (p. 32)
genes (p. 30)
mitosis (p. 32)
nucleus (p. 30)
organelles (p. 30)
osmosis (p. 31)
protoplasm (p. 30)
RNA (p. 30)
selective absorption (p. 32)

Matching Questions

On the blank line, write the letter of the item in column B which is most closely related to the item in column A.

Column A

____ 1. diffusion of water through membrane
____ 2. sacs that store liquids in cell
____ 3. they control cell activity and heredity
____ 4. small bodies that contain chlorophyll
____ 5. outermost part of the nucleus
____ 6. cell parts that produce energy
____ 7. cell material outside of nucleus
____ 8. outermost living part of cells
____ 9. nucleic acid that makes up genes
____ 10. rigid layer of plant and algae cells

Column B

a. mitochondria
b. nuclear membrane
c. cell wall
d. osmosis
e. cytoplasm
f. DNA
g. chloroplasts
h. genes
i. selective absorption
j. cell membrane
k. vacuoles

Multiple-Choice Questions

On the blank line, write the letter preceding the word or expression that best completes the statement or answers the question.

1. All organisms are composed of one or more basic units of life called
a. muscles *b.* cells *c.* bones *d.* cell walls 1 ____

2. All the living material in a cell is called the
a. cytoplasm *b.* protoplasm *c.* ribosome *d.* chromosome 2 ____

3. The living cells of all plants and animals have
a. cell membrane and cytoplasm
b. centrioles and vacuoles
c. centrioles and ribosomes
d. nucleus and chloroplasts 3 ____

4. Which three parts of a cell are alive?
a. cell membrane, cytoplasm, nucleus
b. cell wall, chlorophyll, nucleus
c. vacuole, cell wall, cell membrane
d. cell membrane, vacuole, chlorophyll 4 ____

5. A living thing that consists of only one cell is the
a. ameba *b.* dog *c.* human *d.* lettuce plant 5 ____

6. A cell structure found in plant cells, but not in animal cells, is the
a. nucleus *b.* cytoplasm *c.* cell wall *d.* cell membrane 6 ____

7. All the activities of a cell are controlled by its
a. ribosomes *b.* vacuoles *c.* nucleus *d.* cytoplasm 7 ____

8. Chromosomes are composed of
a. nuclei *b.* genes *c.* chloroplasts *d.* nucleoli 8 ____

9. A part of the cell that sends messages from the nucleus to the cytoplasm is the
a. ribosome *b.* mitochondrion *c.* nucleolus *d.* cell membrane 9 ____

10. Cytoplasm is best described as the type of mixture called a
a. colloid *b.* compound *c.* molecule *d.* transparent solution 10 ____

11. The part of plant cells that is naturally green is the
a. nucleus *b.* cytoplasm *c.* chloroplast *d.* chromatin 11 ____

12. Chromatin threads change to dense rods during the process of
a. osmosis *b.* diffusion *c.* mitosis *d.* absorption 12 ____

13. Substances pass into and out of a living animal cell through the
a. cell wall *b.* mucous membrane *c.* cell membrane *d.* epidermis 13 ____

14. The passage of water through the cell membrane is called
a. photosynthesis *b.* transpiration *c.* digestion *d.* osmosis 14 ____

15. Which part of a cell is primarily composed of DNA?
a. genes *b.* ribosomes *c.* nuclear membranes *d.* vacuoles 15 ____

16. Mitochondria are essential in providing a cell with
a. oxygen *b.* energy *c.* food *d.* water 16 ____

17. During mitosis there is always a separation of the copied
a. centrosomes *b.* chlorophyll *c.* chromosomes *d.* colloids 17 ____

18. Substances pass from the cytoplasm to the nucleus of a cell through the
a. centrioles *b.* vacuole *c.* cell membrane *d.* nuclear membrane 18 ____

19. The odor of perfume can be detected 6 feet away as a result of
a. osmosis *b.* diffusion *c.* selective absorption *d.* oxidation 19 ____

20. Which group of structures is found in the nucleus of a cell?
a. cell wall, vacuoles, chloroplasts *c.* mitochondria, cytoplasm, vacuoles
b. cytoplasm, cell wall, genes *d.* genes, nucleolus, chromosomes 20 ____

Modified True-False Questions

In some of the following statements, the italicized term makes the statement incorrect. For each incorrect statement, write the term that must be substituted for the italicized term to make the statement correct. For each correct statement, write the word "true."

1. The constant movement of the *mitochondria* distributes materials around a cell. 1 ____________

2. In cells, little pockets filled with fluid are the *centrioles*. 2 ____________

3. Cells parts that help make enzymes and proteins are *ribosomes*. 3 ____________

4. Parts that are usually present in animal cells but not in plant cells are *chromosomes*. 4 ____________

5. Mitosis is a series of steps during cell *division*. 5 ____________

6. Molecules of solids, liquids, and gases spread and mix by the process of *osmosis*. 6 ____________

7. A prune was dropped into water and allowed to stand. After 24 hours, the prune became swollen with water as a result of the process of *secretion*. 7 ____________

8. Selective *absorption* means that a cell can let in some substances and keep others out. 8 ____________

9. Just before it divides, a cell has 14 chromosomes. After mitosis, each new cell has *seven* chromosomes. 9 ____________

10. A hard substance associated with plant cell walls is *cellulose*. 10 ____________

Testing Your Knowledge

1. Explain why the guard cells of a lettuce leaf are green whereas other epidermal cells are almost colorless.

____

____

2. Give three ways in which plant and animal cells are alike.

____

____

3. Give three ways in which plant and animal cells are different.

4. In what ways are osmosis and diffusion alike?

5. In what ways are they different?

6. When a stalk of celery is placed in a glass of salt water, the celery wilts. Explain this change.

7. When wilted celery is placed in fresh water, it regains its crispness. Explain this change.

8. Explain the difference between the two terms in each of the following pairs:
a. cell wall and cell membrane

b. cytoplasm and protoplasm

c. RNA and DNA

d. chromatin and chromosomes

9. Before mitosis can occur, the set of genes in the nucleus' chromatin has to be duplicated. Explain why this is necessary for cell division.

10. Explain the differences between a molecule of DNA and a molecule of RNA in terms of their chemical composition.

4

The Composition of Living Things: Tissues and Organs

LABORATORY INVESTIGATION

EXAMINING ONE-CELLED ORGANISMS

A. Make a slide from a drop of culture of mixed *protozoans* (one-celled animallike organisms) or from a drop of stagnant pond water. Examine the culture under the low power of your microscope.

1. Describe one or two of the organisms that you see.

____

____

2. What evidence can you observe that shows these organisms are carrying out life activities?

____

____

3. How many cells make up each organism?

____

4. Sketch two or three of the organisms that you observe. Label the parts, or organelles, of their cells that you recognize.

Use Fig. 4-1 to help you identify a few common protozoans.

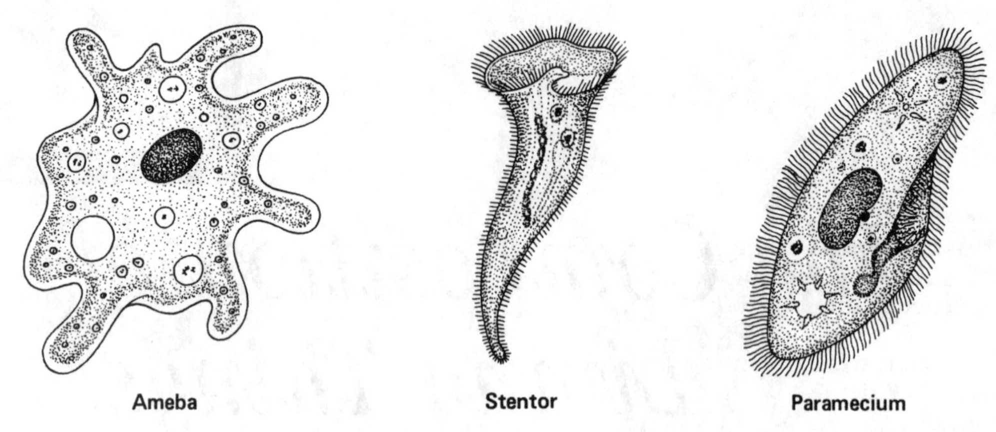

Fig. 4-1. Some common protozoans.

One-Celled and Many-Celled Organisms

ONE-CELLED ORGANISMS

Some organisms, such as the ameba and others in the culture you examined, are composed of only one cell. These organisms are said to be **unicellular.** Living in water, protozoans like these are in close contact with the food, water, and oxygen they need. A one-celled protozoan has to take in its food. Along with this food, the organism also takes in some water. Additional water enters the protozoan's body by osmosis through its cell membrane. Oxygen enters the cell by diffusion. The normal movement of cytoplasm carries the food, water, and oxygen throughout the cell. Waste materials are eliminated directly to the outside of the cell. Most one-celled organisms can survive only in a watery environment.

MANY-CELLED ORGANISMS

Organisms that are composed of many cells are said to be **multicellular.** In relatively large multicellular organisms, most of the cells are deep inside the body. These internal cells, like all others, require food, water, and oxygen. However, the internal cells cannot obtain these substances by themselves. They depend upon other cells to obtain and transport these necessities to them.

Consider the cells in your heart, for example. These cells need food, water, and oxygen to survive and to keep the heart functioning. But they cannot get these things directly from the environment. Instead, cells in the heart depend upon the blood transport system to transport these substances to them.

The cells in multicellular organisms are of many different types. Each type has its own special activity that it performs. Some, like red blood cells, carry oxygen; others, like nerve cells, carry impulses ("messages"). Each type of cell is said to be *specialized,* that is, it has one or more distinctive features that enable it to carry out its special activity.

The specialization of the cells in the body is similar to the division of labor in a factory, where different people perform different jobs. As a result of specialization, many-celled organisms can carry out more complex activities than one-celled organisms.

Tissues

An examination of different multicellular organisms reveals that no living thing is a simple mass of cells. Rather, the cells are arranged in groups. A group of similar cells that works together to perform a specific function is called a **tissue.** Several tissues working together make up an **organ.** Several organs working together make up an **organ system.** All the systems acting together make up the **organism.** In such a complex system, each structure depends upon the others to function properly.

In this chapter, we will consider only plants and animals in detail. We will discuss the other types of organisms in Chapter 5.

ANIMAL TISSUES

Epithelial Tissue. The cells of **epithelial** (or covering) **tissue** lie close together. This tissue covers the body and lines the internal organs. The work of epithelial tissue includes *protection* and *secretion.*

Epithelial tissue protects other cells and tissues. The cells of the epithelial tissue lining the skin and organs fit together like tiles. The close-fitting arrangement of these cells prevents foreign particles from reaching the cells that lie underneath them.

The windpipe (trachea) is lined with a layer of epithelial cells that have tiny vibrating hairs, called **cilia** (Fig. 4-2). As the cilia move, they push foreign particles out of the trachea, protecting the body from infections.

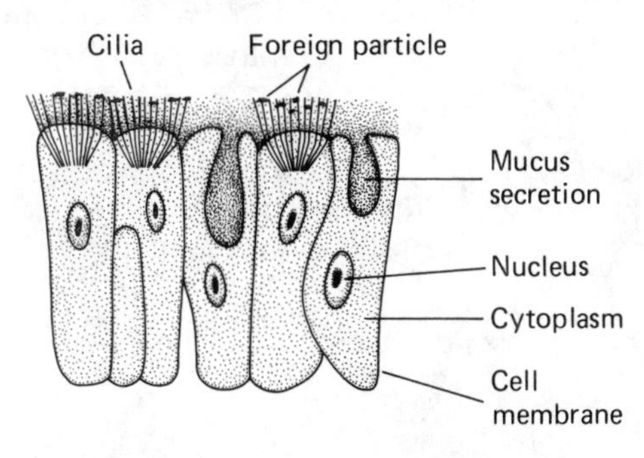

Fig. 4-2. Epithelial cells of the lining of the windpipe.

Epithelial tissue also secretes fluids. *Glands* are lined with epithelial cells that form little pockets. These cells manufacture fluids that collect in the pockets and then are transported to where they are needed in the body. Such fluids, like saliva, help an organism to function properly.

Connective Tissue. Bone and cartilage are made up of dense **connective tissue,** which is found in the skeleton. *Bone* supports and protects soft organs and provides levers to which many muscles are attached. The cells of bone tissue secrete salts of calcium and phosphorus. These mineral salts make up the hard material that lies between the bone cells. *Cartilage* is a similar, but more flexible type of tissue. Another type of fibrous connective tissue binds skin to the tissues below it. (See Fig. 4-3.)

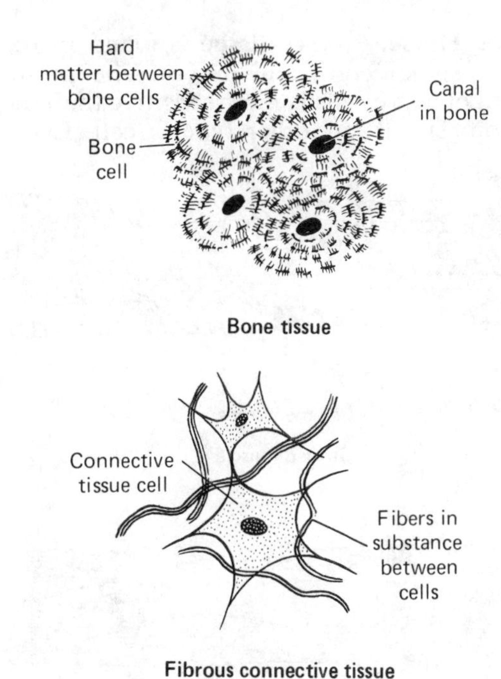

Fig. 4-3. Bone tissue and fibrous connective tissue.

Muscle Tissue. The cells of most **muscle tissue** are long. As muscle cells contract and relax, they cause some part of the body to move. Fig. 4-4 shows the three varieties of muscle tissue.

1. *Striped* (*Striated*) *Muscle.* Stripes are visible in the long cells of this type of muscle tissue. Striped muscles carry out acts that are controlled by the will. Examples of such acts, called *voluntary acts,* are turning the head and bending the fingers. Most striped muscles are attached to two or more bones of the skeleton.

2. *Smooth Muscle.* Stripes are not visible in the short cells of this type of muscle tissue. Smooth muscles carry out acts that are not controlled by the will. Examples of such acts, called *involuntary acts,* are movements of the stomach and the intestines. Smooth muscle is located in the walls of the food tube, the blood vessels, and some other organs.
3. *Heart (Cardiac) Muscle.* Although the cells of heart muscle tissue are striped, they carry out only the involuntary act of pumping blood. Heart muscle makes up the walls of the heart. When this muscle relaxes, the heart, which is hollow, fills with blood. When heart muscle contracts, it forces the blood out of the heart and into the blood vessels.

Nerve Tissue. **Nerve tissue** is found in the brain, spinal cord, and nerves. The protoplasm of nerve cells is more sensitive than the protoplasm of any other type of cell. Every nerve cell has several long, threadlike branches (Fig. 4-5). The sensitive protoplasm within these branches enables nerve cells to carry "messages," or *impulses,* to all parts of the body. Nerve cells are arranged so that messages travel only in one direction along a nerve.

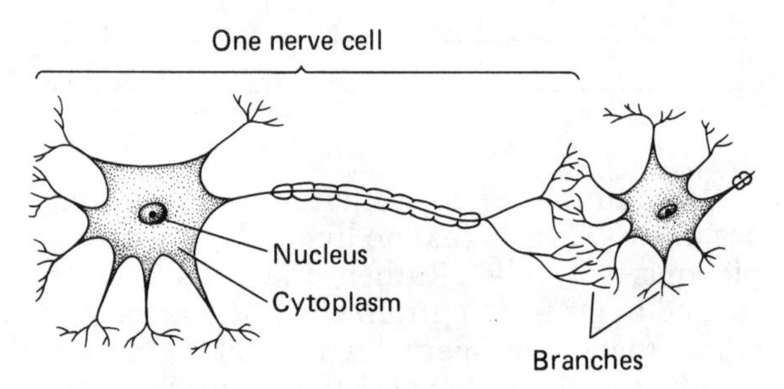

Fig. 4-5. Cells of nerve tissue.

Blood Tissue. There are three main types of cells in **blood tissue**—*red blood cells, white blood cells,* and *platelets* (Fig. 4-6). These cells are carried in a fluid material, the *plasma.*

Blood is present in all the blood vessels. *Hemoglobin,* which is a compound of iron and a protein, is found in red blood cells. The hemoglobin enables these cells to carry out their special job of absorbing and transporting ox-

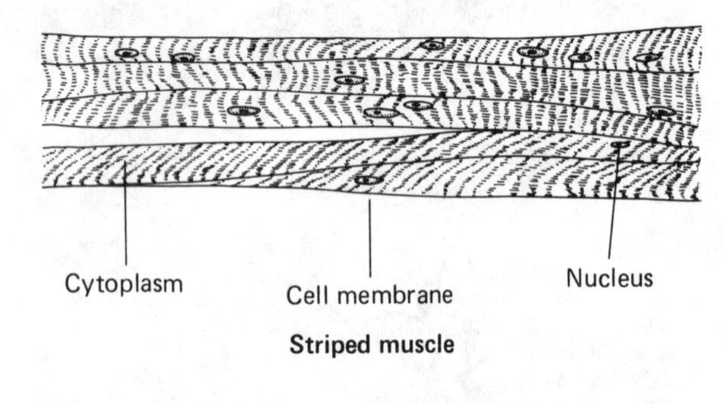

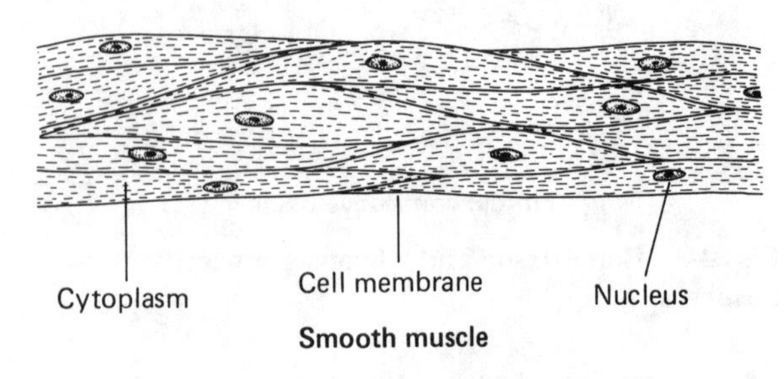

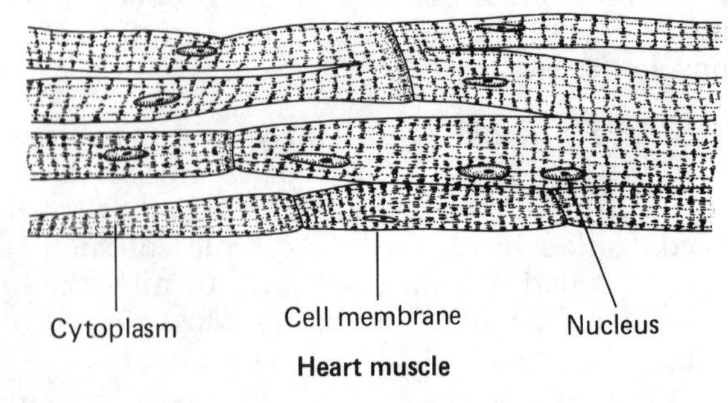

Fig. 4-4. The three varieties of muscle tissue.

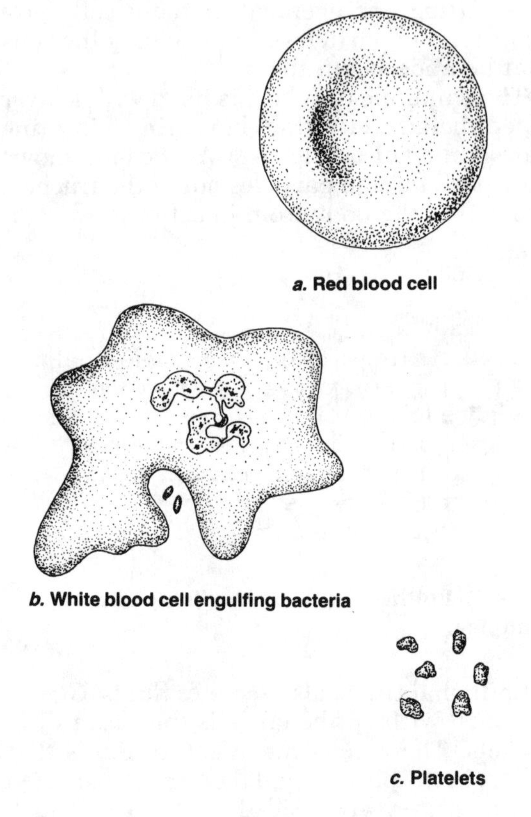

Fig. 4-6. Cells of blood tissue.

ygen to all parts of the body. White blood cells are able to destroy foreign particles that cause infections by taking in the particles and decomposing them. Platelets aid in blood clotting. The plasma, being a liquid, can carry dissolved foods and wastes in addition to carrying the blood cells. (You will learn more about blood tissue in Ch. 16.)

PLANT TISSUES

The cells of plant tissues, like those of animal tissues, vary in form according to the work they do. There are several types of plant tissue.

Epidermal Tissue. **Epidermal tissue** makes up the outermost covering of roots, stems, and leaves. The cells of the *epidermis* lie close together and interlock like pieces of a jigsaw puzzle. Epidermal tissue of plants, like epithelial tissue of animals, protects other tissues that lie below it. (Fig. 3-1, on page 29, shows the epidermal tissue of the lettuce leaf you observed in that laboratory investigation.)

Supporting Tissue. Although plants lack bone, most plants grow upright. Such plants possess **supporting tissue,** which is composed of long cells that have thick, hard, strong cell walls. Supporting tissue is present inside roots, stems, and leaves.

Other Plant Tissues. Other tissues present in plants include food-making tissue, conducting tissue, storage tissue, and growing tissue. These are discussed more fully in Chapter 13.

Organs, Organ Systems, Organisms

ORGANS

A group of tissues that acts together in carrying out a particular job is called an *organ.* Some examples of animal organs are the small intestine, lungs, and heart. Plant organs include the leaf, flower, and root.

The small intestine, for example, is composed of several tissues that, working together, aid in digesting and absorbing food. The muscle tissue in the wall of the small intestine pushes food along the food tube and mixes the food with juices secreted by the epithelial tissue of the intestinal glands. Nerve tissue controls the contraction of the muscles of the small intestine and the secretions of some of its glands. Other tissues also function in the small intestine.

ORGAN SYSTEMS

A group of organs that acts together in performing a special life activity is called an *organ system* (or *system*). See Table 4-1 to review the major parts and jobs of the organ systems of humans and other complex animals.

ORGANISMS

Every living thing, regardless of the number of cells comprising it, is an *organism.* All organisms share the following: they are highly organized, use energy, have a limited life span, and grow and respond to environmental changes.

ORGANIZATION OF MATTER

Matter in the universe is organized, that is, arranged in some systematic, orderly way. Atoms are organized into elements. Elements are organized into the compounds and mixtures that make up living and nonliving matter. Compared to the organization of living matter, the organization of nonliving matter is simple. The organization of living matter is very complicated. Elements, compounds, and mixtures make up the protoplasm of cells. Cells are organized into tissues; tissues into organs; organs into systems; and systems into organisms.

TABLE 4-1. ORGAN SYSTEMS

Organ System	*Major Organs*	*Major Tissues*	*Major Jobs*
Skeletal-muscular	Arms, legs, skull, ribs, spinal column	Bone, blood, striped muscle, nerve	Motion and support
Digestive	Mouth, stomach, intestine, glands	Epithelial, blood, smooth muscle, nerve	Breaking down food to simpler form
Circulatory	Heart, arteries, veins, capillaries	Blood, heart muscle, smooth muscle, nerve, epithelial	Transport of materials
Respiratory	Lungs, diaphragm rib cage	Bone, blood, epithelial, striped muscle, nerve	Exchange of oxygen and carbon dioxide
Excretory	Kidneys, urinary bladder, ureter	Epithelial, blood, nerve, smooth muscle	Excretion of liquid wastes
Nervous	Brain, spinal cord, nerves	Nerve, blood	Sensitivity and carrying impulses
Reproductive	Ovaries, oviducts, testes, sperm ducts	Epithelial, blood nerve	Production of eggs and sperm
Endocrine	Ductless glands	Pituitary, adrenal cortex, thyroid, pancreas	Proper functioning of glands and secretion of hormones

CHAPTER REVIEW

Science Terms

The following list contains all of the boldfaced scientific terms found in this chapter and the page on which each appears.

blood tissue (p. 40)
cilia (p. 39)
connective tissue (p. 39)
epidermal tissue (p. 41)
epithelial tissue (p. 39)
multicellular (p. 38)
muscle tissue (p. 39)
nerve tissue (p. 40)
organ (p. 39)
organ system (p. 39)
organism (p. 39)
supporting tissue (p. 41)
tissue (p. 39)
unicellular (p. 38)

Matching Questions

On the blank line, write the letter of the item in column B which is most closely related to the item in column A.

	Column A	Column B
____	1. group of tissues working together	*a.* muscle tissue
____	2. tiny vibrating hairs in windpipe	*b.* connective tissue
____	3. meaning composed of many cells	*c.* organ
____	4. group of organs working together	*d.* platelets
____	5. covers the body and lines organs	*e.* multicellular
____	6. makes up bone and cartilage	*f.* epidermal tissue
____	7. enables movement of body parts	*g.* organ system
____	8. fluid material in the blood	*h.* unicellular
____	9. outer covering of roots, stems, leaves	*i.* cilia
____	10. aid in blood-clotting process	*j.* epithelial tissue
		k. plasma

Multiple-Choice Questions

On the blank line, write the letter preceding the word or expression that best completes the statement or answers the question.

1. The organ system to which the mouth belongs is the
a. digestive *b.* respiratory *c.* circulatory *d.* muscular — 1 ____

2. A tissue is made up of a group of
a. organs *b.* organisms *c.* cells *d.* systems — 2 ____

3. Protection and secretion are two activities carried out by
a. cilia *b.* epithelial tissue *c.* bone tissue *d.* blood plasma — 3 ____

4. The secretion of body fluids is carried out by
a. bones *b.* blood *c.* nerves *d.* glands — 4 ____

5. The type of tissue that helps support the body is
a. epithelium *b.* bone *c.* muscle *d.* nerve — 5 ____

6. Cells that can contract are part of the tissue called
a. nerve *b.* epithelial *c.* muscle *d.* bone — 6 ____

7. The type of muscle tissue in the wall of the small intestine is
a. smooth *b.* striped *c.* voluntary *d.* cardiac — 7 ____

8. Heart muscle tissue
a. has no stripes and is controlled by the will
b. has stripes and is controlled by the will
c. has stripes and is not controlled by the will
d. has no stripes and is not controlled by the will — 8 ____

9. The shape of a cell of nerve tissue is suited to its ability to
a. carry messages *b.* contract *c.* move *d.* change in weight 9 ____

10. Which tissue has a fluid between its cells?
a. supporting tissue of plants *b.* epidermal *c.* bone *d.* blood 10 ____

11. Epidermal tissue of a lettuce leaf has a job similar to which animal tissue?
a. nerve *b.* epithelial *c.* bone *d.* blood 11 ____

12. Trees are able to grow to great heights because they contain the tissue called
a. storage *b.* food-making *c.* supporting *d.* epidermal 12 ____

13. Cells are to tissues as tissues are to
a. systems *b.* organisms *c.* organs *d.* compounds 13 ____

14. A unicellular organism is composed of
a. many tissues *b.* one cell *c.* many cells *d.* no cells 14 ____

15. Lungs belong to the body system called the
a. digestive *b.* circulatory *c.* respiratory *d.* excretory 15 ____

Modified True-False Questions

In some of the following statements, the italicized term makes the statement incorrect. For each incorrect statement, write the term that must be substituted for the italicized term to make the statement correct. For each correct statement, write the word "true."

1. Body cells that bear *cilia* are found in the windpipe. 1 ____________

2. The division of labor in an auto factory is similar to the *reproduction* of plant and animal cells. 2 ____________

3. Voluntary acts are carried out by *smooth* muscle tissue. 3 ____________

4. The iron compound that can absorb oxygen in the blood is *iron oxide.* 4 ____________

5. Organisms such as the *ameba* have no tissues. 5 ____________

6. Cells deep in the body, such as bone cells, get their food and oxygen from the *blood.* 6 ____________

7. Cells that line the organs make up *connective* tissue. 7 ____________

8. *White* blood cells are specialized for carrying oxygen. 8 ____________

9. Geranium and ivy plants have *supporting* tissue. 9 ____________

10. The lettuce plant is a *unicellular* organism. 10 ____________

Testing Your Knowledge

1. Explain the difference between the two terms in each of the following pairs:
a. multicellular and unicellular

____

b. cell and tissue

___

c. organ and system

___

d. organ and organism

___

e. voluntary acts and involuntary acts

___

f. smooth muscle and striped muscle

___

2. Give a brief explanation for each of the following true statements:
a. The ameba, a one-celled organism, can live although it has no tissues or organs.

___

b. The frog, a many-celled organism, cannot live if its blood tissue is drained off.

___

3. Explain how epithelial tissue is important for protecting the organs of the body.

___

4. Why is hard bone tissue important for the body (that is, what is its role in the body)?

___

5. Describe the special jobs performed by red blood cells and white blood cells. How does the plasma help them perform their jobs?

___

___

6. Complete the table of organs and organ systems.

Organ System	*Major Organs*	*Major Jobs*
Skeletal-muscular	*a.* ______	*b.* ______
c. ______	Mouth, stomach, intestines, glands	*d.* ______
e. ______	Heart, arteries, veins, capillaries	*f.* ______
g. ______	*h.* ______	Exchange of oxygen and carbon dioxide
i. ______	Kidneys, urinary bladder, ureter	*j.* ______
Nervous	*k.* ______	Carrying impulses
l. ______	Ovaries, oviducts, testes, sperm ducts	Production of sperm and eggs

Stem Cell Research: Should It Be Banned?

Stem cell research is the field of science that learns how an organism develops from a single cell. The promise that stem cell research holds has enabled scientists to label this branch of research as regenerative or reparative medicine.

Stem cells and its research can be one of the most exciting areas of biology today, but it raises many scientific and ethical questions as it generates new discoveries.

Stem cells have two characteristics that make them stand out in embryonic cell research. First, they exist for long periods of time as they undergo a series of cell divisions and second, they become cells with special functions, such as the beating cells of heart muscle.

Stem cells are important for living organisms for many reasons. In the early embryo (3–5-day stage), the inner cells of the newly formed structure (called the blastula) give rise to hundreds of highly specialized cells. These cells would normally make up the developing fetus. These cells give rise to the bone marrow, muscle, brain cells, and so on. Could these cells be used to replace cells lost through normal wear and tear of the body, injury, or disease? Could these cells replace those cells that cause Parkinson's disease, diabetes, and heart disease?

Stem cells have been studied for more than 20 years in animals, usually mouse embryos. But it is only since 1998 that human embryonic stem cells have been studied. The unique properties of stem cells make them ideal for research; they are capable of dividing and renewing themselves for long periods; they are unspecialized and they can give rise to specialized cell types. Scientists are working on the mechanisms that allow these cells to become specialized, what signals specialization, and how long can they be "usable." Continued research will find these answers, and the possible curing of disease and genetic mutations could be the result.

The ethical dilemma with human stem cell research is the origin of the cell. Many scientists procured the initial cells through aborted fetuses. This upset many people, both from a religious and ethical viewpoint. Those who were opposed to abortion (i.e., the National Right to Life Committee) suggested that the use of fetuses for medical research might encourage women to have abortions. As the potential for stem cell research increased, scientists started to retrieve stem cells from other methods, such as *in vitro* fertilization. If the fertilization was successful, the scientists would be able to harvest these cells from the petri dishes in which they were developed.

Regardless, such work has been limited by a U.S. government ban providing federal funds for such research. This ban has been supported by President Bush and currently remains in effect.

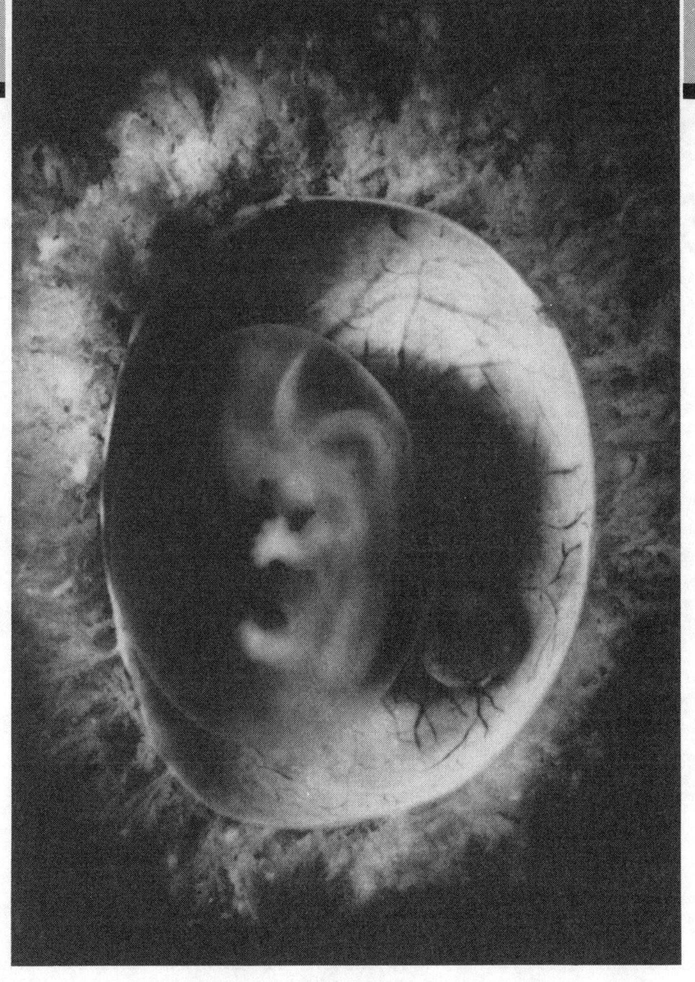

2. What is in vitro *fertilization?*

____

____

____

____

3. On a separate sheet of paper, write a short essay about your feelings on stem cell research and its use. Do you agree or disagree that it is ethical, and why?

1. Why are stem cells important?

____

____

____

____

5

The Great Variety of Living Things

LABORATORY INVESTIGATION

BECOMING ACQUAINTED WITH THE FUNGUS KINGDOM

Your teacher will display a variety of organisms that are all members of the fungus kingdom. Examine the specimens presented and describe some of their characteristics. Record your observations in the space provided.

I. Yeast

a. Where it grows

____

b. What its color is

____

c. Shape or texture

____

II. Mold

a. Where it grows

____

b. What its color is

____

c. Shape or texture

____

III. Mushroom

a. Where it grows

____

b. What its color is

____

c. Shape or texture

____

IV. Lichen

a. Where it grows

____

b. What its color is

____

c. Shape or texture

____

Classification Schemes

In the laboratory investigation, you examined a variety of different types of fungi. In your daily life, you see many different kinds of organisms. Each distinct kind of organism, such as a dog, fly, rose, lobster, mushroom, and maple, is called a **species.** A species is a group of like organisms that can interbreed and produce fertile offspring. More than 1,500,000 species of organisms have been discovered and named by scientists. It is thought that there may be several million more species still unknown to science.

During the millions of years that living things have existed on earth, many species have become extinct. These species died out because they were unable to adapt to changes in their environment. Many other species continued to live because they were able to adapt. Members of these other species managed to carry out their individual life activities and to reproduce their kind before they died. The organisms we see today are their descendants.

In this chapter and the next three chapters, you will study the main groups of living things. You will examine wome of these organisms in your laboratory investigations. Because there are so many organisms, the task of studying them could be very confusing, unless they are classified in some way. In order to understand this huge number of diverse organisms, scientists have classified them according to an established system.

THE FIVE-KINGDOM SYSTEM

Many classification schemes have been used in the past. The scheme that is now used to name and classify organisms is based on the one devised by Swedish naturalist Carolus Linnaeus in the eighteenth century. Linnaeus classified all living things as either plants or animals.

In our modern scheme, we place organisms that are similar in their cell and body structures, and in the way they develop, into the same group. Organisms differ in size, shape, behavior, and environments in which they live. Only a study of their body structures, development, and other biological traits can reveal whether they are related and should be classified together. For example, the whale and the mouse, according to their body structures and development, are more closely related than are the whale and the giant squid.

In this example, size and environment are not as important as some other characteristics used for classifying the animals.

In the scientific classification scheme, the whale and the mouse are both classified as animals that have an internal skeleton with a backbone. Except for size, the structure of their skeletons is very much alike. In addition, since whales and mice share other important characteristics, such as feeding their young with milk, they are classified together in a smaller group, the mammals. The fact that they both feed their young with milk shows that they are more closely related than some other animals that have internal skeletons but do not nurse their young.

Although both the whale and the giant squid are large ocean-dwelling animals, their body structures and the way in which they develop are extremely different. The whale, which possesses a bony skeleton within its body, is classified in one large division of the animal kingdom; the squid, which lacks a bony skeleton inside its body, is classified in another large division.

Some organisms are so different from what we consider to be plants or animals that they are classified in totally separate groups. The euglena, for example, which is unicellular, has some plantlike cell structures and some animallike cell structures. As a result, euglena is classified as a type of alga in the protist kingdom, which consists of mostly one-celled animallike and plantlike organisms. For a long time, scientists classified all organisms into just the animal, plant, and protist kingdoms. See Table 5-1 below, to review the different classification schemes.

Most biologists currently use a **five-kingdom system** for classifying all living things. A **kingdom** is the largest, major group into which similar organisms can be classified. The five kingdoms are: Monerans, Protists, Fungi, Plants, and Animals. The largest divisions of the kingdoms are called **phylums** (or *phyla*). Organisms within a phylum share at least one important characteristic. Each phylum is divided into smaller, more closely related groups, called **classes,** which are further divided into **orders.** Smaller subdivisions include the **family,** the **genus,** and finally the *species.* The scientific name of an organism is its genus and species name. For example, the scientific name for the human species is *Homo sapiens; Homo* is the genus and *sapiens* is the species name.

TABLE 5-1. DIFFERENT CLASSIFICATION SYSTEMS

	Number of Kingdoms			
Kingdom	*Two*	*Three*	*Four*	*Five*
Animal	Animals, protozoa	All multicellular animals	All multicellular animals	All multicellular animals
Plant	Plants, algae, fungi, slime molds	Plants, algae, fungi, slime molds	All multicellular plants and all fungi	All multicellular plants
Protist		Unicellular organisms and colonial protozoa	Most unicellular organisms (algae and protozoans)	Most unicellular organisms (algae and protozoans)
Moneran			Blue-green bacteria, bacteria, and other microorganisms that lack nuclei	Blue-green bacteria, bacteria, and other microorganisms that lack nuclei
Fungi				All fungi (unicellular and multicellular)

Simple Organisms

VIRUSES AND MONERANS

Viruses. A **virus** is a tiny biological particle that contains a core of genes (DNA or RNA) encased in a protein coat. Viruses have a variety of shapes; some have a "head" and a "tail" region (see Fig. 5-1). Although viruses are composed of the same substances found in living cells, they are not living organisms. Viruses have no nucleus, cytoplasm, cell membrane, or other organelles. Furthermore, viruses cannot reproduce unless they are inside an organism's cell, using the cell parts to make copies of the viral genes. Viruses actually inject their genes into living cells. As a result, viruses are harmful because they take over the life activities of the cells that they infect.

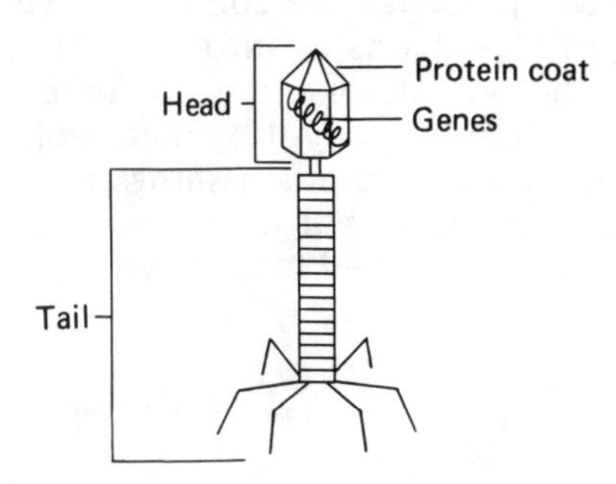

Fig. 5-1. Model of a virus.

Monerans. The **moneran** kingdom consists of simple one-celled microscopic organisms that have no nuclei or any other membrane-bound organelles. Some monerans, known as *blue-green bacteria,* make their own food by photosynthesis. However, most monerans—commonly called **bacteria** (singular, *bacterium*)—absorb nutrients from their environment. There are about 5000 species of monerans.

Bacteria. All bacteria are one-celled. High-powered microscopes are needed to see them. Bacteria are often classified by their shape. There are three main types of bacteria: ball-shaped bacteria, called *cocci;* rod-shaped bacteria, called *bacilli;* and corkscrew-shaped bacteria, called *spirilla* (see Fig. 5-2).

Each bacterium consists of a small mass of protoplasm surrounded by a cell membrane and cell wall. No nucleus is present, but there are genes in the protoplasm. Bacteria can live nearly any place where there is some moisture, food, warmth, and darkness. However, except for the blue-green bacteria, which have chlorophyll and carry out photosynthesis, bacteria cannot manufacture their own food. Some species get their food from dead organic matter. Some get their food from the tissues of living organisms. Others get their food by living in partnership with a green plant.

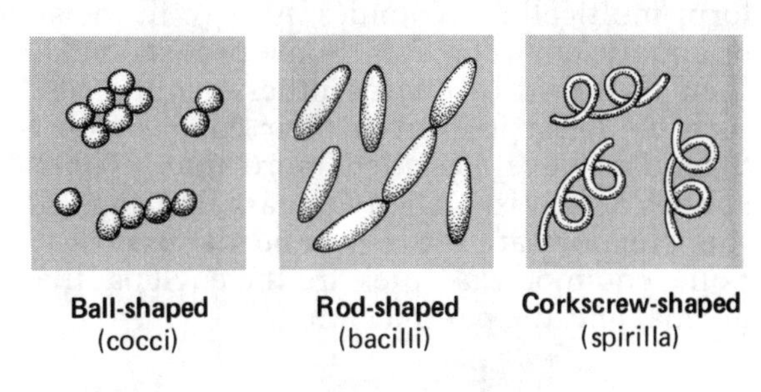

Fig. 5-2. Three types of bacteria.

Bacteria reproduce by *binary fission.* In this process, one organism splits into two by dividing up the contents of the parent cell. When some kinds of bacteria are exposed to unfavorable conditions, their cytoplasm loses water, contracts, and becomes round. A tough coating then forms around the mass of cytoplasm. The organism can remain in this inactive phase for many years, until favorable conditions are restored. Then the coating breaks off and the bacterium becomes active again, resuming its normal life activities.

Importance of Monerans. There are both helpful and harmful species of monerans. Many bacteria are essential to the existence of humans and other organisms. Bacteria of decay break down organic matter, thus restoring minerals to the soil and keeping it fertile. Some bacteria live in the gut of other organisms, aiding in the digestion and absorption of nutrients. People use other strains of bacteria to make cheeses, yogurt, linen, and antibiotic medicines, among other products.

Some bacteria are very troublesome to humans. Unless food is preserved by refrigeration or some other method, bacteria of decay can spoil our foods. Other bacteria cause serious diseases such as diphtheria, tuberculosis, typhus, pneumonia, and tetanus.

PROTISTS

The **protist** kingdom is composed of simple organisms that have nuclei and membrane-bound organelles, but lack special tissues. Most protists are unicellular, although some form multicellular colonies. All live in moist or aquatic environments. Some protists make their own food by photosynthesis; others absorb or ingest food from their environment. Scientists have classified more than 50,000 species as protists. The two main divisions of this kingdom are the *algae* and the *protozoa*. Some common examples are the euglena, the ameba, and the paramecium.

ALGAE

Characteristics. **Algae** are plantlike organisms that live in water or in very moist places. Algae exist either as microscopic single cells or as colonies of various sizes. The colonies vary in shape, from small clusters of cells to large flat sheets of cells. All contain chlorophyll and manufacture their own food by photosynthesis. Most algae have cell walls, similar to those of plant cells. Examples are spirogyra, chlorella, and pleurococcus (Fig. 5-3). Spirogyra is usually found as a stringlike colony, while chlorella cells live separately, and pleurococcus cells stick together in colonies of just a few cells each.

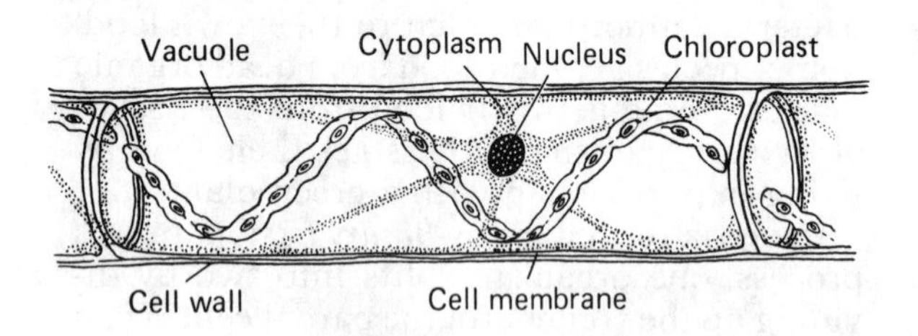

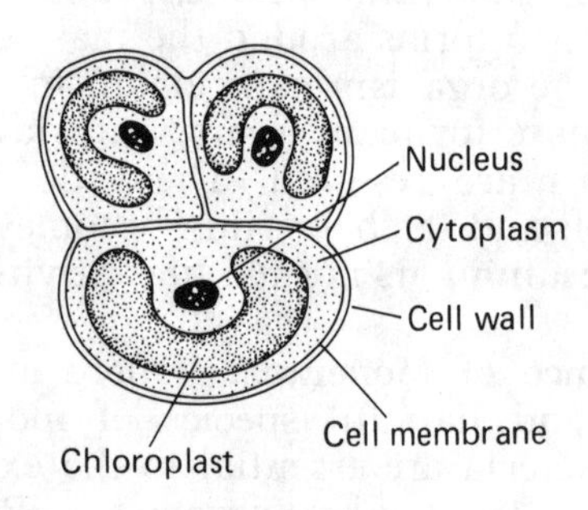

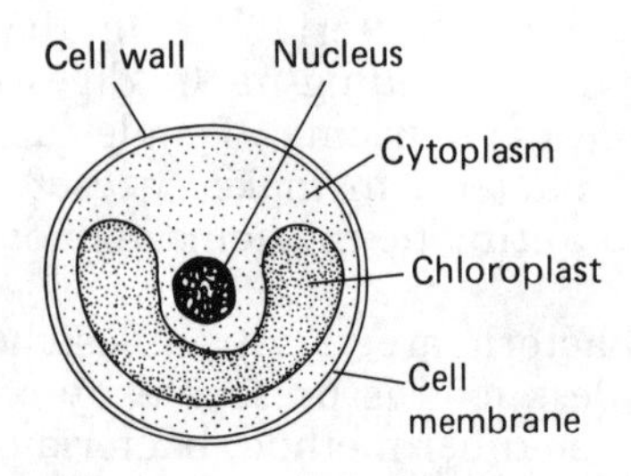

Fig. 5-3. Three examples of algae.

Euglena. The euglena (Fig. 5-4) is an unusual type of freshwater alga, because it has animallike as well as plantlike traits. It is composed of a single cigar-shaped cell that has a whiplike hair projecting from its front end. As the hair whips around, the euglena moves (an animallike trait). The euglena contains chloroplasts and so can make its own food by photosynthesis (a plantlike trait). A nucleus, cytoplasm, cell membrane, and contractile vacuole are present. The contractile vacuole discharges liquid wastes into a channel that leads to the outside of the cell. The euglena can detect changes in light by means of a reddish *eyespot.* This organism shows a positive response to light, in that it moves toward the source of light detected by the eyespot.

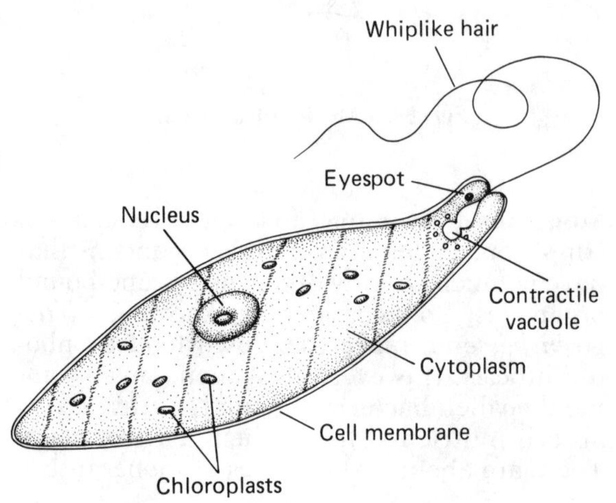

Fig. 5-4. Euglena—a type of alga.

Importance of Algae. Since all algae make their own food, they are important as a direct source of food for water-dwelling animals and, indirectly, important as a food supply for people. Some algae, like seaweeds, are eaten by people, and are a good source of iodine, protein, mineral salts, and vitamins. Other types of algae are used in making food (like ice cream) and cosmetic products (like shampoo), and as animal feed. Agar, which comes from seaweed, is a jelly used for growing bacteria in laboratories. However, there are some harmful algae, such as the species that causes "red tide." This is an overgrowth of toxic (poisonous) algae, referred to as an *algal bloom*.

PROTOZOANS

Characteristics. **Protozoans** are tiny one-celled animallike organisms. Like the other protists, they live in moist or aquatic environments. Most protozoans reproduce by binary fission. Protozoans do not have a cell wall.

Ameba. The ameba (Fig. 5-5) is the simplest protozoan known. Along with certain bacteria, the ameba most probably resembles the first organisms that existed on earth billions of years ago. It inhabits ponds and streams, living on the surface of underwater plants. The ameba consists of a nucleus, a cell membrane, and cytoplasm containing several food vacuoles and one contractile vacuole (Fig. 5-5*a*). As the ameba moves, it constantly changes its shape. The movement of an ameba resembles crawling—the ameba pushes out part of its cell membrane, into which its cytoplasm flows. This "false foot" that forms is called a *pseudopod.*

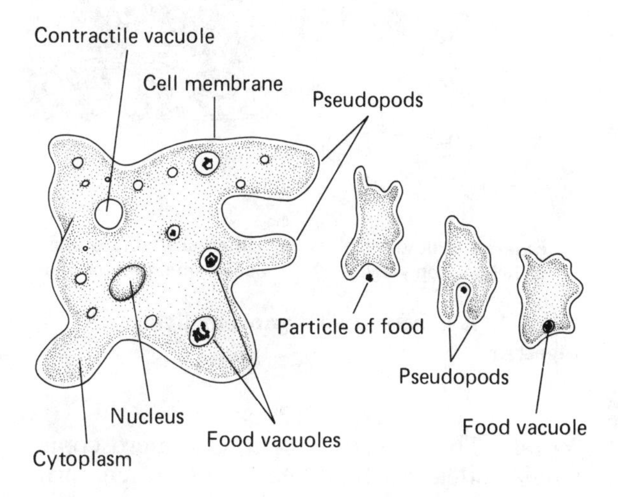

Fig. 5-5. Ameba—a simple protozoan.

Amebas eat tiny food particles by engulfing them, that is, by surrounding them with their pseudopods (Fig. 5-5*b*). Upon being eaten, the food particles, along with a drop of water and part of the cell membrane, become a *food vacuole.* Food is digested inside the food vacuole. As the cytoplasm streams around the cell, food vacuoles are circulated. Digested food diffuses out of the food vacuoles into the cytoplasm. Undigested food (solid waste) is eliminated from a food vacuole by a process that is the opposite of engulfing.

Dissolved wastes and excess water gather in the *contractile vacuole.* When full, this vacuole contracts and discharges the liquid wastes to the outside of the ameba. Waste carbon dioxide leaves the ameba, and oxygen enters into it, by diffusion through the cell membrane.

The ameba is sensitive to stimuli in its environment and responds to them in its movements. It shows a negative response to intense heat or light and irritating substances, and a positive response to moderate light, food, and oxygen.

Paramecium. Unlike the ameba, the paramecium (Fig. 5-6) has a definite shape—it resembles a slipper. The paramecium has numerous specialized structures and is more complex than the ameba. The paramecium swims rapidly by means of *cilia,* tiny hairs that beat in a definite rhythm and move the organism through the water. Cilia also sweep food into its mouth. The food of the paramecium consists of bacteria and other organisms of similar small size. These organisms, like the paramecium, inhabit stagnant water, such as ponds.

Food that has entered the paramecium's oral groove is swept down the gullet to its base, where food vacuoles form. The food is digested and circulated in the same manner as in the ameba. Water diffuses into the cytoplasm, and oxygen and carbon dioxide are also exchanged by diffusion through the cell

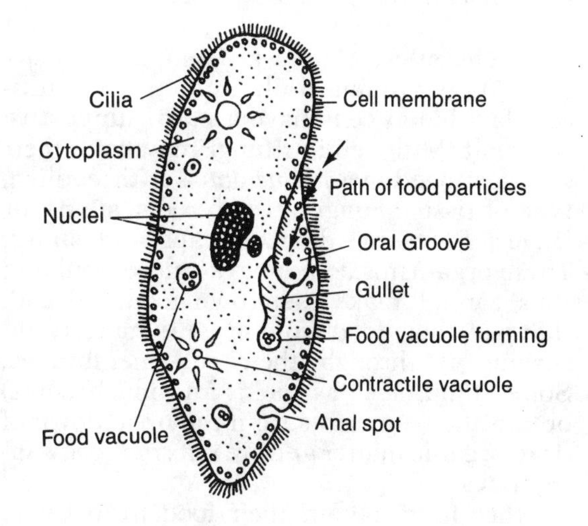

Fig. 5-6. Paramecium—a more complex protozoan.

membrane. Solid wastes are eliminated at the anal spot. Liquid wastes are discharged by means of two contractile vacuoles.

Like the ameba, the paramecium adjusts to changes in its environment by means of responses. The paramecium avoids irritating chemicals and bright light. It moves toward dimly lit, moderately warm areas and toward acids. Bacteria, which serve as food for the paramecium, are often found in such areas. Tiny rodlike structures lie under the cell membrane of the paramecium. When stimulated, these structures are discharged, forming long threads that the paramecium uses for defense or for anchoring itself while feeding.

Importance of Protozoans. Many larger organisms that live in water feed upon the microscopic protozoans. Some protozoans possess hard shells composed of chalk (limestone). When these organisms die and their shells accumulate, chalk deposits may be formed. Many other protozoans are harmful because they cause diseases in people and other animals. An example of such a disease is malaria, which affects millions of people worldwide.

FUNGI

The **fungus** kingdom consists of anywhere from 65,000 to 100,000 species of simple organisms. Fungi are found everywhere on earth: in forests, deserts, oceans, and even in your home. They live in and on organic and inorganic substances, growing best in warm, damp areas. While most fungi are microscopic, there are many larger species that look almost plantlike. Examples of fungi include the yeasts, molds, and mushrooms.

Characteristics. Fungi are simple in structure. Many are unicellular; others are multicellular. Fungi cells have a hard, supportive cell wall. Multicellular fungi may have their cells arranged in a *filamentous,* or threadlike, type of tissue. Fungi have no roots, stems, or leaves; but some do have caps and stalks. These organisms do not have chlorophyll and thus cannot make their own food. Instead, they get their food from organic matter, absorbing it through their cell membranes. Some fungi (such as molds and mushrooms) obtain their nutrients from the breakdown of dead organic matter and are referred to as **saprophytes.**

Other fungi obtain their food from living things and are referred to as **parasites.** The organism on which a parasite feeds is called the **host.** Parasites live either in the tissues or on the surface of their hosts, doing harm as they take nutrients from host cells. Some examples of parasitic fungi are the moldlike wheat rust and corn smut.

Some fungi actively prey on living things and are referred to as *predatory* fungi. For example, the oyster fungus obtains its nutrients by secreting a substance that slows down roundworms, enabling the fungus to capture the worm and absorb nutrients from it. In this type of interaction, the fungus is the *predator* and the roundworm is the *prey,* or victim.

Another unusual interaction of a fungus and another living thing is that of the **lichen** (Fig. 5-7). The organisms that make up a lichen are actually an alga (protist) and a fungus living in close association with one another. The fungus provides inorganic substances that the alga uses in photosynthesis. As a result of this association, the fungus obtains food from the alga. This process, whereby two different organisms live together and benefit each other, is known as *symbiosis.* The two organisms involved are referred to as *symbionts.*

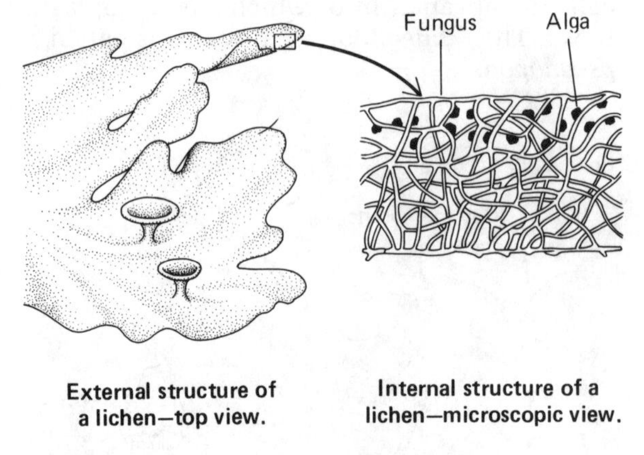

External structure of a lichen—top view.

Internal structure of a lichen—microscopic view.

Fig. 5-7. A lichen—two organisms living together.

Yeast. The yeast (Fig. 5-8*a*) is a microscopic, single-celled fungus. Yeast cells often form small colonies. Some species of yeast get their food from the tissues of living organisms. Many yeasts use sugar as their food. They obtain energy by oxidizing the sugar to carbon dioxide and alcohol. This process is called *fermentation.* Yeast usually reproduces by *budding,* a form of reproduction in which a small bulge develops on the parent cell, grows, and breaks off to become a new individual cell. Yeasts grow well in warm, moist places.

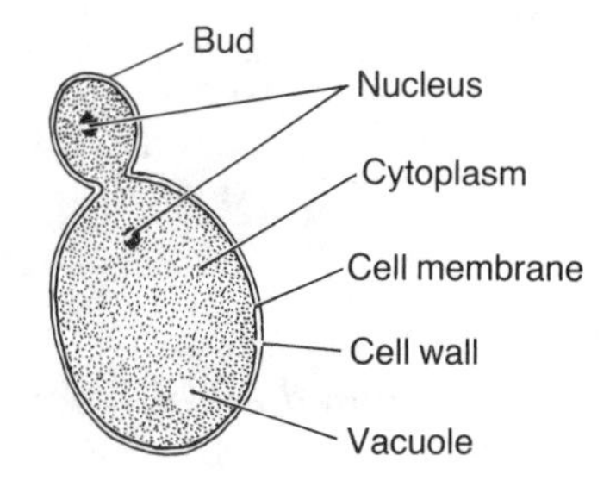

a. **Yeast (much enlarged)**

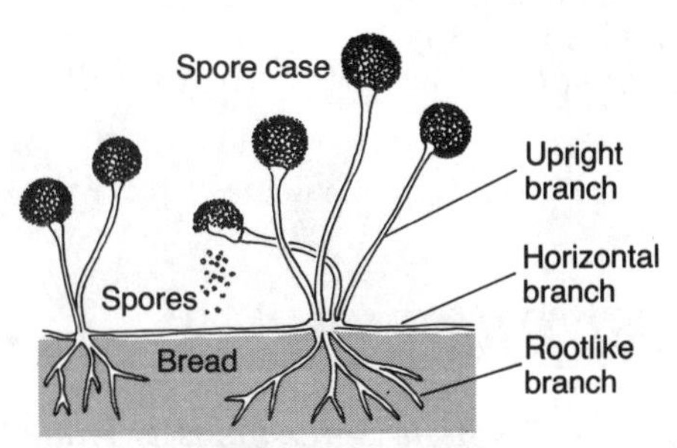

b. **Bread mold (enlarged)**

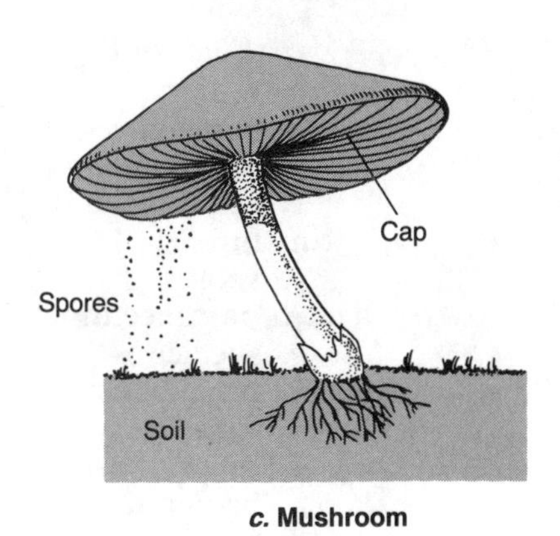

c. **Mushroom**

Fig. 5-8. Three examples of fungi.

Bread Mold. The bread mold (Fig. 5-8*b*) that grows on moist bread is usually visible to the unaided eye as a fuzzy patch. The mold's body actually consists of numerous branching threads that spread over the surface of the bread, grow down into it, and absorb nutrients from it. A mold usually reproduces asexually by producing **spores,** tiny reproductive cells protected by hard walls. These spores spread through the air and then land on a food source (such as bread that is left out on a counter), where they take hold and grow into new molds.

Mushroom. The mushroom (Fig. 5-8*c*), like the bread mold, is a multicellular fungus. Mushrooms live in moist soil that is rich in decaying material, from which they absorb their food. These fungi also grow on trees, absorbing nutrients from them. Mushrooms reproduce by forming spores, which they release from their caps. The main threadlike body of mushrooms is found underground.

Importance of Fungi. Because yeast produces carbon dioxide gas, yeast is used in baking to make dough rise. In addition, yeast produces alcohol, and so it is used in making alcoholic beverages. Yeast is also one of the organisms most widely used in studies on heredity.

Like many bacteria, some kinds of mold spoil our foods and cause diseases. Ringworm and athlete's foot are diseases caused by molds. Other kinds of molds produce valuable antibiotic medicines, like penicillin and cheeses such as Roquefort. Some mushrooms are valued as food; others are extremely poisonous to eat. Fungi that obtain their food from decaying or dead material (or organisms) help return simple nutrients to the soil. This enables new plants to absorb the nutrients and grow. These fungi are called decomposers.

CHAPTER REVIEW

Science Terms

The following list contains all of the boldfaced scientific terms found in this chapter and the page on which each appears.

algae (p. 52)
bacteria (p. 51)
class (p. 50)
family (p. 50)
five-kingdom system (p. 50)
fungus (p. 54)
genus (p. 50)
host (p. 54)
kingdom (p. 50)
lichen (p. 54)
moneran (p. 51)
order (p. 50)
parasites (p. 54)
phylum (p. 50)
protist (p. 52)
protozoans (p. 53)
saprophytes (p. 54)
species (p. 49)
spores (p. 55)
virus (p. 51)

Matching Questions

On the blank line, write the letter of the item in column B which is most closely related to the item in column A.

Column A

____ 1. largest division of a kingdom
____ 2. distinct group of organisms
____ 3. unusual type of algae
____ 4. slipper-shaped protozoan
____ 5. a "false foot" on an ameba
____ 6. most common monerans
____ 7. tiny biological particle
____ 8. reproduction by splitting in half
____ 9. tiny reproductive cells with walls
____ 10. group that includes mushrooms

Column B

a. euglena
b. pseudopod
c. budding
d. fungus
e. spores
f. paramecium
g. species
h. phylum
i. virus
j. binary fission
k. bacteria

Multiple-Choice Questions

On the blank line, write the letter preceding the word or expression that best completes the statement or answers the question.

1. Bacteria are classified in the kingdom called
a. viruses *b.* monerans *c.* fungi *d.* protists 1 ____

2. Which one-celled organism has a structure that can detect light?
a. bacteria *b.* paramecium *c.* ameba *d.* euglena 2 ____

3. Fermentation of sugar is the result of the life activities of
a. amebas *b.* yeast *c.* alcohol *d.* algae 3 ____

4. Rod-shaped bacteria are called
a. spirilla *b.* cocci *c.* bacilli *d.* yeast 4 ____

5. Plantlike organisms that live in water are called
a. spores *b.* algae *c.* protozoans *d.* fungi 5 ____

6. Which is an example of algae?
a. yeast *b.* bacterium *c.* spirogyra *d.* ameba 6 ____

7. Algae are most unlike protozoans in the way they
a. reproduce *b.* assimilate *c.* get oxygen *d.* get food 7 ____

8. The group that contains only fungi is
a. mushrooms, molds, yeast *c.* mushrooms, potatoes, rice
b. molds, potatoes, mushrooms *d.* molds, mushrooms, algae 8 ____

9. Useful substances that are produced by molds include
a. fungicides *b.* antibiotics *c.* spores *d.* peat moss 9 ____

10. Molds reproduce by forming hard-walled cells called
a. cocci *b.* spores *c.* seeds *d.* algae 10 ____

11. Which group contains only protozoans?
a. paramecium, ameba *c.* bacterium, ameba
b. ameba, virus *d.* virus, euglena 11 ____

12. Scientists classify organisms in order to
a. make jobs for scientists *c.* find useful plants and animals
b. find missing links among viruses *d.* make it easier to study organisms 12 ____

13. The smallest subdivision for classifying an organism is
a. phylum *b.* species *c.* class *d.* order 13 ____

14. The shape of an ameba is described as
a. changeable *b.* slipperlike *c.* starlike *d.* round 14 ____

15. The ameba eats food particles by
a. engulfing them *c.* catching with tentacles
b. stinging them *d.* catching with cilia 15 ____

16. The paramecium and ameba both use food vacuoles to
a. store food and excrete wastes *c.* store and digest food
b. oxidize food and excrete water *d.* excrete carbon dioxide 16 ____

Modified True-False Questions

In some of the following statements, the italicized word makes the statement incorrect. For each incorrect statement, write the term that must be substituted for the italicized term to make the statement correct. For each correct statement, write the word "true."

1. Since the beginning of life on earth, many species of organisms have become *extinct*. 1 ____________

2. Our modern scheme of classification is based on the one devised by *Darwin*. 2 ____________

3. Organisms are classified in the same group if they have very similar *body structures*. 3 ____________

4. A paramecium swims by means of its *pseudopods*. 4 ____________

5. The euglena can make its own food because it has several *nuclei*. 5 ____________

6. Bacteria reproduce by means of *binary fission*. 6 ____________

7. Many diseases are caused by bacteria and *protozoans*. 7 ____________

8. Ball-shaped bacteria are called *spirilla*. 8 ____________

9. Pleurococcus is a type of colonial *virus*. 9 ____________

10. A type of fungus that helps us make cheese is *yeast*. 10 ____________

11. Many foods are spoiled by molds and *bacteria* of decay. 11 ____________

12. Some foods and cosmetics are made with products from *algae*. 12 ____________

13. The simplest protozoan known is the *paramecium*. 13 ____________

14. Dissolved wastes are discharged from an ameba's *food* vacuole. 14 ____________

15. A lichen is an alga and a *virus* living together. 15 ____________

16. The organism on which a parasite feeds is the *host*. 16 ____________

17. Fungi cannot make their own food because they have no *chlorophyll*. 17 ____________

18. Organisms that get their nutrients from the breakdown of dead organic matter are called *saprophytes*. 18 ____________

Testing Your Knowledge

1. Draw and name the three main types of bacteria.

____

2. Draw a labeled diagram of an ameba.

3. Describe how an ameba performs each of the following life activities:

a. gets food

____

b. gets oxygen

____

c. gets rid of wastes and excess water

____

4. Explain why a whale is classified in the same large group as a mouse, and not in the same large group as a giant squid.

____

____

5. What activity of yeast makes it valuable for use in baking?

____

____

6. Explain how algae are important to the food supply of humans, both directly and indirectly.

____

7. Explain the difference between the two terms in each of the following pairs:
a. parasite and host

____

b. parasite and saprophyte

____

c. budding and binary fission

____

8. Explain the importance of fungi to future generations of plants.

____

____

9. Why do scientists need a system of classification for animals and plants?

____

____

10. Why was the classification system expanded from two kingdoms to five?

____

____

6

The Great Variety of Plants

LABORATORY INVESTIGATION

BECOMING ACQUAINTED WITH THE PLANT KINGDOM

In this investigation you will examine specimens of plants that belong to different divisions of the plant kingdom. Record your observations in the space provided.

I. Nonvascular Plants—(Plants Without Conducting Tissue)

1. *Mosses and Liverworts*
a. Examples seen ______________________________
b. Where they live ______________________________
c. Characteristics ______________________________

II. Vascular Plants—(Plants Having Conducting Tissue)

2. *Ferns and Horsetails* (*Spore Plants*)
a. Examples seen ______________________________
b. Where they live ______________________________
c. Characteristics ______________________________

3. *Conebearers, or Conifers* (*Nonflowering Seed Plants*)
a. Examples seen ______________________________
b. Where they live ______________________________
c. Characteristics ______________________________

4. *Flowering Seed Plants*
a. Examples seen ______________________________
b. Where they live ______________________________
c. Characteristics ______________________________

The Plant Kingdom

A few characteristics distinguish the plant kingdom from other kingdoms of organisms. Unlike most animals and protozoans, plants lack organs of locomotion and are usually stationary. The outermost part of plant cells is the cell wall (made of cellulose), which is a fairly stiff, nonliving structure. The outermost part of animal cells, by contrast, is the cell membrane, a flexible living structure.

Unlike the fungi, plants are green—they have chlorophyll and make their own food by the process of photosynthesis. Plants are complex multicellular organisms. They probably evolved from algae, which also use chlorophyll to photosynthesize their food. Scientists have classified about 350,000 species of living plants.

Plants are classified into two main groups—**vascular** and **nonvascular.** Vascular plants have true roots, stems, leaves, and special tissue for conducting water; nonvascular plants lack these specialized structures and conducting tissue. Study Table 6-1 below, which outlines the major divisions into which all plants are classified.

TABLE 6-1. MAJOR DIVISIONS OF THE PLANT KINGDOM

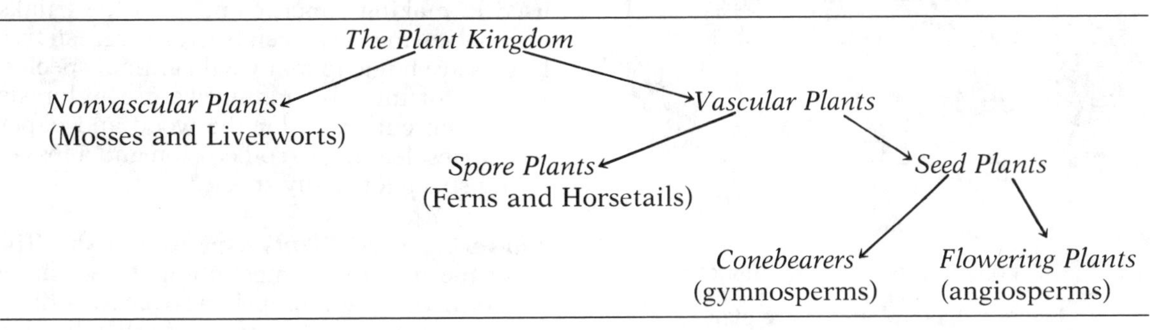

NONVASCULAR PLANTS: THE MOSSES AND LIVERWORTS

Characteristics. The division of nonvascular plants includes the **mosses** and **liverworts** (Fig. 6-1). The most abundant species of this division are the mosses, which are found worldwide. Mosses are low-growing tiny green plants that live in very moist environments. These plants are more complex than the algae and fungi. A moss has leaflike and stemlike structures. The leaflike structure performs photosynthesis. Instead of roots, a moss has stringy rootlike extensions that grow down into the soil. These structures anchor the moss in the surface of the soil, absorbing water and minerals from it. Unlike the larger plants, mosses lack vascular and supporting tissues and thus cannot grow tall. Mosses reproduce by means of spores, found in spore cases at the top of stalks. (You will learn about spore formation in Ch. 9.)

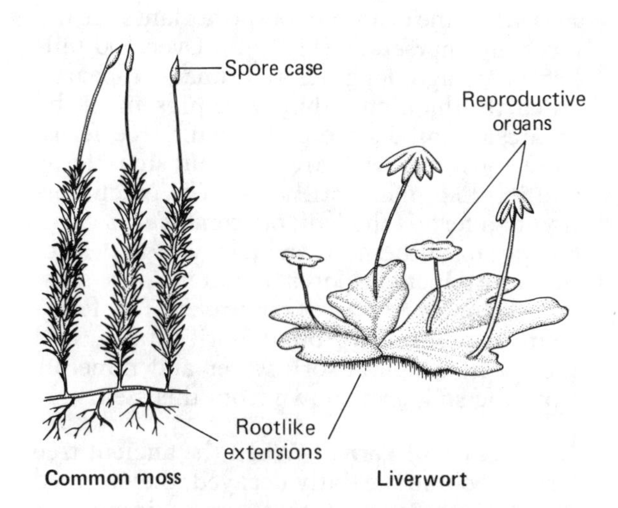

Fig. 6-1. Examples of bryophytes.

Importance of Mosses. Mosses provide a source of food for wild animals. Mosses also help to form soil and protect it from erosion. *Peat moss* is a spongy moss that grows in swampy areas. After being dried, peat moss is often mixed with soil by gardeners and florists because this moss absorbs water and keeps it in the soil. Dried, partially decayed peat moss is also used as fuel in some northern European countries.

VASCULAR PLANTS: PLANTS HAVING CONDUCTING TISSUE

Characteristics. All the plants in this group have true roots, stems, and green leaves. These structures, which transport food and water throughout the plant's body, are made of such tissues as xylem, phloem, and supporting tissue. There are three major groups of vascular plants—the *spore plants* (or *ferns*), the *conebearers* (or *conifers*), and the *flowering plants.*

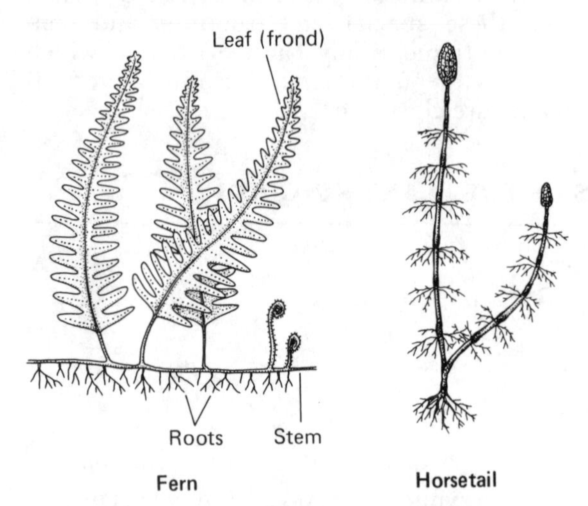

Fig. 6-2. Examples of spore plants.

Ferns. Some examples of spore plants are the **ferns** and **horsetails** (Fig. 6-2). Over 300 million years ago, long before humans appeared on earth, the plants that were present in the greatest numbers were the giant tree ferns. Now, most tree ferns are gone, but small ferns and fernlike plants still exist. The green portion of a fern is the leaf, or **frond.** Ferns carry out photosynthesis in their leaves, and form tiny reproductive spores (found in spore sacs) on the undersides of their leaves. The fern's stem grows underground horizontally, and the roots, which absorb water and minerals from the soil, grow down from this stem.

Importance of Ferns. When the ancient tree ferns died and partially decayed, they formed the deposits of coal that are so important today as a source of fuel. The leaves of present-day ferns are used chiefly for decorative purposes.

Conebearers **(Gymnosperms).** The **conifers**, or **conebearers**, are the nonflowering seed plants that bear *cones.* These cones are the reproductive organs that produce fertilized seeds. The leaves of conifers are usually needlelike and evergreen. Examples of this division are the pine, spruce, and fir tree (Fig. 6-3).

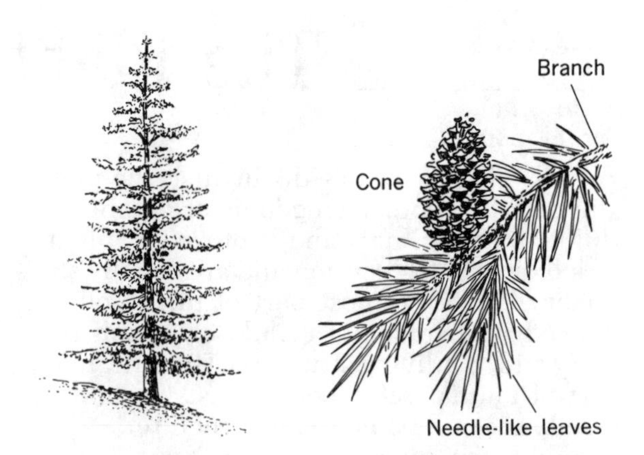

Fig. 6-3. A conebearer—the pine tree.

Importance of Conebearers. Most of the lumber used as building material, and the pulp used in making paper, come from the trunks of conebearing evergreen trees. Evergreen tree forests are home to many wild animal species, too. Unfortunately, many evergreen forests have been cut down for the wood and paper industries, leading to soil erosion and a loss of living space for many species.

Flowering Seed Plants **(Angiosperms).** The **angiosperms** are the most abundant group of plants alive today, comprising over two-thirds of all plant species. The plants in this division bear flowers, which produce seeds that are enclosed in a *fruit* (Fig. 6-4). Flowering plants

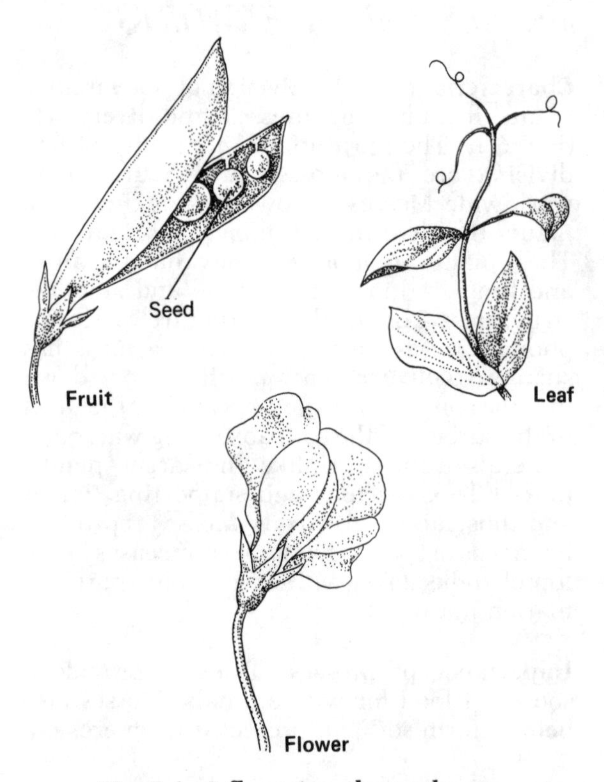

Fig. 6-4. A flowering plant—the pea.

have leaves that are usually flat and much broader than the needlelike leaves of conifers. Some members of this group are the pea, rose, violet, carrot, cherry, elm, oak, ivy, bamboo, and grasses.

Importance of Flowering Plants. The flowering plants supply thousands of species of insects, birds, and other animals with food and shelter. Flowering plants are important to humans, both as a source of foods and fibers and for decorative purposes. Among the flowering plants that are of economic importance are rice, corn, peanuts, soy, lettuce, cotton, maple, flax, apples, oranges, coffee, chestnut, celery, tulips, and wheat. In addition, the ovary protects the developing ovule (seeds) as maturation takes place. These seeds will disperse, forming new plants when conditions for seed germination are proper.

CHAPTER REVIEW

Science Terms

The following list contains all of the boldfaced scientific terms found in this chapter and the page on which each appears.

angiosperms (p. 62)
conebearers (p. 62)
conifers (p. 62)
ferns (p. 62)
frond (p. 62)
gymnosperms (p. 62)
horsetails (p. 62)
liverworts (p. 61)
mosses (p. 61)
nonvascular (p. 61)
vascular (p. 61)

Matching Questions

On the blank line, write the letter of the item in column B which is most closely related to the item in column A.

Column A

___ 1. ferns
___ 2. conebearers
___ 3. mosses
___ 4. flowering plants
___ 5. fronds
___ 6. nonvascular plants
___ 7. formed coal deposits
___ 8. peat moss
___ 9. vascular plants
___ 10. cones

Column B

a. produce seeds inside a fruit
b. include mosses and liverworts
c. have true roots, stems, leaves
d. have needlelike leaves
e. vascular spore plants
f. tiny plants in moist areas
g. reproductive organs of conifers
h. the leaves of ferns
i. found in spore cases
j. ancient tree ferns
k. used for gardening and for fuel

Multiple-Choice Questions

On the blank line, write the letter preceding the word or expression that best completes the statement or answers the questions.

1. Mosses are most closely related to
a. ferns *b.* horsetails *c.* liverworts *d.* grasses 1 ____

2. Ferns are most closely related to
a. horsetails *b.* mosses *c.* ivy *d.* palm trees 2 ____

3. Mosses cannot grow tall because they lack
a. leaflike parts
b. supporting tissue
c. rootlike parts
d. spore cases 3 ____

4. Ferns form reproductive spores on their
a. stems *b.* roots *c.* leaves *d.* stalks 4 ____

5. Which group contains conifers only?
a. pine trees and spruce trees
b. pine trees and maple trees
c. maple trees and cactus
d. spruce trees and ferns 5 ____

6. Which group contains angiosperms only?
a. corn, apple, ferns
b. bamboo, lettuce, moss
c. cotton, pine, maple
d. violet, maple, oak 6 ____

7. The vascular plant group includes all the following *except*
a. horsetails
b. liverworts
c. flowering plants
d. conebearers 7 ____

8. When they reproduce, broad-leaved plants such as the oak have
a. flowers *b.* cones *c.* spores *d.* needles 8 ____

9. The following plants reproduce by means of spores *except* for
a. mosses *b.* liverworts *c.* ferns *d.* pines 9 ____

10. Deposits of coal were formed by the decay of
a. mosses *b.* tree ferns *c.* conifers *d.* liverworts 10 ____

Modified True-False Questions

In some of the following statements, the italicized term makes the statement incorrect. For each incorrect statement, write the term that must be substituted for the italicized term to make the statement correct. For each correct statement, write the word "true."

1. Peat comes from a type of *fern* plant. 1 ____________

2. Carrots belong to the group of *conebearer* plants. 2 ____________

3. Unlike most animals, plants lack organs for *locomotion*. 3 ____________

4. Plants probably evolved from *fungi*. 4 ____________

5. True roots, stems, and leaves are found in *vascular* plants. 5 ____________

6. Ferns carry out photosynthesis in their *stems*. 6 ____________

7. Needlelike leaves are usually found in the *angiosperms*. 7 ________

8. The most abundant group of plants today are the *liverworts*. 8 ________

9. Most lumber comes from the trunks of *tree ferns*. 9 ________

10. Flowering plants enclose their seeds inside a *fruit*. 10 ________

Testing Your Knowledge

1. Explain why mosses and liverworts are small and lie close to the ground.

____

2. Describe two functions performed by a fern's leaves.

____

3. Give several reasons why it is important to replant evergreen (conifer) forests after they have been cut down.

____

____

4. Name five flowering plants and give a useful product that is derived from each one.

____

____

5. Explain the difference between the two terms in each of the following pairs:

a. angiosperm and gymnosperm ____

____

b. vascular and nonvascular ____

____

c. seed plants and spore plants ____

____

6. Explain the difference between a plant ovary and an ovule. What does each part develop into in a mature plant?

____

____

Tropical Deforestation: One of the Greatest Tragedies?

"The devastation is unbelievable. It is one of the greatest tragedies of all history." So said then-Senator Albert Gore after returning from a visit to Brazil's Amazon rain forest.

Former Vice President Gore was referring to deforestation in a tropical region whose area is 90% that of the continental United States. It is a dense forest that covers parts of nine countries in South America and, from east to west, is 3500 kilometers across. The watery artery that gives its more than a million animals and plants species life is the world's second longest river, the Amazon.

Could former Vice President Gore be correct about this huge forest's destruction? And why should we, who are so far away, even care? Furthermore, who would want to "devastate" such a place? And for what reasons?

Let's start with the last question first. The Amazon rain forest is a largely untapped resource for human beings. It, and the other rain forests of the world, have already provided us with substances for about one fourth of all medicines prescribed in the United States, including some used to treat cancer and high blood pressure. Yet, according to some experts, for every one tropical rain forest plant that has been tested for its medicinal properties, 99 others still have not.

Rain forests have also given us all sorts of important foods, including bananas, chocolate, cinnamon, coffee, cola, corn, oranges, and sugarcane. But the forests have even more to give, according to some people. They say the forests hold undeveloped land for farming, vast numbers of trees for construction, minerals like gold and iron, and grazing land for cattle.

So bulldozers clear the land for farms and cattle ranches. And sometimes the forest is burned to clear it, producing smoke clouds that block out the sun for weeks at a time. Loggers come with chain saws to cut down trees for wood and paper. And miners come with their drilling and shoveling equipment to scrape away the land.

Conservationists estimate that each year these activities destroy areas of rain forest the size of the state of Washington! However, many leaders of the countries in which tropical rain forests flourish—mostly in South America, Africa, and Southeast Asia—say the land is needed for these uses to help their countries prosper. Although there may be some truth to this position, many economists and scientists point out that more money can be made by keeping the forests alive than by killing them.

For example, cutting down trees and selling their lumber gives loggers $400 per acre. But tapping the trees and plants in the same area for rubber, medicines, and food items like fruits and nuts provides $2400 per acre. What's more, say scientists, you can get the rubber, medicines, and food without killing the trees and plants. Otherwise, once the trees and plants are gone, their value goes with them. The soil of tropical rain forests is so poor in nutrients that these areas tend to become permanent deserts after the trees are destroyed.

And according to many scientists, when a tropical rain forest is destroyed, thousands

of plant and animal species we have never even identified are lost forever—their potential values unknown. Sadly, some experts estimate that about 130 species of plants, animals and insects currently do become extinct every day due to such destruction.

Does any of this affect you personally? It well might, if the world's climate were to be affected. And many scientists say that the loss of rain forests may lead to such large-scale changes.

So, was former Vice President Gore's statement about the devastation on target or an exaggeration? What do you think?

1. *The percentage of rain forest plants tested for their medicinal properties is:*

   a. 0.1 b. 1.0 c. 10 d. 100

2. *The world's major tropical rain forests are found in*

   a. Australia, Africa, and South America.
   b. North America, South America, and Africa.
   c. Southeast Asia, Africa, and South America.
   d. Southeast Asia, South America, and Australia.

3. *Products we can get from the rain forest without killing trees include all but the following:*

   a. fruits and nuts
   b. wood and paper
   c. medicines
   d. rubber

4. *A headline in* Science News *said: "Have your rain forest and eat it too." Based on the feature you have just read, interpret the meaning of the headline.*

   ____

   ____

5. *Some scientists have predicted that destruction of tropical rain forests will change the world's climate due to a possible warming caused by the "greenhouse effect." Do library research on this topic and write a short essay explaining why you do, or do not, agree with these scientists.*

7

Animals Without Backbones: The Invertebrates

LABORATORY INVESTIGATION

BECOMING ACQUAINTED WITH THE INVERTEBRATES

Your teacher will display a variety of organisms that represent the major invertebrate phyla in the animal kingdom. Examine the specimens and record your observations in the space provided.

I. Sponges (Poriferans)

a. Examples seen

___

b. Where they live

___

c. Characteristics

___

II. Jellyfish, Hydra, and Corals (Coelenterates)

a. Examples seen

___

b. Where they live

___

c. Characteristics

___

III. Flatworms (Platyhelminthes)

a. Examples seen

____

b. Where they live

____

c. Characteristics

____

IV. Roundworms (Nematodes)

a. Examples seen

____

b. Where they live

____

c. Characteristics

____

V. Segmented Worms (Annelids)

a. Examples seen

____

b. Where they live

____

c. Characteristics

____

VI. Hard-Shelled Animals (Mollusks)

a. Examples seen

____

b. Where they live

____

c. Characteristics

____

VII. Spiny-Skinned Animals (Echinoderms)

a. Examples seen

____

b. Where they live

____

c. Characteristics

____

VIII. Jointed Animals (Arthropods)
a. Examples seen

____

b. Where they live

____

c. Characteristics

____

The Invertebrates

In the laboratory investigation, you observed a variety of animals that are all known as **invertebrates.** This group has many different types of animals in it, and many of its members do not at first even appear to be animals. However, the greatest variety and largest number of animals alive on earth today are classified as invertebrates. In spite of their great variety, these animals are grouped together because all of them lack an internal backbone.

SPONGES: PHYLUM PORIFERANS

Characteristics. Some sponges inhabit fresh water, but most species of sponges live in the ocean. All sponges are composed of many cells that are arranged in two main layers around a hollow cavity (Fig. 7-1). The cell layers are pierced by thousands of pores, hence the name **poriferans,** or *porebearers.* Imbedded among the cells is a tough supporting material that forms the skeleton of the sponge. Some sponges have a hard skeleton composed of lime, while others are of glasslike silica; a third type of sponge has a tough but flexible skeleton composed of *spongin.* Adult sponges do not move around; they grow attached to solid underwater objects. Sponges can reproduce asexually by forming a *bud,* which is a small bulge that breaks off to become a new sponge. Sponges can also reproduce sexually, by forming sperm and egg cells.

Vase-Shaped Sponge. The attached (bottom) end of the sponge is closed; the unattached end is open. Certain cells of the inner layer bear whiplike hairs similar to cilia. The beating of these hairs causes a current of water to flow into the pores, to the hollow cavity, and out through the open end of the sponge. Some cells of the inner layer engulf microscopic food particles from the water and digest them. All the cells absorb dissolved oxygen and release carbon dioxide. The current of water carries wastes out through the open end of the sponge.

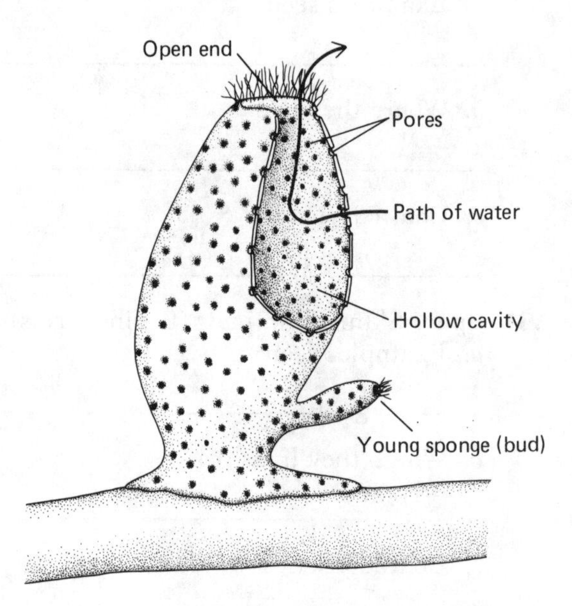

Fig. 7-1. Sponge (partly cutaway view).

Importance of Sponges. When sponges are brought to the surface by sponge divers, and placed in the air to dry out, the living cells die and rot. The tough supporting material remains. When washed and dried, these spongin skeletons become the "natural sponges" used for bathing. Living sponges are an important part of coral reef communities.

JELLYFISH, HYDRA, CORALS: PHYLUM COELENTERATES

Characteristics. This phylum comprises a large group of simple animals that includes the *jellyfish, hydra, corals,* and *sea anemones* (Fig. 7-2). A few species of this phylum inhabit fresh water; most live in salt water. Like the sponges, the **coelenterates** are many-celled and are composed of two main cell layers. All coelenterates have a layer of jellylike material between the two cell layers. Coelenterates lack hard internal body parts, and are either bell-shaped or vase-shaped. Musclelike cells, nerve cells, and stinging cells are found in this group.

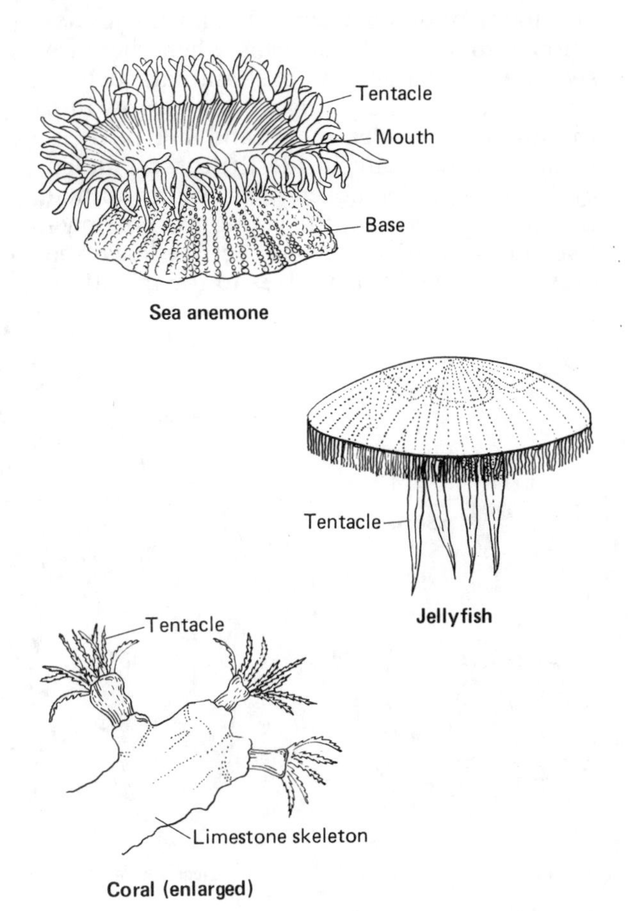

Fig. 7-2. Examples of coelenterates.

All members of the coelenterate group have mouths surrounded by long flexible arms, called **tentacles.** These tentacles have stinging cells for capturing prey. The mouth leads directly into a central, hollow digestive cavity. In fact, the name *coelenterate* means "hollow gut." The jellyfish and hydra are delicate and transparent. Their bodies consist of about 95% water. Some jellyfish and hydra reproduce by forming *buds*. Others form sperm and egg cells to reproduce sexually. The sea anemone looks like a large hydra and is often colorful, resembling a flower. Coral animals are soft and tiny; but each animal secretes a hard limestone skeleton around itself. As a coral animal reproduces by budding, the new individuals remain attached to the parent, forming a *colony*. Colonies made up of millions of attached limestone skeletons form *coral reefs*.

Hydra. Like the sponge, the hydra is usually attached to an underwater object. It is closed at the attached end (foot) and open at the other end (mouth). A hydra can extend itself to a length of about 1 cm. When disturbed, it contracts into a tiny ball. The hydra, unlike the sponge, can move from place to place by slowly sliding along on its base, or foot. Sometimes it inverts itself, turning its foot upward. It then moves along on its arms or in very slow somersaults. Its musclelike cells make these movements possible.

The hydra captures food by means of its arms, which sting the prey and then push it through the mouth into the digestive cavity. Here, special cells of the inner layer secrete enzymes and digest the food. Solid wastes are discharged through the mouth. Since every cell of the animal is close to water, each cell can take in oxygen directly from the water by diffusion and, similarly, can excrete liquid wastes directly into the water.

By means of a network of nerve cells, the hydra receives stimuli and responds to them by muscular action or by discharging its stinging cells. When stimulated, the stinging cells shoot out hooked, poisonous threads that can entangle and paralyze very small animals. In this manner, the hydra protects itself against enemies and captures animals smaller than itself for food.

Importance of Coelenterates. Large jellyfishes, with poisonous stinging cells in their tentacles, endanger swimmers who are stung by them. Corals are important because accumulations of their skeletons form underwater reefs. When ships run into such reefs, both the ship and the reef may be damaged. Large colonies of corals also form islands that can be inhabited by humans. A few kinds of coral are used in making ornaments; however, some corals are now becoming too rare to be used this way.

FLATWORMS: PHYLUM PLATYHELMINTHES

Characteristics. Some **flatworms** live in fresh water, but most of them live inside other animals as parasites. The name **platyhelminthe** means "flatworm." The animals in this group are definitely flat in shape, resembling a piece of ribbon or tape. Flatworms are soft-bodied, consisting of three layers of cells—outer, middle, and inner layers—arranged around a pouchlike, branched food tube. Special tissues and some simple organ systems, which develop from the middle cell layer, are present. Among these are the muscular, nervous, and reproductive systems. There is a definite head region in which a small brainlike mass of nerve tissue is located. Branches from this tissue project backward through the body, forming two solid nerve cords. Reproduction occurs both by binary fission and by the mating of males and females.

There are three main types of flatworms: the *planaria*, the *tapeworm*, and the *liver fluke* (Fig. 7-3).

Planaria. The planaria lives in fresh water under solid objects, such as rocks. It swims slowly by means of cilia, which cover its body, and by muscular contractions. It is usually about 1 cm. in length and about 1/2 cm. in width. Two *eyespots,* dark areas in the head region, can detect differences in light. The planaria moves away from bright light. The mouth of this worm is at the end of a tube located on the underside of its body.

Tapeworm. This flatworm is a parasite that lives in the intestines of such hosts as humans, cattle, and pigs. The tapeworm has no means of locomotion. It lacks a mouth and other digestive organs. With *suckers* and *hooks* found on its "head" region, the worm fastens itself to the intestinal wall of its host. Through its own body wall, the worm absorbs the digested food that is present in the host's intestine. Although seldom more than about 1 cm. in width, some tapeworms may reach a length of over 3 meters.

The body of a tapeworm consists of a series of identical, separate sections, each of which produces sperm and egg cells. When "ripe," each section contains thousands of fertilized eggs. Ripe sections pass out of the host's body with solid wastes. There are two stages in the development of tapeworms. First, the eggs develop into young tapeworms when they are swallowed by a grazing host animal, such as a cow or a pig. The young worms burrow into the animal's muscles (meat), and then lie there in an inactive state inside little sacs, called **cysts.** If the undercooked meat of this animal is eaten by a person, the young worms, inactive but still alive, become active, leave the cysts, and attach themselves to the intestinal

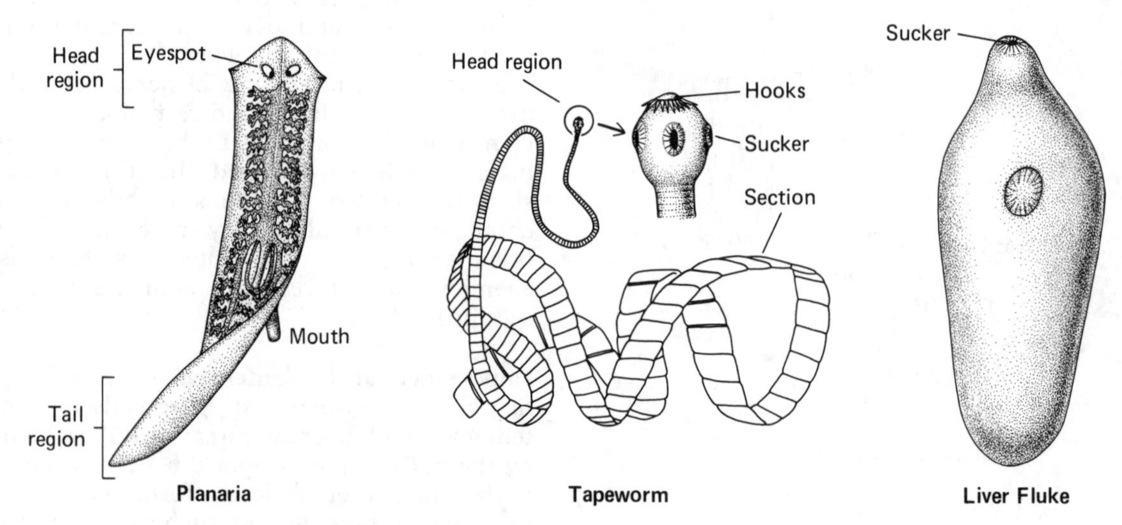

Fig. 7-3. The three types of flatworms.

wall of the new human host. There they develop into adult tapeworms, capable of reproduction.

Liver fluke. This flatworm is a parasite that inhabits the liver of such hosts as humans and sheep. Flukes may reach a length of about 2 cm. and a width of about 1 cm. The fluke (which has a sucker, but no mouth) attaches to the host's liver, and feeds upon its blood. The fluke produces eggs that reach the intestine of the host by way of the bile duct and leave the host's body with solid wastes. If the eggs are swallowed by a certain kind of snail, they hatch within the snail. The young flukes then leave the body of the snail and enter the body of certain fishes. When people eat such fish uncooked, the young flukes pass into their livers, where they grow into adult flukes.

Importance of Flatworms. Tapeworms and liver flukes cause serious diseases in their hosts, leading to physical weakness and, sometimes, death. These diseases are easily prevented by thoroughly cooking all red meat and fish until heat has penetrated the food. Worm parasites cannot withstand temperatures above boiling for more than about 20 minutes. Better sanitation of water supplies in developing nations can also help cut down flatworm infections in people.

ROUNDWORMS: PHYLUM NEMATODES

Characteristics. Some species of **roundworms** live in soil; some live in water; some are parasites of animals; others are parasites of plants. Most free-living roundworms are very small; the parasitic roundworms can be over a meter long. The body of a **nematode** is slender, rodlike in shape, soft, and also composed of three layers of cells. The middle layer of cells encloses a body cavity. Tissues and simple organ systems are present. Roundworms, which are pointed at both the head and tail ends, possess a mouth and an anus. Among the most important roundworms are the *hookworm* and the *trichina worm* (Fig. 7-4). Other examples of roundworms are the *threadworm* and the *pinworm.*

Hookworm. Adult hookworms, less than 2 cm. long, are often found attached to the intestinal wall of humans who live in warm regions. The worms feed upon the host's blood, causing the person to feel weak. The eggs of

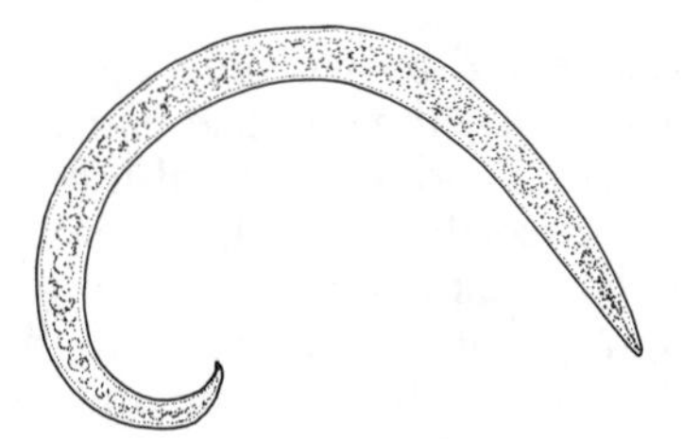

Hookworm (about twice natural size)

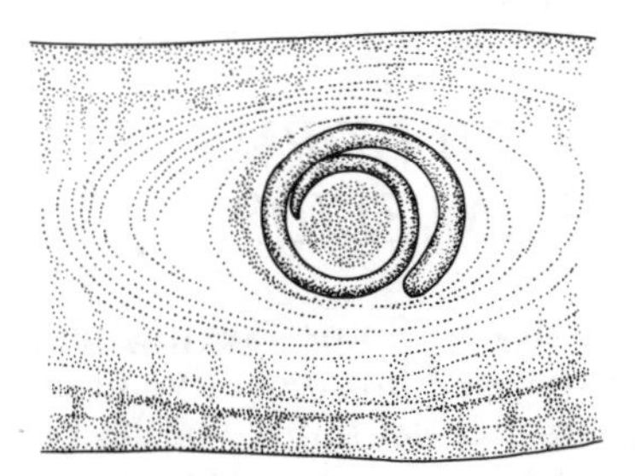

Trichina worm in muscle tissue
(microscopic view)

Fig. 7-4. Examples of roundworms.

the hookworm pass out of the host's body with the solid wastes. If these eggs reach moist soil, they hatch into young hookworms. These worms can enter the body of a new host by boring through the skin of the feet. The worm then enters the blood and is carried to the lungs or another organ. Eventually, the worm arrives in the intestine, fastens itself to the intestinal wall, and bores into a blood vessel there. Feeding on blood, the hookworm grows to adulthood and reproduces. Hookworms also infect the intestines of dogs, cats, and horses.

Trichina Worm. This tiny (1/2 cm.) worm is a parasite in the human, the pig, and the rat. Pigs that eat rats infected with the trichina worm become infected themselves. The young trichina worms burrow into the pig's muscles and, like tapeworms, lie inactive there inside cysts. If a human eats infected pork that has not been thoroughly cooked, the worms leave the cysts, enter the person's small intestine, mature and reproduce. The new generation of young worms then enters the muscles of the same human to form more cysts, causing intense muscular pain.

Importance of Roundworms. *Hookworm disease* results in a loss of blood and general body weakness. In this condition, a person is likely to catch other diseases. Hookworm disease can be prevented by the sanitary disposal of human wastes and by wearing shoes in rural areas. *Trichinosis,* the disease caused by the trichina worm, is not only painful but may, if untreated, cause death. This disease can be prevented by cooking pork and other meats long enough and at temperatures high enough to kill the parasites that may be present. Some of the parasitic roundworms also destroy tomatoes and other crops by sucking out the sap from their roots.

SEGMENTED WORMS: PHYLUM ANNELIDS

Characteristics. Some **segmented worms** live in fresh water, but most live in salt water. A few others live on land. The body of these worms is long and tubelike in form. Unlike the surface of roundworms, which is smooth, the surface of these worms is divided into many sections, or **segments.** Like the two other worm phyla and, in fact, like all higher animal phyla, segmented worms, or **annelids,** have three layers of cells.

The organ systems are complex. The food tube extends from mouth to anus, having specialized regions, such as an intestine, between these openings. A complete circulatory system is present. It is made up of arteries, capillaries, and veins. The nervous system is composed of a solid mass of nerve tissue in the head region (a "brain") and a solid nerve cord that runs along the lower side of the body. Although some segmented worms reproduce by splitting in half, all are capable of sexual reproduction. Every earthworm has both male and female reproductive organs. The *clitellum* is the body segment that plays a role in the earthworm's mating process. Examples of segmented worms are *sandworms, leeches, giant tube worms,* and *earthworms.* The earthworm is the one most familiar to people (see Fig. 7-5).

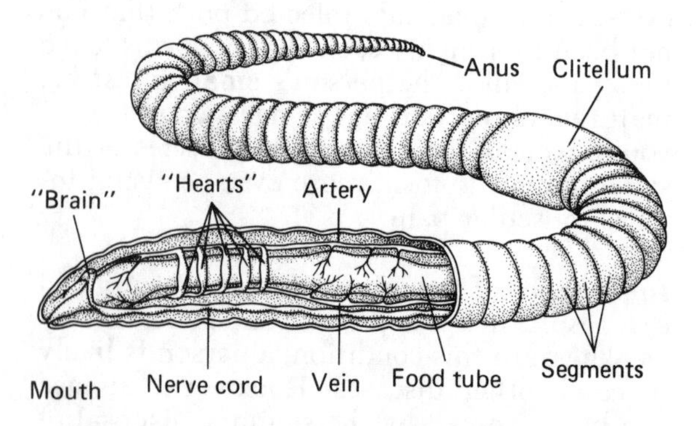

Fig. 7-5. Earthworm (partly cutaway view).

Earthworm. The earthworm lives in moist soil. This worm burrows into the soil by eating it and the organic matter that it contains. As the soil passes through the worm's food tube, the organic matter is digested and absorbed. The remaining waste material is discharged through the anus.

The circulatory system of the earthworm consists of two major blood vessels—an artery that runs above the food tube and a vein that lies below the food tube. About 2 1/2 cm. behind the front end of the worm, these blood vessels are joined by five pairs of more muscular blood vessels (the "hearts"). These five pairs of hearts continually pump blood. Elsewhere in the body, branches from the artery join the vein by way of capillaries. The worm takes in oxygen and gives off carbon dioxide through its moist skin.

The earthworm responds to stimuli by means of reflex actions that are controlled by its nervous system. (You will learn about reflexes in Ch. 20.)

Importance of Segmented Worms. Segmented worms serve as food for many animals. The earthworm is probably most important to people, because this worm's activities make farming possible. As earthworms burrow, they aerate the soil, bring deep soil to the surface, and enrich the soil with their wastes. Earthworms and sandworms are frequently used as bait in fishing. The leech (or bloodsucker), which is a temporary parasite on humans, turtles, and fish, is also important to people. Because the leech's saliva contains a substance that prevents blood clotting, leeches are now being used to aid blood flow and healing in some types of surgery.

HARD-SHELLED ANIMALS: PHYLUM MOLLUSKS

Characteristics. There are over 100,000 species of **mollusks,** divided into three distinct classes based on their shells. Some animals in this group live in salt water; some live in fresh water; and others live on land. The body of all mollusks is soft and enclosed by a soft tissue called the **mantle.** The outer side of the mantle produces a hard shell material. Some of the animals in this phylum, like the *snail,* have a single coiled shell; others, like the *clam* and *oyster,* have two shells connected by a *muscular hinge;* and a few, such as the *squid,* have a shell that is not visible because it is small

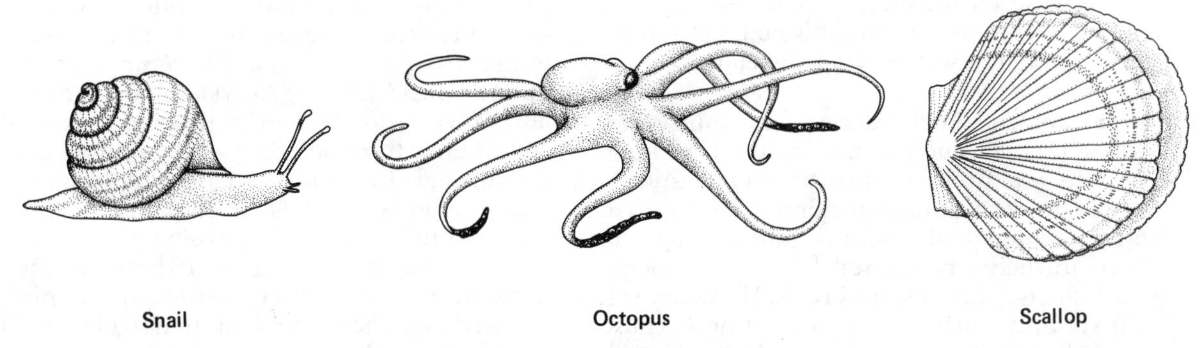

Fig. 7-6. Examples of mollusks.

and imbedded in the soft tissues. (The *octopus* has no shell at all.) Clams and snails move by means of a muscular foot. Squids and octopuses move by squirting jets of water. See Fig. 7-6 for some examples of common mollusks.

Clam. Many kinds of clams burrow in sand in shallow water. The central mass of the clam's body contains all the major organ systems (see Fig. 7-7). Muscles open and close the shells of the clam. Folds of the mantle form two passageways through which water flows in and out, and passes over the *gills*. The flowing water brings in oxygen and food. The oxygen is taken up by blood inside the gills. The food, which consists of microscopic organisms, is directed to the animal's mouth by cilia.

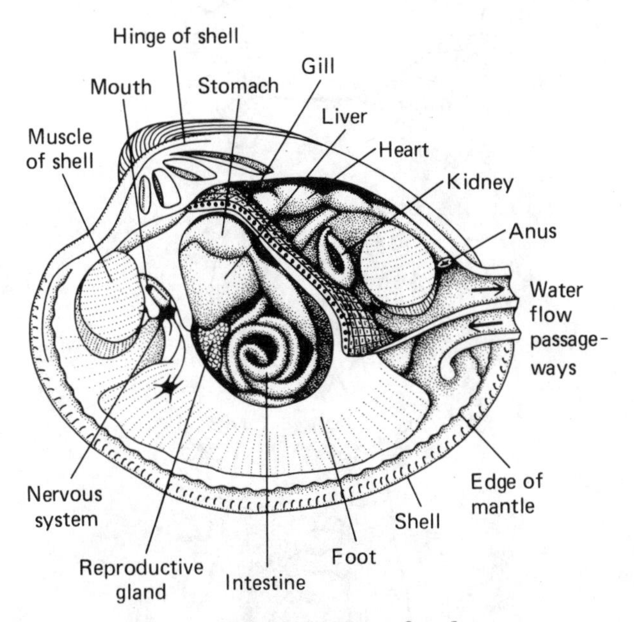

Fig. 7-7. Cutaway view of a clam.

Importance of Mollusks. Among the mollusks that are used as food by people and other animals are oysters, clams, snails, octopuses, and squids. Some freshwater and saltwater clams and oysters produce pearls when particles of sand become lodged between the mantle and the shell. The shells of some mollusks are used to make buttons, jewelry, and even money. Animals of this group that are harmful include the "shipworm" (which bores through the wood of ships and piers), snails that transmit parasitic diseases, and snails and slugs that feed on our crops.

SPINY-SKINNED ANIMALS: PHYLUM ECHINODERMS

Characteristics. There are over 5000 different kinds of **echinoderms.** All the animals in this group live in the ocean. Most echinoderms have a body shape that either resembles a star or is round. The body is covered with numerous sharp projections, called *spines,* hence the name *echinoderm,* which means "spiny skinned." Echinoderms have no head; but they do have a mouth. These animals have special organs of locomotion, called **tube feet.** All of the major organ systems are present. Common examples of echinoderms are the *starfish, sea urchin, brittle star, sea lily,* and *sea cucumber* (Fig. 7-8).

Starfish. The starfish lives in shallow offshore water. Its body consists of five arms, or **rays,** that extend from a central disc. The undersurface of each arm bears hundreds of tube feet lined up in a groove. The tube feet are used not only in locomotion, but in capturing and feeding upon such animals as oysters and clams. The starfish wraps its arms around a clam and attaches its tube feet to the shells of the clam. By exerting a steady, strong pull upon the shells, the starfish tires the clam until the mollusk relaxes its muscles and opens its shells. Then, the starfish pushes its stomach out of its small mouth and into the space between the clam's shells. Before it is pulled back into the starfish's body, the stomach digests the soft parts of the clam.

Importance of Echinoderms. In some Asian countries, sea cucumbers are used as food. Certain sea urchins have poisonous spines, which are very painful if stepped on. Starfish often damage and destroy clams and oysters that are being cultivated for market. Some starfish also destroy coral reefs. However, scientists find starfish very interesting because starfish can grow back lost body parts, such as an arm.

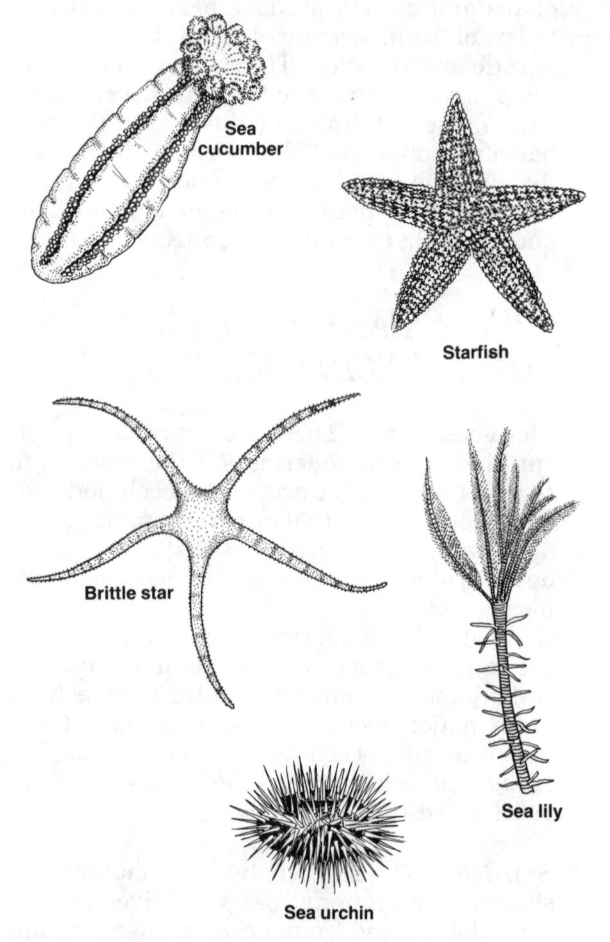

Fig. 7-8. Examples of echinoderms.

JOINTED ANIMALS: PHYLUM ARTHROPODS

Characteristics: All **arthropods** have bodies that are divided into either two or three sections and are covered by an outer shell, or **exoskeleton.** Hingelike joints between the body sections give the animals flexibility. Even the legs and *antennae* (feelers) of these animals are jointed. In fact, the name *arthropod* means "jointed feet." The organ systems are complex and well developed. The hard exoskeleton that covers an arthropod does not grow along with the animal. In a process called **molting**, the exoskeleton is shed. Then, the animal grows rapidly until a new hard exoskeleton forms.

There are more kinds of arthropods than there are of all other species of animals counted together. The four major classes of jointed animals are *crustaceans, myriapods, arachnids,* and *insects.*

Class I: Crustaceans. Examples of this class are the *lobster*, the *shrimp*, and the *crab* (Fig. 7-9). There are over 25,000 species of **crustaceans.** These animals usually have two body sections containing two pairs of antennae and five pairs of legs. Crustaceans are mainly aquatic and they breathe by means of gills. Many animals in this group are prized as food by people all over the world. Crustaceans such as krill (a shrimp) are an important food source for whales, too.

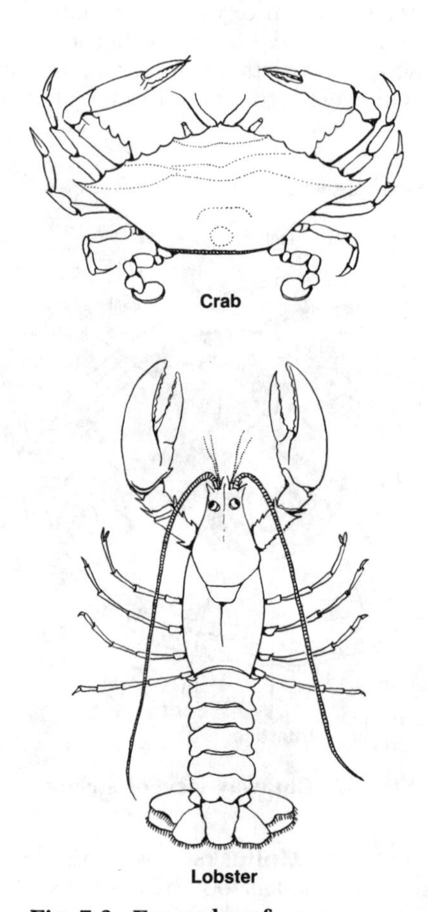

Fig. 7-9. Examples of crustaceans.

Class II: Myriapods. Millipedes and centipedes (Fig. 7-10) are the main examples of this class. **Myriapods** (meaning "a great many feet") live in the ground, usually in dark, moist places. *Millipedes* (meaning "thousand feet") have two pairs of legs on each body segment; *centipedes* (meaning "hundred feet") have one pair of legs per body segment. Some centipedes are beneficial—they eat insect pests. Other centipedes are harmful—their poisonous bite is painful to humans. Millipedes eat decaying plants.

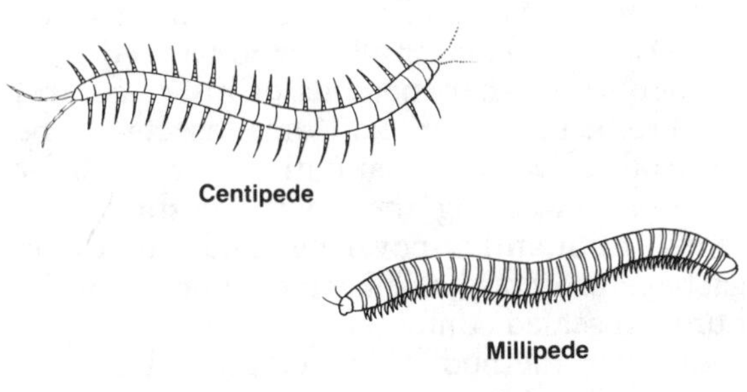

Fig. 7-10. Myriapods.

Class III: Arachnids. Examples of the **arachnids** are *spiders, scorpions, ticks,* and *mites.* (See Fig. 7-11.) These animals have two body sections and four pairs of walking legs attached to their front body section. There are over 100,000 species of arachnids. Garden spiders, which eat insects, are useful to us. The scorpion, which has a poisonous sting, is harmful to humans. Also harmful are some ticks and mites, which frequently carry diseases. Spiders are well-known for the variety of intricate webs they weave to catch prey.

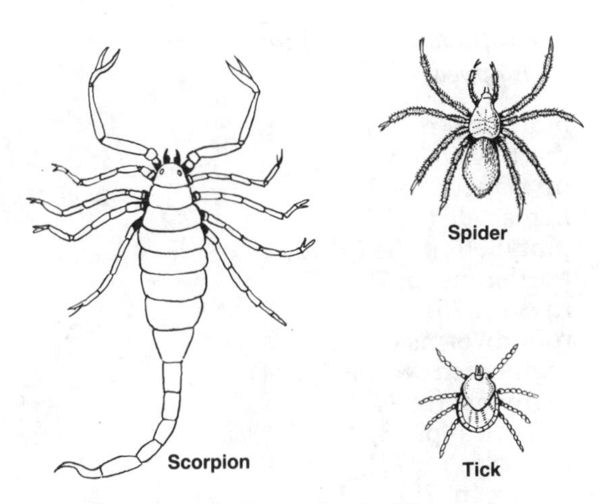

Fig. 7-11. Arachnids.

Class IV: Insects. Most **insects** have three body sections, or regions, with three pairs of legs and two pairs of wings, which may be unequal in size. There are more than 600,000 different species of insects. Included among these are the *wasp, damselfly, aphid, ant, silverfish, potato beetle,* and *grasshopper.* See Fig. 7-12 for some examples.

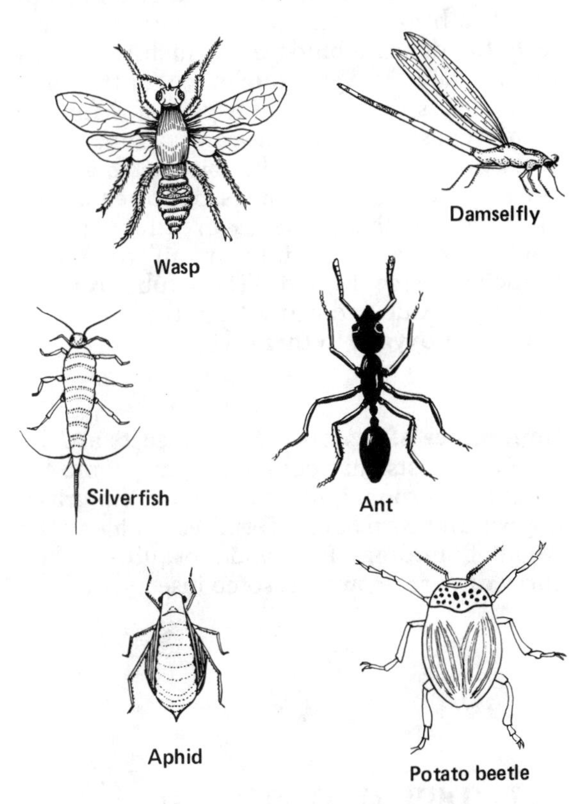

Fig. 7-12. Examples of insects.

Grasshopper. The head of a grasshopper (Fig. 7-13) bears two large **compound eyes,** which detect moving objects. Each compound eye consists of thousands of tiny eyes. Also present

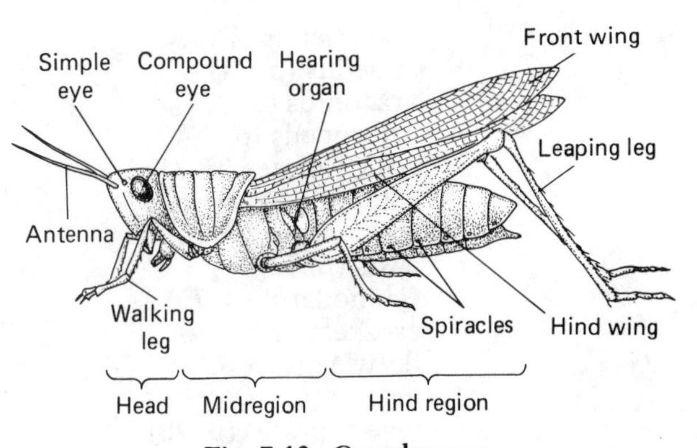

Fig. 7-13. Grasshopper.

on the head are three small simple eyes that distinguish light and dark. Projecting from the head region are two *antennae*. Imbedded in each antenna are nerve endings for touch and smell. The mouth has two lips and two pairs of jaws, which cut and grind vegetation. The *midregion* (**thorax**) of the body bears three pairs of legs and two pairs of wings. Four of the legs are used in walking and two in leaping. The front pair of wings covers and protects the delicate hind pair, which is used in flying. The *hind region* (**abdomen**) of the body bears on its first segment a pair of drumlike membranes, which are hearing organs. Most of the other segments of the abdomen bear two **spiracles.** These are openings, one on each side of the body, that serve as breathing pores. Each spiracle opens into an air tube that branches inside the body. These tubes remove carbon dioxide and water from the body cells and bring oxygen to them.

Importance of Insects. Many insects are destructive pests that compete with people for their food crops. Among these are the grasshopper and corn borer. Termites eat away the wood of buildings. Flies and mosquitoes often carry disease. However, some insects are beneficial. Among these are the honeybee, which pollinates flowers and produces honey; the silkworm moth, which produces natural silk; and the praying mantis, which eats insect pests. In addition, insects are a source of food for a great many animals.

How We Control Insects. Most of us think of chemical sprays as the method for controlling insect reproduction. The most common sprays (chlordane, nicotine, arsenic) are very effective but are also harmful to other organisms, especially humans. When using these sprays in the home, follow the directions on the container very carefully. Some chemicals (such as DDT) are no longer allowed to be used.

Therefore, other methods of insect control need to be implemented. Home insects can be controlled by certain sanitation measures, for example, sweeping the floors, washing surfaces frequently, covering and removing garbage promptly, and keeping all food in the home in sealed containers.

Another method of insect control that is receiving a lot of recent attention is the encouragement of the reproduction of insects' natural enemies, which would then naturally control the number of harmful insects in the general population.

CHAPTER REVIEW

Science Terms

The following list contains all of the boldfaced scientific terms found in this chapter and the page on which each appears.

Matching Questions

On the blank line, write the letter of the item in column B which is most closely related to the item in column A.

Column A	Column B
____ 1. related to hydra	*a.* crab
____ 2. has ribbonlike shape	*b.* centipede
____ 3. porebearer	*c.* earthworm
____ 4. a type of roundworm	*d.* clam
____ 5. has spines and tube feet	*e.* jellyfish
____ 6. has five pairs of "hearts"	*f.* grasshopper
____ 7. has five pairs of walking legs	*g.* flatworm
____ 8. has four pairs of walking legs	*h.* hookworm
____ 9. has a double shell and gills	*i.* sponge
____ 10. has three pairs of legs and four wings	*j.* spider
	k. starfish

Multiple-Choice Questions

On the blank line, write the letter preceding the word or expression that best completes the statement or answers the question.

1. An animal whose body is characterized by thousands of pores is the
a. hydra *b.* sponge *c.* coral *d.* starfish 1 ____

2. Sponges can reproduce asexually by
a. forming a bud *c.* producing spores
b. splitting in half *d.* releasing seeds 2 ____

3. The number of cell layers in both corals and sponges is
a. one *b.* two *c.* four *d.* six 3 ____

4. An animal that has stinging cells is the
a. sponge *b.* starfish *c.* jellyfish *d.* tapeworm 4 ____

5. Soft-bodied, ribbonlike animals are classified as
a. roundworms *b.* jellyfish *c.* flatworms *d.* seafood 5 ____

6. The liver fluke feeds upon
a. intestines *b.* blood *c.* decayed meat *d.* water plants 6 ____

7. A worm that can enter the body through the soles of the feet is the
a. hookworm *b.* tapeworm *c.* trichina *d.* leech 7 ____

8. The hookworm feeds upon blood inside the human
a. intestine *b.* heart *c.* leg *d.* liver 8 ____

9. A worm that lives inside human muscle tissue is the
a. tapeworm *b.* fluke *c.* trichina *d.* hookworm — 9 ____

10. Thorough cooking of pork can help prevent infection by
a. malaria and trichina
b. trichina and tapeworm
c. hookworm and tapeworm
d. tapeworm and malaria — 10 ____

11. The number of cell layers in a segmented worm is
a. two *b.* three *c.* four *d.* six — 11 ____

12. Examples of segmented worms include the
a. earthworm, tapeworm, liver fluke
b. earthworm, sandworm, tapeworm
c. earthworm, sandworm, leech
d. sandworm, leech, hookworm — 12 ____

13. The circulatory system of the earthworm includes
a. arteries, veins, and capillaries
b. only arteries and veins
c. only capillaries
d. only veins and capillaries — 13 ____

14. Which worm normally lives all its life in moist soil?
a. tapeworm *b.* pinworm *c.* earthworm *d.* hookworm — 14 ____

15. In respiration, the earthworm takes in oxygen through its
a. skin *b.* gills *c.* lungs *d.* air tubes — 15 ____

16. The animal most closely related to the clam is the
a. starfish *b.* snail *c.* trichina *d.* shrimp — 16 ____

17. The body of mollusks is covered with
a. scales *b.* spines *c.* a mantle *d.* hinged joints — 17 ____

18. The body of a starfish is covered with
a. sharp spines
b. short cilia
c. tough scales
d. a smooth shell — 18 ____

19. The echinoderms carry out locomotion by means of
a. tentacles *b.* pseudopods *c.* small tails *d.* tube feet — 19 ____

20. Which animal possesses an exoskeleton with joints?
a. snail *b.* lobster *c.* earthworm *d.* coral — 20 ____

21. The process of molting refers to
a. a reaction to warmth
b. shedding a body covering
c. a way of reproducing
d. use of stinging cells — 21 ____

22. Crustaceans, such as the shrimp, breathe by means of
a. air tubes *b.* its skin *c.* gills *d.* lungs — 22 ____

23. The animal most closely related to the scorpion is the
a. spider *b.* sea urchin *c.* octopus *d.* starfish — 23 ____

24. The grasshopper breathes through its
a. lungs *b.* spiracles *c.* gills *d.* mouth — 24 ____

25. The two shells of clams and oysters are connected by
a. a muscular hinge
b. an exoskeleton
c. special joints
d. overlapping scales — 25 ____

Modified True-False Questions

In some of the following statements, the italicized term makes the statement incorrect. For each incorrect statement, write the term that must be substituted for the italicized term to make the statement correct. For each correct statement, write the word "true."

1. Animals with many pores and a central hollow cavity are known as *poriferans*. 1 ____________
2. In capturing its food, a jellyfish uses its *tentacles*. 2 ____________
3. Hydra can reproduce by forming little *spores*. 3 ____________
4. Animals whose skeletons form reefs are known as *sponges*. 4 ____________
5. The tapeworm is a type of flatworm that is a *host*. 5 ____________
6. Hookworms and trichina worms are two types of *flatworms*. 6 ____________
7. An annelid's body is divided into many *segments*. 7 ____________
8. An octopus is a type of *arthropod*. 8 ____________
9. Segmented arthropods having many legs are called *myriapods*. 9 ____________
10. Most insects have wings and *four* body sections. 10 ____________

Testing Your Knowledge

1. Explain each of the following true statements:

   a. A type of coelenterate can wreck a ship.

   ____

   b. Tapeworm disease can be prevented in the kitchen.

   ____

   c. A starfish uses its tube feet for feeding as well as for moving.

   ____

   ____

2. Why is hookworm disease rare among people living in cities?

   ____

3. Why will a grasshopper *not* drown if only its head is below water?

   ____

   ____

4. Why is the spider classified as an arachnid and not an insect?

____

____

5. Describe how earthworms make soil suitable for farming.

____

____

6. In each of the following groups, one term does not belong with the others. Write that term in the blank space at the right.

a. earthworm, hookworm, trichina worm, pinworm ________

b. leech, sandworm, earthworm, tapeworm ________

c. oyster, starfish, snail, squid ________

d. wasp, beetle, scorpion, butterfly ________

e. sea urchin, brittle star, coral, sea lily ________

7. When spraying house plants for insects, it is necessary to avoid spraying foods. Why?

____

____

8. Many residential yards have fruit trees. Many of these trees never bear a successful crop. Can you list five reasons for this?

____

____

9. What kind of picture does a grasshopper see through its compound eyes?

____

____

10. Why is the grasshopper an enemy of the farmer?

____

____

8

Animals With Backbones: The Vertebrates

LABORATORY INVESTIGATION

BECOMING ACQUAINTED WITH THE VERTEBRATES

In this laboratory investigation, you will study examples of animals belonging to the different vertebrate classes. Record your observations in the spaces below.

I. Fishes
a. Examples seen ______________________
b. Type of body covering ______________________
c. Type of limbs/locomotion ______________________
d. Where they live ______________________

II. Amphibians
a. Examples seen ______________________
b. Type of body covering ______________________
c. Type of limbs/locomotion ______________________
d. Where they live ______________________

III. Reptiles
a. Examples seen ______________________
b. Type of body covering ______________________
c. Type of limbs/locomotion ______________________
d. Where they live ______________________

IV. Birds

a. Examples seen ______

b. Type of body covering ______

c. Type of limbs/locomotion ______

d. Where they live ______

V. Mammals

a. Examples seen ______

b. Type of body covering ______

c. Type of limbs/locomotion ______

d. Where they live ______

The Vertebrates

In the laboratory investigation, you examined a variety of animals that are classified as **vertebrates.** Most of the animals that you encounter in your daily life are probably vertebrates (such as dogs and birds). Yet, these animals make up only about five percent of all animal species.

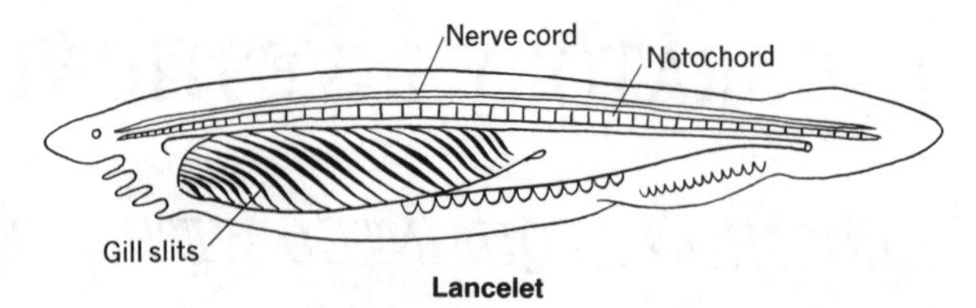

Fig. 8-1. Lancelet—a lower chordate.

Characteristics. Vertebrates have two very important features that the invertebrates lack: an *internal skeleton* (**endoskeleton**) and a *backbone.* The vertebrate group is actually a division of the **chordate** phylum. At some point in their development, all chordates have several pairs of *gill slits* (even if they do not breathe through gills later), a *notochord* (a flexible support rod), and a hollow *nerve cord.* There are two other divisions of chordates besides the vertebrates. These consist of simple marine animals that are referred to as the *lower chordates.* (See Fig. 8-1.)

The Classes of Vertebrates

All vertebrates have a backbone, or **spinal column,** that is composed of a series of separate bones, the **vertebrae.** The backbone is part of the internal supporting skeleton.

Nearly all vertebrates have two pairs of limbs that are used in locomotion. Vertebrates also have a body cavity that contains the major organs, and head that has a fairly well-developed brain. All vertebrates reproduce sexually, by producing sperm and egg cells. The vertebrate subphylum is divided into the following seven classes: the jawless fishes; the cartilaginous fishes; the bony fishes; amphibians; reptiles; birds; and mammals.

The three classes of fishes, the amphibians, and the reptiles are all referred to as **cold-**

blooded animals. This means that the body temperature of the animal changes as the temperature of the environment changes. The birds and mammals are called **warmblooded animals,** because they can maintain a warm body temperature, regardless of the temperature of the environment.

THE FISHES

Characteristics. There are three main classes of fishes: the *jawless, cartilaginous,* and *bony.* Nearly all fishes have a streamlined body, which aids in swimming. All fishes breathe by means of gills. In addition, all fishes have a two-chambered heart.

JAWLESS FISHES

The members of this class have no jaws, no scales, no paired fins, and a skeleton made of **cartilage** (not bone). These fish have a sucking mouth equipped with sharp teeth, which they use to cut into the bodies of other fish. **Jawless fishes** are found in both fresh water and salt water, where they live as parasites, sucking the blood from living fish. Examples of this class are the *lamprey* and *hagfish* (Fig. 8-2).

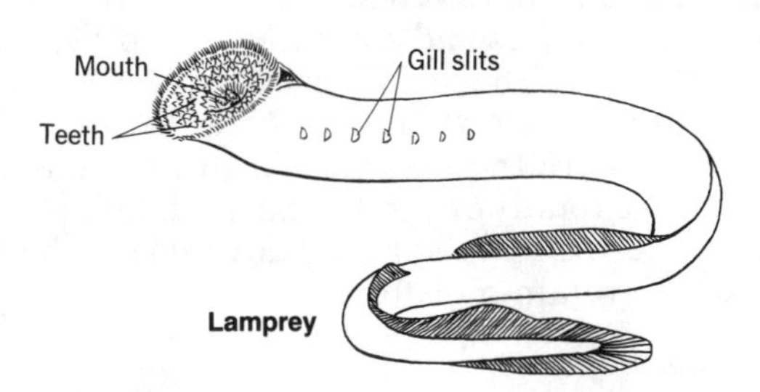

Fig. 8-2. Lamprey—a jawless fish.

CARTILAGINOUS FISHES

The members of this class have jaws, which in many species are equipped with rows of sharp teeth. **Cartilaginous fishes** are so named because they have skeletons that are made of cartilage. They also have gill slits, paired fins, and tough scales. These fishes are found in fresh water and salt water. Cartilaginous fishes are usually active predators, feeding on fishes and other aquatic forms of life. Examples of this class are the *sharks* and *rays* (Fig. 8-3).

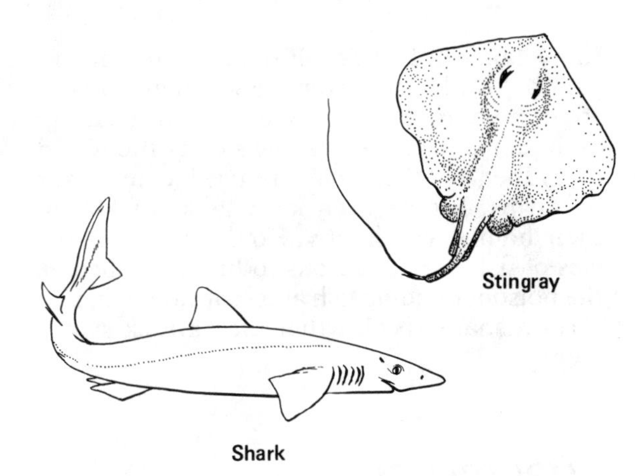

Fig. 8-3. Shark and ray—cartilaginous fish.

BONY FISHES

These are the fishes with which you are most familiar. The **bony fishes** have a skeleton made of bone, smooth overlapping scales, gills, paired fins, and a swim bladder. There are thousands of species of bony fishes. Examples include the *cod, eel, trout, sturgeon,* and *salmon* (Fig. 8-4).

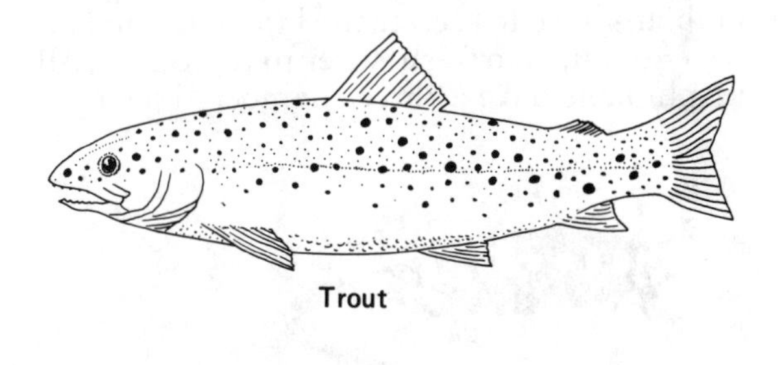

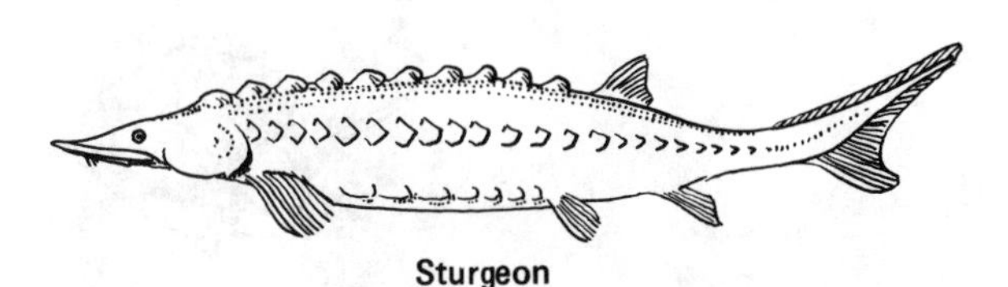

Fig. 8-4. Examples of bony fishes.

Importance of Fishes. Fish are an important part of coral reef systems and other aquatic environments. Many species of fish make up an important part of people's diets the world over. Inedible fish parts are used as fertilizer on farms. People even keep tanks of fish in their homes as a hobby. However, some species of fish are dangerous to humans, such as the poisonous stone fish and scorpion fish, and certain shark species that may attack swimmers.

AMPHIBIANS

Characteristics. Examples of this class are the *frog, toad,* and *salamander* (Fig. 8-5). Unlike fishes, **amphibians** lack scales. In their early stages of development, amphibians live in the water, breathing through their gills and thin skin. In their adult stage, most amphibians come out of the water onto land, and breathe air through their lungs and skin. The early aquatic stage is fishlike in appearance; as adults, amphibians have four limbs for moving on land. Adult frogs and toads have no tails, whereas salamanders maintain their long tails when adult. Most land-dwelling amphibians have to keep their skin moist, and all have to return to fresh water to reproduce. All amphibians have a three-chambered heart.

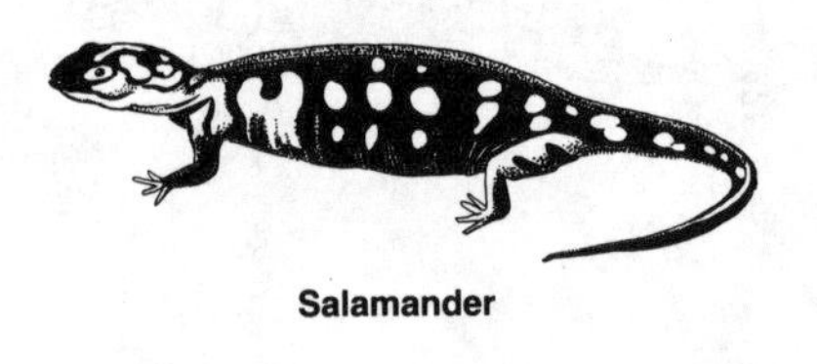

Fig. 8-5. Amphibians.

Frog. The frog starts its life as a **tadpole,** which is an aquatic fishlike animal that has a tail and no limbs. As it develops, the tadpole loses the tail and grows two sets of limbs. The long hind limbs are particularly well-adapted for jumping on land. Once its legs and lungs are fully developed, the young frog lives mostly on land, staying close to water in order to keep moist and reproduce. Frogs eat insects, which they catch with their long sticky tongues. Toads resemble frogs, but they have rough, bumpy skin and live mostly on land.

Importance of Amphibians. Because they eat so many insects, amphibians are an important part of land and water environments. Frogs and toads eat many insects, such as grasshoppers, that could be agricultural pests. Unfortunately, the number of amphibians is becoming increasingly rare worldwide, as the natural areas in which they live are being destroyed, and as many frogs are captured for food and research purposes.

REPTILES

Characteristics. Examples of this class are the *lizard, snake, crocodile,* and *turtle* (Fig. 8-6). These may seem like very different types of animals, but all **reptiles** have skin that is dry and covered by close-fitting scales. This scaly covering holds moisture inside the animal's body. Although many reptiles spend most of their time in water, all of them breathe air by means of lungs. Most reptiles have a three-chambered heart; alligators and crocodiles have a more advanced four-chambered heart. Like fishes and amphibians, reptiles are coldblooded. They often bask in the sun to warm up, and rest in the shade to cool off. Reptiles were the first vertebrates to be able to survive totally on land because of their scaly skin covering and their shell-covered egg, both of which retain moisture.

Lizards and Snakes. Lizards are fast-moving insect eaters that usually live in warm regions. They have two pairs of limbs, with claws on their toes, a long tail, and are usually small (some tropical species, however, grow up to 4 meters long). Snakes, which are thought to have evolved from lizards, have no limbs. They have extra-wide scales on their undersides; with strong muscle movements, snakes use these scales to slide along the ground and other surfaces. Snakes are predators; they have small teeth, so they swallow their prey whole. Some snakes have poisonous fangs, which they use to protect themselves and to catch prey. Snakes live worldwide: on the ground, in trees, in rivers, and even in the ocean. They vary in size, some reaching lengths of over 10 meters.

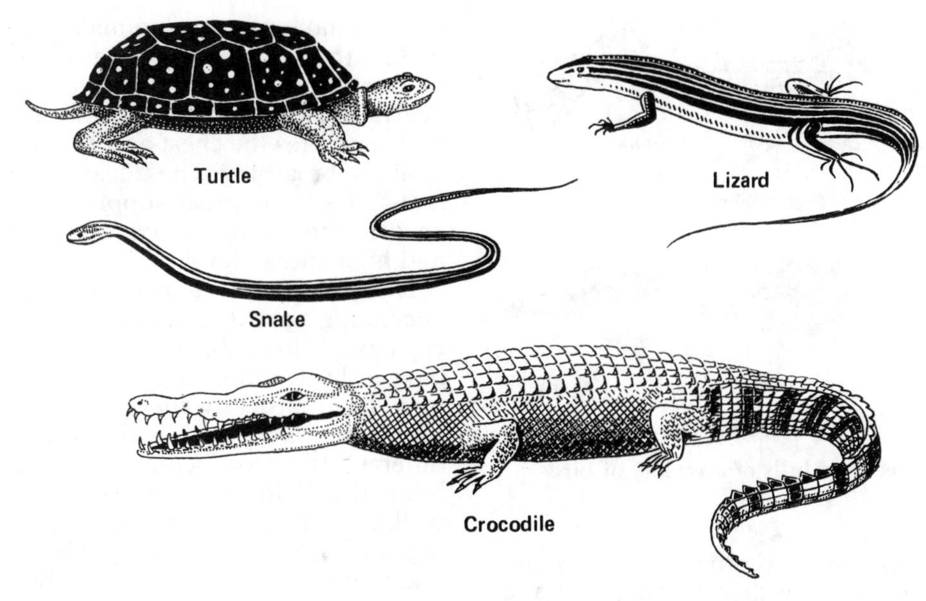

Fig. 8-6. Reptiles.

Alligators and Crocodiles. Although they look like big lizards (having four limbs with claws and a long tail), alligators and crocodiles are actually more "advanced" reptiles—some species have been observed caring for their young after they hatch in their nest. These reptiles live in warm regions around the world. They are found in rivers and swamps, feeding on fish and small animals. Some species grow quite large, reaching over 10 meters in length.

Turtles. This group includes land *tortoises,* freshwater *terrapins,* and *sea turtles.* All turtles have a horny beak (without teeth), which they use to catch food, and an upper and lower shell made up of flattened bony plates. Although some turtles are strictly aquatic, all breathe air and all return to land to lay their eggs. The largest species include the sea turtles and the giant land tortoises. The front limbs of sea turtles have evolved from legs into flippers suited for swimming.

Importance of Reptiles. Reptiles eat many insects and rodents (such as mice and rats) that may carry diseases and destroy crops. Snakes especially keep rodent populations down. Many species of reptiles are hunted for food and for their skins (to make leather). Now most of these are protected by law, because their populations have decreased due to overhunting and loss of habitat. Numerous snake species, like rattlesnakes and cobras, are dangerous to people because of their poisonous bite. Some alligator and crocodile species have been known to attack and eat humans.

BIRDS

Characteristics. There are nearly 9000 species in this class. Examples of birds are the *robin, eagle, ostrich,* and *penguin.* The body of a bird is covered with stiff, lightweight feathers. Birds are warmblooded animals, and feathers help retain the high body heat that is needed for flight energy. Other features that show that birds are suited to flight are a streamlined body; a strong four-chambered heart; powerful lungs; a pair of wings (modified front limbs) with strong flight muscles; air pockets (*air sacs*) throughout the body; hollow lightweight bones; tail feathers (no tail); and a lightweight **bill** (beak), rather than a heavy jaw with teeth.

Birds eat mainly high-energy foods, such as insects and seeds; some birds also eat a variety of vertebrates, including other birds. Very keen eyesight and good hearing are two characteristics that help birds locate their food. The feet and bills of birds are specialized for eating different types of food and for living in various habitats (Fig. 8-7). For example, the sparrow's feet are suited for perching; its short bill for cracking seeds. The heron's feet are suited for wading; its long bill for catching fish. The hawk's strong feet are suited for grasping prey; its hooked bill for tearing it apart. The duck's webbed feet are suited for swimming; its bill for straining food from the water. And the woodpecker's feet are suited for walking up trees; its pointed bill for digging insects out of tree trunks.

Because birds are warmblooded, they have been able to survive in a variety of environments. Some species of birds are even flight-

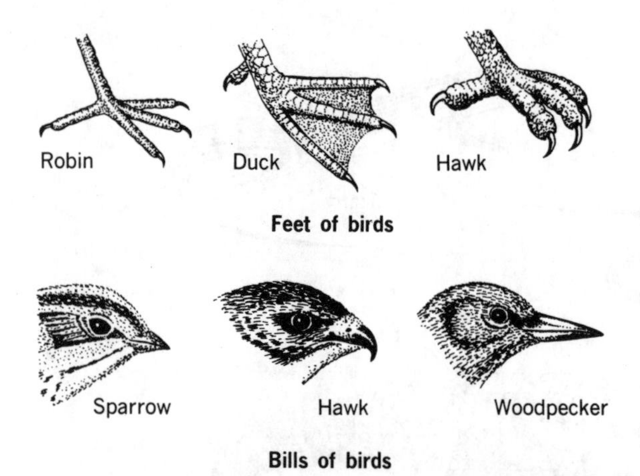

Fig. 8-7. Feet and bills of a variety of birds.

less: penguins waddle on the icy ground and "fly" underwater with their webbed feet and flipperlike wings; ostriches have long legs and strong feet for running fast.

Birds incubate their hard-shelled eggs in nests, and feed and protect their young once they hatch. Some birds are known for their spring and fall migrations, which show that they have an excellent ability to navigate.

Importance of Birds. Many species of birds are beneficial to people because they eat insects, rodents, and seeds of weeds that destroy crops. Birds are important in nature because they spread seeds and pollinate plants as they feed on nectar. People use some birds, like chickens, and their eggs as a source of food. Many people also enjoy birdwatching and keeping birds as pets. Unfortunately, the trade in wild birds for the pet industry has endangered some species of parrots. Many birds are losing their habitats as shorelines, wetlands, and forests are cleared and built upon. Certain birds, such as crows, may be considered pests when they eat our crops. Pigeons are sometimes considered a health hazard, because they can transmit a respiratory disease to people.

MAMMALS

Characteristics. There are about 5000 species in this class. Examples are the *cow, cat, whale, rabbit, monkey,* and *human.* Because **mammals** are warmblooded, they have been able to live in all kinds of environments throughout the world. They are the most abundant group of land vertebrates today.

There are several key features that have helped mammals do so well. The body covering found on most mammals is hair which, like feathers, helps to retain body heat. All mammals have large lungs, a four-chambered heart, and a **diaphragm**—a sheet of muscle that separates the chest and abdomen regions to aid in breathing. These features help mammals obtain the great supply of oxygen they need to maintain a warm body temperature and high energy level.

The digestive system of mammals varies, depending upon the animal's diet. Plant-eaters have a long digestive tract; meat-eaters have a short, simple digestive tract. All mammals have basically the same skeletal structure, with variations in the limbs reflecting different types of locomotion. Mammals exhibit the following variety of movements: walking, running, swimming, flying, swinging, and climbing. Perhaps most importantly, mammals have a large, highly developed brain. The cerebrum—the part that controls thinking and learning—is particularly well-developed; it is larger than that of any other type of animal. Mammals care for their young, and they form complex social groups. Like birds, some species of mammals migrate.

Most mammals develop within the **uterus** (or **womb**), which is a special organ of the female reproductive system. While in the uterus, the unborn mammal receives food and oxygen from the mother through a special organ called the **placenta.** Newborn mammals are all fed on milk, which is produced by special glands called **mammary glands** (hence the class name *mammals*).

Mammals are classified into about a dozen divisions, called *orders.* These groupings are based mainly upon important differences in feet (or limbs) and teeth. As in the birds, these differences reflect a variety of habitats and diets. Major differences in the development of the young separate the mammals into three main groups: the *egg-laying,* the *pouched,* and the *placental* mammals. (See Fig. 8-8, which shows mammals from each of the three main groups.)

Egg-laying Mammals. The **egg-laying mammals** make up the most unusual order of mammals. Its members—the platypus and the spiny anteater—lay eggs that have a leathery shell. In this way, they are similar to some reptiles. However, after the young hatch, they are fed on milk, like all mammals.

Pouched Mammals. The **pouched mammals** are also called **marsupials;** examples are the kangaroo, opossum, koala, and wombat. Marsupial females carry their still-developing

Fig. 8-8. Examples of the different types of mammals.

young inside a pouch on their abdomen, in which the mammary (milk) glands are found. Most marsupials live in Australia.

Placental Mammals. The **placental mammals** include several orders.

1. *Insect-eaters.* This group includes the hedgehogs, moles, and shrews. These are very small animals that hunt insects that live in the ground. Most insect-eaters are active only at night.
2. *Bats (flying mammals).* These are the only mammals that can fly. Their front limbs have evolved into wings, with skin stretched between the elongated finger bones, and back to the hind leg on each side. Most bats eat insects; some eat fruits.
3. *Toothless mammals.* This order includes the anteaters, tree sloths, and armadillos. Anteaters have no teeth; the other species just have molars. These animals have strong front claws (to open insect nests) and sticky tongues to eat insects. Tree sloths eat mainly leaves. Some anteater species live in the trees; others live on the ground.
4. *Rodents.* These are the gnawing mammals; **rodents** have large front gnawing teeth, which they use to chew through wood and other foods. This order includes the rat, mouse, squirrel, and beaver. Rabbits, which have a different type of gnawing teeth, are in their own separate order.
5. *Hoofed mammals (ungulates).* Most hoofed animals are plant-eaters (*herbivores*). The hoof is actually a very large, specialized toenail. Most hoofed animals have long legs and are fast runners. There are two groups of hoofed mammals—the odd-toed and the even-toed. The odd-toed order includes the horse, zebra, rhino, and tapir. The even-toed order includes the cow, pig, camel, deer, antelope, giraffe, and goat. All these browsers and grazers have specialized teeth that grind up the plants they eat.
6. *Meat-eaters (carnivores).* This order includes the dog, cat, bear, lion, fox, raccoon, weasel, and seal. All carnivores have sharp claws and sharp teeth for catching, killing, and eating their prey.
7. *Elephants.* Elephants are the trunk-nosed mammals. There are two types, the African and the Asian elephant, and they are the largest land animals alive today. Elephants are herbivores; they eat leaves, fruits, bark, and small trees.
8. *Whales.* This group includes the whales and dolphins. Although they breathe air, they are totally aquatic animals. Whales and dolphins eat fish and other sea life. Another type of aquatic mammal, the *manatee,* eats aquatic vegetation. The limbs of aquatic mammals are actually *flippers* and tails that are suited to swimming.
9. *Primates.* The **primates** include the lemur, monkey, ape, and human. Most primates live in or around trees, eating a variety of vegetation. Primates have good color vision (with eyes set in front of the face); hands that can grasp and handle objects; the ability to stand upright on two legs; and a highly developed cerebrum, making them among the most intelligent of all the mammals. In addition, they have grasping fingers and nails instead of claws (on the fingers and toes).

Importance of Mammals. Wild and domesticated mammals provide us with food, clothing, and companionship. Some mammals, like horses and oxen, are used to carry heavy loads and for transportation. Dogs are trained to aid blind people and to work in rescue operations. Some monkeys are trained to aid paralyzed people who are in wheelchairs. Other mammals, such as rats and mice, are considered pests because they carry diseases and raid our crops. Unfortunately, many mammals are abused in captivity, and many others are overhunted by people in the wild, either for their meat or for their skins and horns.

CHAPTER REVIEW

Science Terms

The following list contains all of the boldfaced scientific terms found in this chapter and the page on which each appears.

amphibians (p. 86)
bill (p. 87)
bony fishes (p. 85)
cartilage (p. 85)
cartilaginous fishes (p. 85)
chordate (p. 84)
coldblooded animals (p. 85)
diaphragm (p. 88)
egg-laying mammals (p. 88)
jawless fishes (p. 85)
mammals (p. 88)
mammary glands (p. 88)
marsupials (p. 88)
placenta (p. 88)
placental mammals (p. 89)
pouched mammals (p. 88)
primates (p. 89)
reptiles (p. 86)
rodents (p. 89)
spinal column (p. 84)
tadpole (p. 86)
uterus (p. 88)
vertebrae (p. 84)
vertebrates (p. 84)
warmblooded animals (p. 85)
womb (p. 88)

Matching Questions

On the blank line, write the letter of the item in column B which is most closely related to the item in column A.

Column A

____ 1. early stage of frog
____ 2. mammals with wings
____ 3. dry, scaly skin covering
____ 4. toothless mammals
____ 5. totally aquatic mammals
____ 6. have upper and lower bony plates
____ 7. feathers and lightweight bones
____ 8. mammals with pouches
____ 9. gnawing mammals
____ 10. scales, gills, and swim bladder

Column B

a. turtles
b. whales and dolphins
c. bony fishes
d. rodents
e. bats
f. tadpole
g. reptiles
h. anteaters
i. lower chordates
j. traits for flight
k. marsupials

Multiple-Choice Questions

On the blank line, write the letter preceding the word or expression that best completes the statement or answers the question.

1. The two main features of *all* vertebrates are the backbone and the
a. gills *b.* internal skeleton *c.* scales *d.* milk glands 1 ____

2. A fish having a cartilaginous skeleton and jaws with many teeth is the
a. lamprey *b.* hagfish *c.* shark *d.* trout 2 ____

3. Gills and a swim bladder are both found in the
a. whales *b.* tadpoles *c.* bony fishes *d.* manatees 3 ____

4. Two coldblooded animals are the
a. trout and penguin
b. lizard and antelope
c. toad and eagle
d. frog and turtle 4 ____

5. Most amphibians spend their lives
a. first in water and then underground
b. first in water and then above ground
c. first underground and then above ground
d. first above ground and then under water 5 ____

6. A four-chambered heart is found among all
a. birds and mammals
b. fish and mammals
c. birds and amphibians
d. reptiles and amphibians 6 ____

7. The body temperature of warmblooded animals
a. may vary with the outside temperature
b. always remains nearly the same
c. rises 10°F during hot weather
d. rises 10°F during cold weather 7 ____

8. The feet of a duck are best suited for
a. walking *b.* running *c.* flying *d.* swimming 8 ____

9. The hawk's hooked bill is best suited for
a. cracking seeds
b. digging for insects
c. tearing into flesh
d. eating soft fruits 9 ____

10. The following animals lay eggs that have shells:
a. fish and birds
b. fish and reptiles
c. birds and reptiles
d. amphibians and birds 10 ____

11. A large, well-developed brain is characteristic of the
a. reptiles *b.* mammals *c.* bony fishes *d.* birds 11 ____

12. *Most* young mammals develop within
a. a pouch *b.* an egg *c.* a uterus *d.* a mammary gland 12 ____

13. The various groupings of mammals are based mainly upon differences in their
a. body size
b. limbs and teeth
c. brain size
d. types of fur 13 ____

14. Which of the following is an egg-laying mammal?
a. opossum *b.* hedgehog *c.* platypus *d.* fruit bat 14 ____

15. Which of the following is a pouched mammal?
a. koala *b.* anteater *c.* tree sloth *d.* squirrel 15 ____

16. Humans are included in which order of mammals?
a. gnawing *b.* toothless *c.* primates *d.* carnivores 16 ____

17. The three classes of fishes include all *except* the following:
a. jawless *b.* pouched *c.* cartilaginous *d.* bony 17 ____

18. The trunk-nosed mammals are the
a. whales *b.* insect-eaters *c.* elephants *d.* bats 18 ____

Modified True-False Questions

In some of the following statements, the italicized terms makes the statement incorrect. For each incorrect statement, write the term that must be substituted for the italicized term to make the statement correct. For each correct statement, write the word "true."

1. A chordate's flexible support rod is known as the *notochord.* 1 ____________

2. Jawless fishes that live as parasites are the lamprey and the *hagfish.* 2 ____________

3. Unlike fishes and reptiles, amphibians lack *skin.* 3 ____________

4. Most hoofed mammals are *carnivores.* 4 ____________

5. Most reptiles and all amphibians have a *3-chambered* heart. 5 ____________

6. Rats and mice are classified as *insect-eaters.* 6 ____________

7. Sharks and rays have skeletons that are made of *bone.* 7 ____________

8. The largest land animals alive today are the *primates.* 8 ____________

9. Air sacs and a strong heart and lungs are traits that aid birds in *flight.* 9 ____________

10. The most abundant group of land vertebrates today are the *reptiles.* 10 ____________

Testing Your Knowledge

1. Both bats and robins are warmblooded animals that have four-chambered hearts. Why are they not classified together?

____

____

2. Why do scientists classify the dolphin as more closely related to the gorilla than to the tuna?

____

____

3. Why aren't 'shellfish' (lobsters, crabs, etc.) classified as fish?

____

____

4. Name two or three ways in which the "advanced" reptiles—the crocodiles and alligators—are similar to birds.

____

5. In what way do the feathers of birds and hair (fur) of mammals serve a similar function?

____

6. Explain why reptiles have to bask in the sun to warm themselves.

____

7. Explain how the skin of frogs and scales of reptiles differ, and how these features help the animals survive in their different environments.

____

____

8. In each of the following groups, one term does not belong with the others. Write that term in the blank space at the right.

a. frog, toad, snake, salamander ____________

b. horse, deer, cow, elephant ____________

c. eagle, chicken, trout, robin ____________

d. fox, lion, wolf, bat ____________

e. lemur, opossum, monkey, ape ____________

9. How do whales differ from fishes?

____

____

10. How do mammals differ from birds and reptiles?

____

____

Endangered Species: Can They Be Protected?

Bulldozers roar. Chain saws whine. Gunshots ring out. Traps clang shut.

To many people, these are the sounds of progress. The bulldozers clear away trees and shrubs to make room for growing populations and farms. These powerful machines also scrape away the land, so people can more easily get at valuable natural resources such as coal and copper. The chain saws bring down huge 200-year-old trees whose wood will be used to build houses and make furniture. The gunshots and traps will kill animals to provide furs for clothing, ivory for jewelry, excitement and recreation for hunters, and "medicines" for the superstitious.

In the process of doing these things, people change the environment. Farms, towns, and mines replace forests. Once-thriving populations of animals are reduced from millions to a few thousand or less as their habitats are destroyed or their members hunted down. A dramatic and typical example is the rhinoceros, which has a 60-million-year history on our planet. It is a descendant of the largest land mammal ever to thunder across the Earth; an animal whose muscular shoulders rose twice as tall as a one-story house, and from tail to nose was longer than a school bus.

Conservationists tell us that if nothing is done to save the rhino, its long history will soon end. They point out that only 90 years ago, there were one to two million of these giant vegetarians living in Africa and Asia. Today, the five existing species number only about 8000 in total.

The story of the decline of the rhino is similar to that of other endangered species, including tigers, elephants, sea turtles, chimpanzees, mountain gorillas, snow leopards, whales, even many frogs and toads. As a matter of fact, today there are well over 1000 animal species that have been declared endangered by various government agencies and private groups. And at least 550 of these species live in the United States. *Endangered* means on the verge of becoming extinct. Many other animals are listed as *threatened,* which means on the verge of becoming endangered.

Back to the story of the rhino. Growing human populations in Africa and Asia continue to clear land for houses, farms, and other human uses. Much of this land was once the home of the rhino. In addition, this great mammal was hunted for sport and is still hunted illegally for its horns. Some people in the Far East believe that the horns possess magical or medicinal powers. Other people in the Middle East still fashion the horns into dagger handles and other ornaments.

In the United States, and in many other countries, laws have been passed to protect animals that are declared endangered. The rhino is among those listed for protection. (This means that people cannot import or export products made from rhinoceros.) In 1973, the U.S. Congress passed the Endangered Species Act. This law requires certain government agencies to devise habitat conservation plans to protect endangered animals and then to carry out recovery plans to prevent their numbers from decreasing.

Unfortunately, devising and enforcing such plans require much time and money. In addition, conflicts sometimes arise between

plans for animal protection and plans for land use and development. As a result, many endangered species are still not adequately protected. And, according to some, they shouldn't be, if the animals' needs conflict with those of people. What do you think?

1. *Animal species are threatened by all the following except*

   a. farmers.
   b. hunters.
   c. miners.
   d. conservationists.

2. *The numbers of rhinoceroses 90 years ago was about*

   a. 11,000
   b. one to two million.
   c. one to two thousand.
   d. a few hundred.

3. *In 1973, the United States Congress passed the*

   a. Threatened Species Act.
   b. Extinct Species Act.
   c. Endangered Species Act.
   d. Save the Species Act.

4. *The two final sentences of the feature above include a provocative statement and a question. Write a response to the question. If you need more space continue on a separate sheet of paper.*

   ____

   ____

   ____

   ____

5. *Recently, conservationists succeeded in getting the northern spotted owl listed as an endangered species. This bird lives only in the old-growth timber forests of America's northwest coast, where many people work as lumberjacks. To protect the owls, the trees have to be left standing. Some people ask: "Don't we need the wood and lumberjack jobs more than we need the spotted owl? After all, what's the loss of one species of owl worth anyway?" Other people ask: "Can't we protect the spotted owl and still have enough wood and jobs too?" Write a short essay in which you address each of the questions posed above.*

9

Asexual Reproduction

LABORATORY INVESTIGATION

VEGETATIVE PROPAGATION: GROWING PLANTS FROM PLANT PARTS

This laboratory investigation will take about ten days before observations can be made. Your teacher will present the class with the following plant parts: *bryophyllum* or begonia leaves; coleus or geranium cuttings; onion bulbs; carrot tops or sweet potatoes.

A. Propagation by Leaves. Obtain one of these plant parts and proceed as follows:

1. Place a *bryophyllum* leaf flat on a wet cellulose sponge in a dish. Keep the sponge moist. After seven to ten days, observe your leaf. Describe the changes that have occurred.

____

2. Plant the tiny stalk (*petiole*) of a begonia leaf in a small pot of soil. Keep the soil moist. After seven to ten days, observe your leaf cutting.

a. Describe the changes that you see.

____

b. What parts of a plant body do you now observe besides the leaf?

____

B. Propagation by Stems. Obtain one of these plant parts and proceed as follows:

3. Place a geranium or coleus plant cutting in a glass of water, or plant it in a small pot of sand or soil. Keep moist. After ten days, observe the cutting. (If necessary, uproot and then replant it.)

a. Describe the plant growth that has occurred.

____

b. What plant parts did the cutting originally have?

___

c. On which part of the cutting did the new growth occur?

___

C. Propagation from Bulbs. Obtain an onion bulb and proceed as follows: (See Fig. 9-1.)

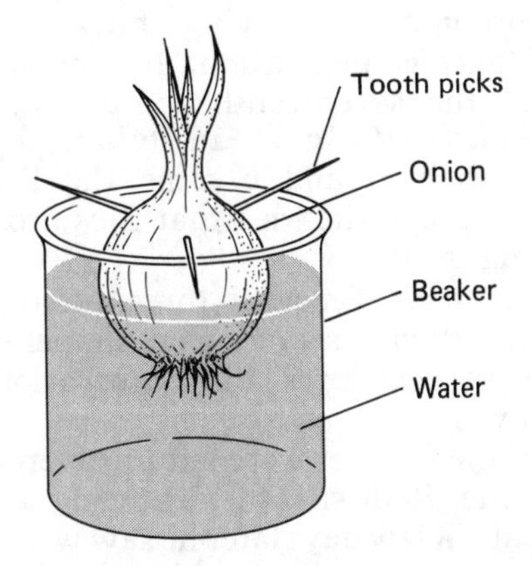

Fig. 9-1. Vegetative propagation from an onion bulb.

4. Poke three toothpicks into the top half of an onion. Place the onion in a beaker of water so that the toothpicks hold the onion up, with just the bottom half of it submerged in the water. Be sure to refill the beaker as needed to keep the bottom of the onion in water. After about ten days, observe your onion.

a. Describe the kinds of plant growth that you see.

___

b. What part of the plant do you think the bulb of an onion is?

___

D. Propagation from Roots. Obtain one of these plant parts and proceed as follows:

5. Take the top 2 cm of a carrot or an end piece of a sweet potato and place it, cut end facing down, in a dish of water. Replace water as needed. After about one week, observe your carrot or sweet potato.

a. Describe the changes that you see.

___

b. What part of the plant is the carrot?

___

c. The sweet potato?

___

d. Describe what you would expect to see happen if these plant parts were left in the water for a few more weeks.

___

Reproduction

People once believed that new generations of living things arose from nonliving matter. For example, it was thought that frogs and fish arose from river mud, and that worms and flies arose from decaying meat. This false idea about the production of living things was called **spontaneous generation.** It was not until late in the seventeenth century that *Francesco Redi* (1626–1697), an Italian physician, experimented and became the first scientist to disprove the idea that flies arose from decaying meat.

Redi placed slices of meat in three different sets of jars and then observed what happened (Fig. 9-2). His procedures and observations were as follows:

The jars of the first set were left uncovered. In a short time, Redi saw flies entering and leaving the jars. A few days later he saw worm-like organisms (*maggots*) crawling on the meat. Within one month, he saw these organisms develop into flies.

The jars of the second set were tightly sealed with a paper covering. Neither flies nor air could enter the jars. And Redi found neither flies nor maggots in or around these jars.

Redi reasoned that the odor of the decaying meat in the open jars attracted adult flies. In the first set of jars, these flies laid eggs on or near the meat. The eggs developed into maggots and then into flies. Since no odor escaped from the sealed jars, flies did not come near them; nor were the flies able to lay their eggs in them. Redi's conclusion was that each generation of flies comes from eggs laid by a preceding generation like itself.

Some scientists of the time did not believe Redi's results; they thought the second set of jars did not hatch maggots and flies because air could not get into them. So Redi experimented again, placing meat in a third set of jars covered with a cloth netting that let air in and out. Although flies could smell the meat and land on the jars, they could not enter to lay their eggs on the meat. The result was that neither maggots nor flies hatched in these jars—again supporting Redi's idea that living things come only from other living things.

Today we know that all living things give rise to offspring like themselves by the process of *reproduction.*

In the course of living, every organism develops and grows, fights disease, adjusts to its environment, escapes enemies, and competes for food with other organisms. In other words, organisms struggle to exist. Eventually, every organism loses the struggle and dies. If, before their death, members of a species did not reproduce, the entire species would die out. Although an individual need not reproduce in order to live, some individuals must reproduce for their species to continue.

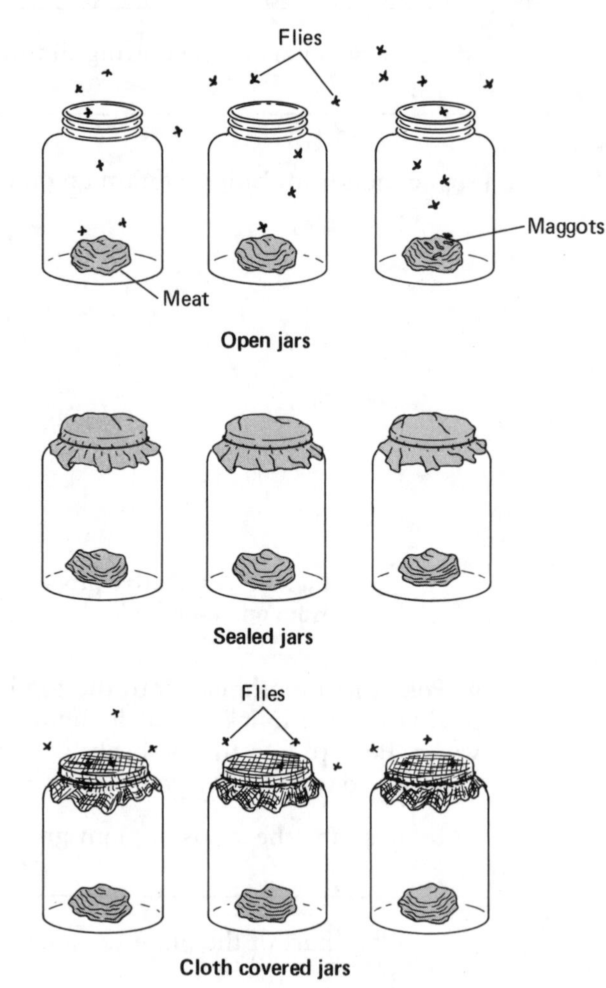

Fig. 9-2. Redi's experiment.

During the process of reproduction, genes, which control the traits of organisms, pass from the old generation to the new one. Before this occurs, the genes, which are located on the chromosomes of a cell, are copied, forming another set of genes. Depending upon the type of reproduction, either half a set from each parent, or one whole set of genes from one parent passes into, and controls the development and traits of, the new individual.

There are two types of reproduction: *asexual* and *sexual.*

Asexual Reproduction

When offspring are produced by only one parent, the process is called **asexual reproduction.** There are several types of asexual reproduction, but all have one characteristic in common—a set of genes from only one parent passes into the next generation. Consequently, the offspring usually have traits exactly like those of the parent and like those of one another.

BINARY FISSION

Binary fission is the splitting of an organism into two equal parts. It is the simplest type of reproduction, and it occurs in simple organisms like the ameba, paramecium, bacteria, and euglena.

Binary Fission in the Ameba. After an ameba feeds, grows, and reaches full size, its nucleus divides into two equal parts (see *mitosis*, page 32). At the same time, the cytoplasm thins out in the middle, pinches in, and separates into two equal parts. Each of these parts has one nucleus in it (Fig. 9-3). Each of the two new individuals, or *daughter cells*, is smaller than the original, or *parent cell*. Both daughter cells have the same set of genes.

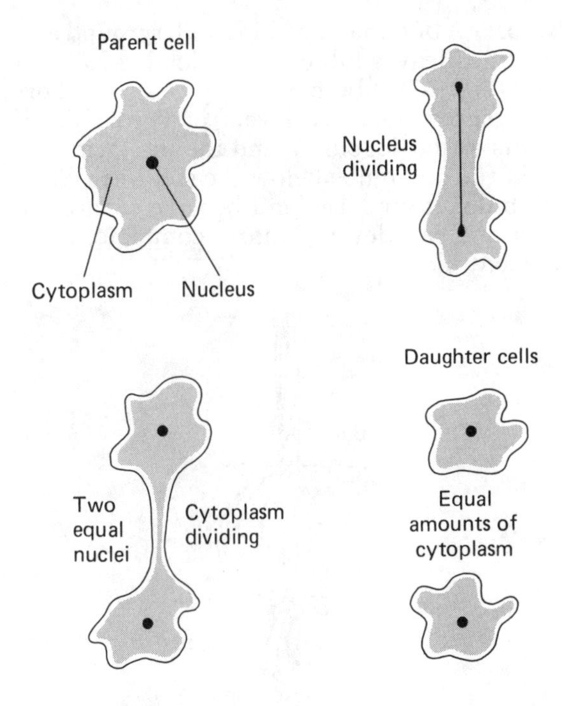

Fig. 9-3. Binary fission in the ameba.

BUDDING

Budding is the splitting of an organism's cytoplasm into two unequal parts, and its nucleus into two equal parts. At the beginning of this process, the new generation grows out of the parent and goes through its early development while still attached to the body of the parent. One-celled organisms, such as yeast, and simple multicellular organisms, like the hydra, often reproduce by budding.

Budding in Yeast. In a yeast cell that is about to bud, the moving cytoplasm carries the nucleus over to one side of the cell. Mitosis takes place and the nucleus divides into two equal parts. The cytoplasm and the cell wall near the nucleus push out and form a small projection, the *bud* (Fig. 9-4). One of the nuclei moves into the bud. The other nucleus stays inside the parent cell. Eventually, the cell wall grows in and separates the bud from the parent. Although the new daughter cell is smaller than the parent cell, both cells have identical sets of genes. A bud may even produce a bud of its own before breaking away from the parent. This is the way in which a yeast *colony* starts.

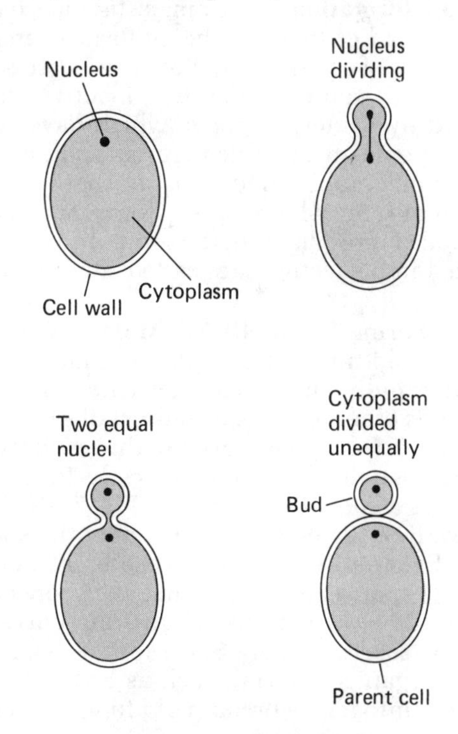

Fig. 9-4. Budding in yeast.

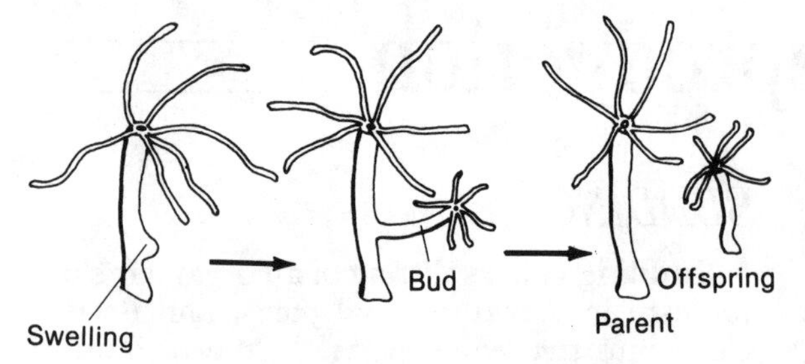

Fig. 9-5. Budding in hydra.

Budding in Hydra. A small swelling, which becomes the bud, forms on the body below the arms of the hydra (Fig. 9-5). The bud consists of numerous cells. As the bud matures, it develops a mouth and small arms. In time, the bud becomes detached from the parent hydra and grows to adult size.

SPORE FORMATION

Spore formation is like binary fission except that instead of two cells being formed, many small cells of equal size are formed. These cells are called **spores.** Like the inactive cells formed by bacteria, spores, which have very little cytoplasm, are able to resist dryness and other unfavorable conditions. In those plants and fungi in which spore formation is a method of reproduction, the spores are usually formed in protective sacs called **spore cases.**

Spore Formation in Bread Mold. When a bread mold fungus has food and moisture, it grows rapidly and spreads over its food. The free ends of many of its upright stalks become swollen. The living material within each swollen tip then divides into many cells. Each cell becomes a spore (refer back to Fig. 5-8*b*). The cell wall of the enlarged tip is now the spore case. Eventually, the spore case breaks open and the spores are freed. Since the spores are tiny and very light in weight, air currents readily scatter them. Spores that land on moist organic material, such as bread, begin to grow into a new bread mold fungus, if conditions are favorable.

VEGETATIVE REPRODUCTION

An organ of a plant that makes, transports, or stores food is called a **vegetative organ.** Leaves, buds, roots, and stems are the major vegetative organs of vascular plants. As you learned in your lab investigation, these non-reproductive organs can give rise to entire new organisms of the same kind. This process is called **vegetative reproduction** (or **vegetative propagation**). Plants produced in this way are identical to their parent plant.

Tubers. Some stems develop underground and become swollen with excess (stored) food. Such a stem that has buds projecting from it is called a **tuber** (Fig. 9-6). The buds on tubers are commonly called "eyes."

A white-potato plant produces several tubers. The tubers, or white potatoes, can have many buds on them. Each bud can eventually produce a new potato plant.

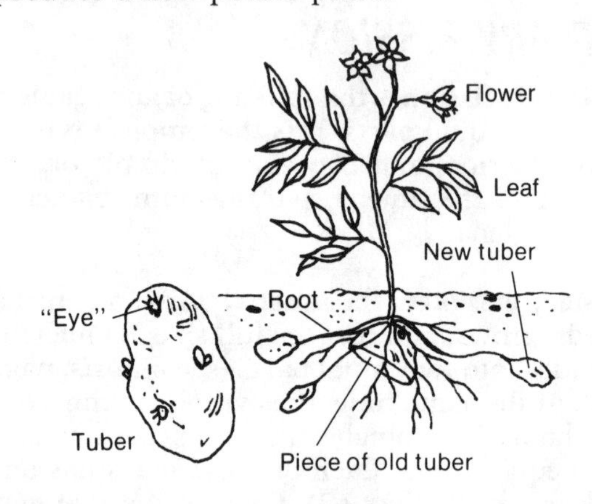

Fig. 9-6. Vegetative reproduction—tuber (white potato).

Bulbs. A bud that develops underground and has thick leaves full of excess food is called a **bulb** (Fig. 9-7). A bulb is actually a small, short stem surrounded by layers of leaves. Such plants as garlic, onion, and the lily form several attached bulbs below the soil line. When the bulbs grow large and become separated, each one can develop into a complete, independent plant.

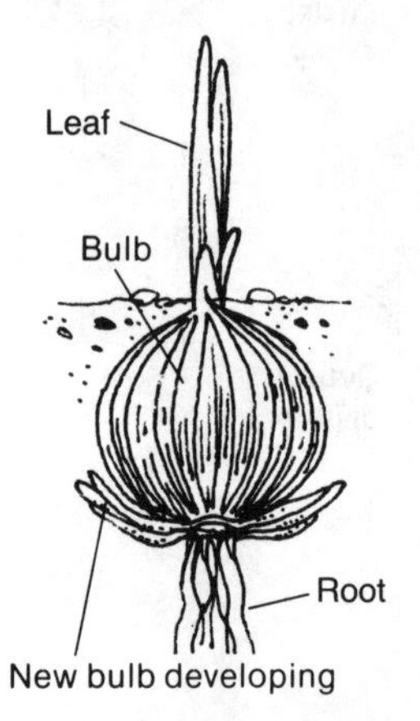

Fig. 9-7. Vegetative reproduction—bulb (onion).

Stem Cuttings. A **stem cutting** (Fig. 9-8) is a piece of stem or branch that has been cut from a growing plant and then planted. After the cutting forms roots and its buds open into leaves, the cutting becomes established as a new plant. Plants such as the lilac, geranium, and begonia are often grown from cuttings.

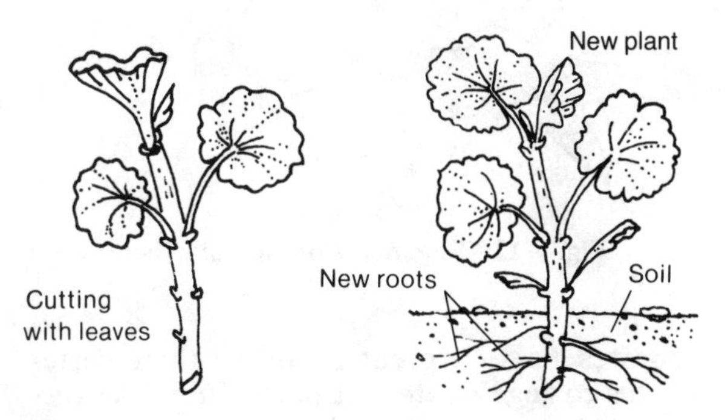

Fig. 9-8. Vegetative reproduction—stem cutting (geranium).

Fleshy Roots. A **fleshy root** is a root that becomes enlarged with stored food (Fig. 9-9). When a fleshy root is planted, it can develop leaves and a stem and form a complete plant. Such plants as the carrot and the sweet potato can be grown from their fleshy roots.

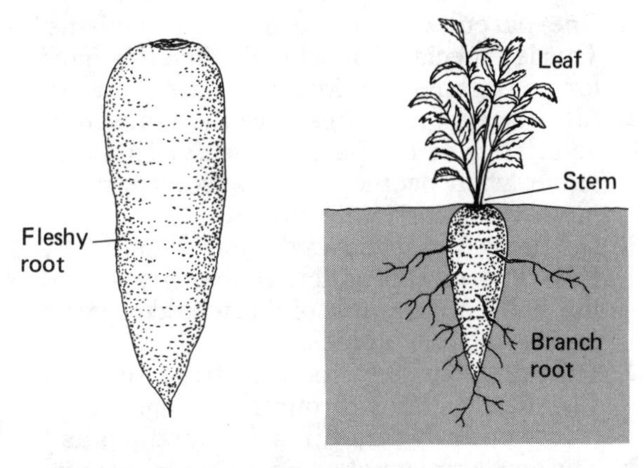

Fig. 9-9. Vegetative reproduction—fleshy root (carrot).

GRAFTING

Grafting is a type of artificial vegetative reproduction that occurs only with the help of people. When grafting is carried out, part of one plant, either a twig, stem, or a bud, is cut off and transferred to another plant. The part of the first plant that has been cut off and transferred to the other plant is called the *scion.* The second plant, which has roots and receives the scion, is called the *stock* (Fig. 9-10). Good grafts are usually obtained when the stock and the scion are of plant species or varieties that are closely related. The closer the relationship between the plants, the better the chance for their conducting tissue and growing tissue to make good contact. Such contact enables minerals, water, and food to pass between the grafted parts. The growing layers of both parts can then grow together and form a permanent attachment.

Ordinarily, grafting cannot be used to produce new breeds of fruits or flowers. Both the scion and the stock continue to grow their parts as they did before being grafted. For example, when a branch from a purple-flowered lilac is grafted to a white-flowered lilac plant, the scion continues bearing purple flowers and the stock, white flowers.

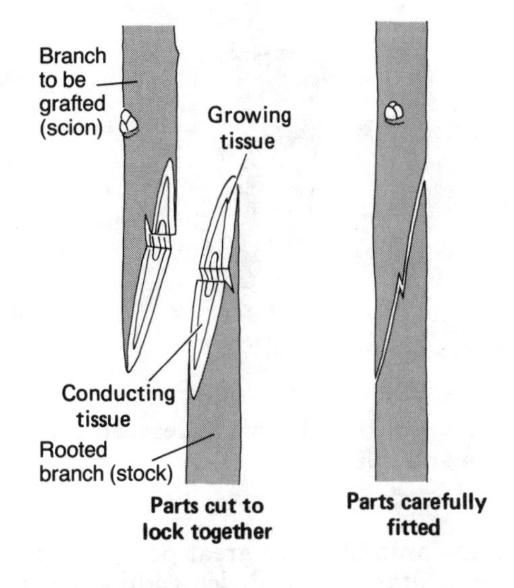

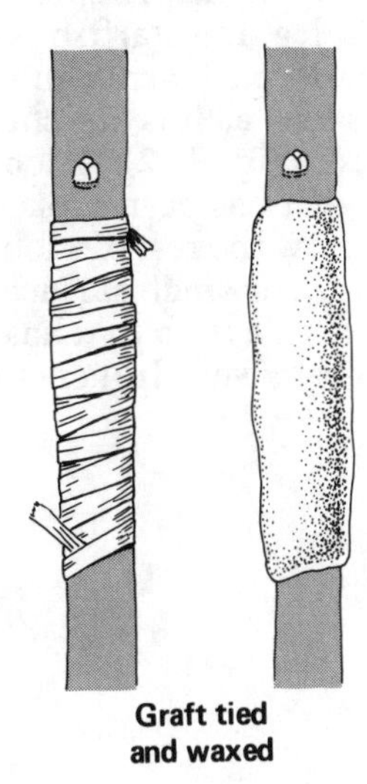

Fig. 9-10. Vegetative reproduction—grafting.

REGENERATION

Regeneration is the process by which an organism grows back a part of its body that has been lost. In plants, regeneration is often the same as vegetative reproduction because complete plants are formed. In animals, regeneration also occurs, but to a limited extent. Hardly any animals reproduce naturally by regeneration. Planaria worms can reproduce this way, either when cut in half or when they split in half lengthwise (see Fig. 9-11). You can even produce a worm with two heads or two tails if you cut planaria the proper way.

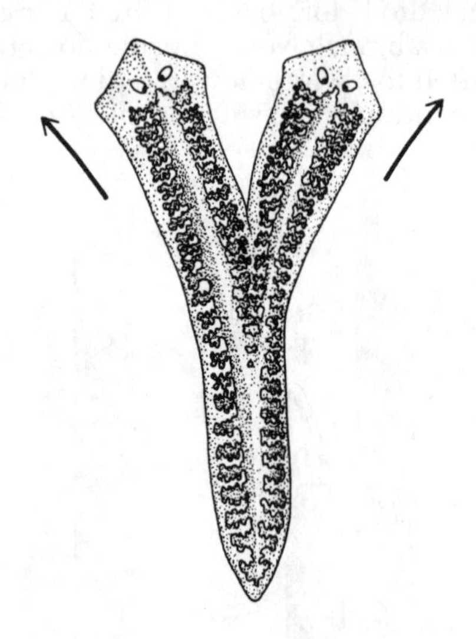

Fig. 9-11. Reproduction by regeneration—a planaria splitting.

Some animals have great powers of regeneration, others very little. Tadpoles can regrow a lost tail or leg. If a starfish is chopped into five equal parts, each arm can grow into a new starfish, as long as it is attached to part of the central disk (Fig. 9-12). A lobster also can regrow a leg that has been broken off, but the leg cannot regrow the rest of a lobster. The human body can repair small surface injuries, but cannot grow back a limb that has been cut off or surgically removed. In general, invertebrates have a greater ability than vertebrates to regenerate lost parts. Very few vertebrates can regenerate lost parts.

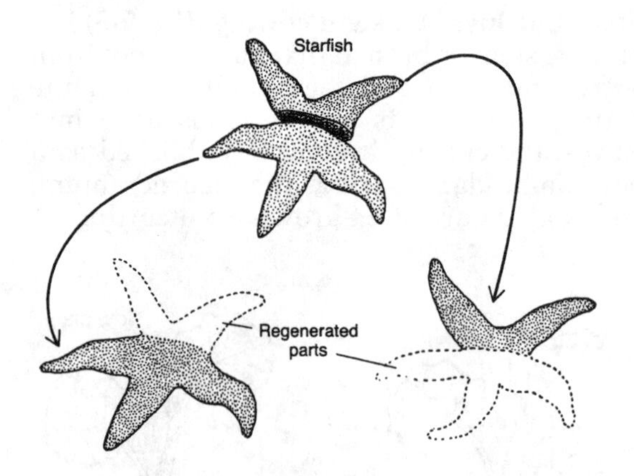

Fig. 9-12. Regeneration in a starfish.

BENEFITS OF ASEXUAL REPRODUCTION

Asexual reproduction benefits organisms and people for the following reasons:

1. One parent can produce new organisms. This is especially beneficial when it is hard for an organism to locate a mate.
2. All the new organisms have the same characteristics as the parent variety. This is of value when people want to produce many plants of a particular breed.
3. Offspring are produced more rapidly by this type of reproduction than by sexual methods. An example of this would be yeast and bacterial colonies.
4. Offspring can be produced from seedless varieties of plants through grafting. This is a very desirable trait in many food plants.
5. The development of an organism by asexual reproduction often takes less time than development from sexual reproduction.

CHAPTER REVIEW

Science Terms

The following list contains all of the boldfaced scientific terms found in this chapter and the page on which each appears.

asexual reproduction (p. 99)
binary fission (p. 99)
budding (p. 99)
bulb (p. 100)
fleshy root (p. 101)
grafting (p. 101)
regeneration (p. 102)
spontaneous generation (p. 98)
spore cases (p. 100)
spore formation (p. 100)
spores (p. 100)
stem cutting (p. 101)
tuber (p. 100)
vegetative organ (p. 100)
vegetative propagation (p. 100)
vegetative reproduction (p. 100)

Matching Questions

On the blank line, write the letter of the item in column B which is most closely related to the item in column A.

Column A

____ 1. splitting of organism into two equal parts
____ 2. offspring from one parent only
____ 3. splitting of organism into two unequal parts
____ 4. reproductive cells formed in cases
____ 5. leaves, buds, roots, and stems
____ 6. growth of new plant from plant organs
____ 7. underground stem with reproductive buds
____ 8. underground stem with layers of leaves
____ 9. carrots and sweet potatoes
____ 10. regrowth of lost body parts

Column B

a. vegetative organs
b. fleshy roots
c. tuber
d. grafting
e. asexual reproduction
f. bulb
g. budding
h. regeneration
i. binary fission
j. spores
k. vegetative reproduction

Multiple-Choice Questions

On the blank line, write the letter preceding the word or expression that best completes the sentence or answers the question.

1. Where do houseflies come from?
a. eggs *b.* garbage *c.* worms *d.* rotten fruit — 1 ____

2. Before a cell divides, the genes in the nucleus
a. are halved
b. are copied
c. move to the surface
d. move into the nucleolus — 2 ____

3. The number of parents that take part in asexual reproduction is
a. none *b.* 2 *c.* 3 *d.* 1 — 3 ____

4. Two amebas that came from one parent are likely to have
a. different genes
b. the same genes
c. the same number but not the same kind of genes
d. the same kind of genes but not the same number — 4 ____

5. Which is the simplest type of reproduction?
a. fertilization *b.* sexual reproduction *c.* binary fission *d.* grafting — 5 ____

6. In binary fission, the nucleus divides equally by the process of
a. cell division *b.* mitosis *c.* sporing *d.* budding — 6 ____

7. Unequal division of an organism occurs during the process of
a. sporing *b.* fission *c.* spontaneous generation *d.* budding — 7 ____

8. Two organisms that reproduce by budding are
a. the spirogyra and the starfish
b. the yeast and the clam
c. the yeast and the earthworm
d. the yeast and the hydra — 8 ____

9. The asexual reproductive parts of a bread mold are called
a. buds *b.* grafts *c.* spores *d.* roots — 9 ____

10. The *stock* and the *scion* are terms for plant parts used in
a. budding *b.* grafting *c.* sporing *d.* seeding — 10 ____

11. A mold growing on bread was probably brought to the bread by
a. an air current *b.* a water current *c.* a bird *d.* cilia — 11 ____

12. A vegetative organ of a plant is ordinarily used by the plant for
a. reproduction *b.* nutrition *c.* locomotion *d.* excretion — 12 ____

13. Tulips and lilies are usually grown by gardeners from
a. slips *b.* grafts *c.* bulbs *d.* tubers — 13 ____

14. When a plant is grown from a stem cutting, the structures that grow back soon are the
a. roots *b.* flowers *c.* spores *d.* seeds — 14 ____

15. The "eye" of a potato is actually a
a. root *b.* stem *c.* bud *d.* seed — 15 ____

16. An example of a fleshy root that can grow into a complete plant is the
a. sweet potato *b.* ivy *c.* geranium *d.* rose — 16 ____

17. The part of a plant that is grafted onto another plant is the
a. tuber *b.* seed *c.* stem *d.* leaf 17 ____

18. The fruits that would be expected to grow on a plum twig that has been grafted to a peach tree are
a. plums *b.* peaches *c.* nectarines *d.* plums and peaches 18 ____

19. Successful plant grafts depend on direct contact between the layers of
a. leaves *b.* roots *c.* bark *d.* growing tissue 19 ____

20. An advantage of asexual reproduction is
a. similarity of offspring *c.* fewer offspring
b. more mates for the offspring *d.* different offspring 20 ____

Modified True-False Questions

In some of the following statements, the italicized term makes the statement incorrect. For each incorrect statement, write the term that must be substituted for the italicized term to make the statement correct. For each correct statement, write the word "true."

1. The life process by which offspring are produced is called *regeneration.* 1 ________

2. The belief that flies and worms come from decaying meat is known as *spontaneous generation.* 2 ________

3. A starfish can regrow a lost arm by the process of *binary fission.* 3 ________

4. More seedless oranges can be produced by *planting seeds.* 4 ________

5. Yeast cells form colonies by producing *bulbs.* 5 ________

6. In Redi's experiment, flies hatched only in the jars that were *uncovered.* 6 ________

7. Offspring formed by binary fission are *unequal* in size. 7 ________

8. When a yeast cell forms a bud, its *nucleus* divides into two unequal parts. 8 ________

9. A white potato is actually an underground *stem.* 9 ________

10. Onions can be produced more rapidly by *vegetative reproduction* than by sexual reproduction. 10 ________

Testing Your Knowledge

1. Explain the difference between the two terms in each of the following pairs:
a. scion and stock

____

____

____

b. bulb and fleshy root

c. tuber and stem cutting

d. binary fission and budding

2. By drawing clearly labeled diagrams, show how budding in yeast differs from binary fission in ameba.

3. When a starfish loses an arm or a lobster loses a claw, these animals regenerate (regrow) the missing parts. Why isn't this ability to regenerate considered a method of reproduction?

4. Explain why and how you would employ grafting in the following case: The roots of European grapevines, which bear sweet fruit, are destroyed by parasites when they are planted in American soil. The roots of the wild American grapevines, which bear sour fruit, are immune to these parasites.

5. Newspaper advertisements often offer for sale a tree that bears McIntosh, Delicious, and other apple varieties. Explain the reason for believing that the advertisement could be true.

6. Describe an experiment that you could perform to tell whether mosquitoes arise from water.

Sexual Reproduction

LABORATORY INVESTIGATION

THE REPRODUCTIVE SYSTEM OF FLOWERING PLANTS

Sexual reproduction is basically the same in both plants and animals: sperm cells fertilize egg cells to produce offspring. Your teacher will present you with a flower that contains both the male and female reproductive organs of the plant. Use a hand lens, if necessary, to closely examine the flower parts. Record your observations in the space provided.

A. Place your flower on a sheet of white paper and examine it carefully. Push aside, or remove, the petals to get a better view of the flower's reproductive parts. Refer to Fig. 10-1 on the following page. Draw each part, as you observe and identify it.

1. Identify the *stamens*. How many stamens does your flower have? ______________________

____

2. *a.* Identify the *anthers*. Where are they located? ______________________

____

b. Describe some characteristics of the anthers. ______________________

____

3. Identify the *anther stalks*. Where are they found? ______________________

____

4. Identify the *pistil*. How many pistils are there? ______________________

____

5. *a.* Identify the *stigma*. Where is it located? ______________________

____

b. Describe some characteristics of the stigma. ______________________

____

6. Identify the *style*. Where is it found? ______________________

B. Using a scalpel, carefully slice open the base of the pistil crosswise.

7. Identify the *ovary* and the *ovules*. How many ovules are there? ________________

____

8. Is your flower male and/or female. Explain your answer. ________________

____

Sexual Reproduction

In asexual reproduction, only one parent is needed to produce offspring. In **sexual reproduction,** two parents are needed to produce offspring. One of these parents is female and the other is male. In the reproductive process, each parent contributes a special cell, called a **sex cell.** When the male sex cell, nucleus or **sperm,** unites with the female sex cell, nucleus or **egg,** a *fertilized egg* is formed. The offspring that develops has genetic material from both parents.

SEXUAL REPRODUCTION IN PLANTS

Some simple plants can reproduce asexually. However, most plants have two stages in their life cycles. One is an asexual stage (spore formation), and the other is a sexual stage in which eggs are fertilized by sperm.

In the flowering plants, egg cells and sperm cells are produced within the flower.

Structure of a Flower. Most flowers have four groups of parts—*sepals, petals, stamens,* and *pistils* (Fig. 10-1).

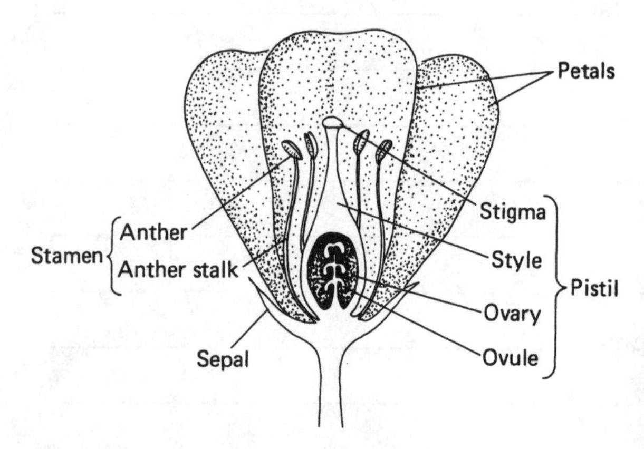

Fig. 10-1. Parts of a flower (cutaway view).

Sepals are the leaflike, outermost structures that are located in a circle at the base of a flower. Sepals are usually green and carry out photosynthesis. The sepals surround and cover the delicate flower before it is fully developed, protecting its inner parts.

Petals are the parts that are usually brightly colored. They are leaflike in structure and are located just within the circle of sepals. Special cells near the bottom of the petals produce a fluid, called *nectar*. The nectar contains sugar and substances that give flowers their characteristic scent. The color, scent, and nectar of flowers are attractive to insects and other animals.

The **stamens** are organs that produce the male sex cells of the plant. Each stamen has two important parts:

1. The *anther* is a sac that contains pollen grains. The **pollen grains** actually mature into the sperm cells of a flower.
2. The *anther stalk* is a slender structure that supports the anther.

The **pistil** is the organ that produces the female sex cells of the plant. Some flowers have several pistils; others, only one. A pistil consists of the following parts:

1. The *stigma* is the topmost part of the pistil. It is sticky and often hairy, so it can catch and hold pollen grains.
2. The *style,* which is below the stigma, connects the stigma to the ovary.
3. The *ovary,* the enlarged base of the pistil, contains one or more round bodies, called **ovules.** Eventually, each ovule produces a mature egg cell.

Pollination of a Flower. The first step in bringing together the eggs and sperm of flowers is **pollination.** In this process, pollen grains are transferred from an anther to a stigma. *Self-pollination* is the transfer of pollen from an anther of one flower to a stigma of that same flower. *Cross-pollination* is the transfer of pollen by some means from an anther of one flower to a stigma of the same kind of flower on another plant. Natural agents of cross-pollination include wind, water, insects, birds, and some small mammals. Pollination of large, colorful flowers is usually carried out by the actions of animals; in small, plain flowers, it is usually done by the wind. *Artificial pollination* is the transfer of pollen done intentionally by people.

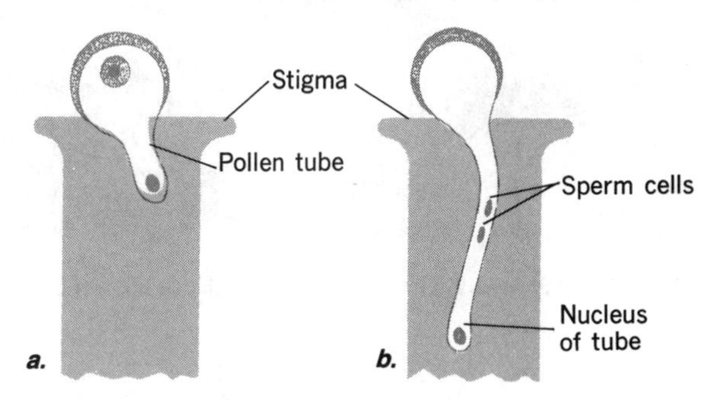

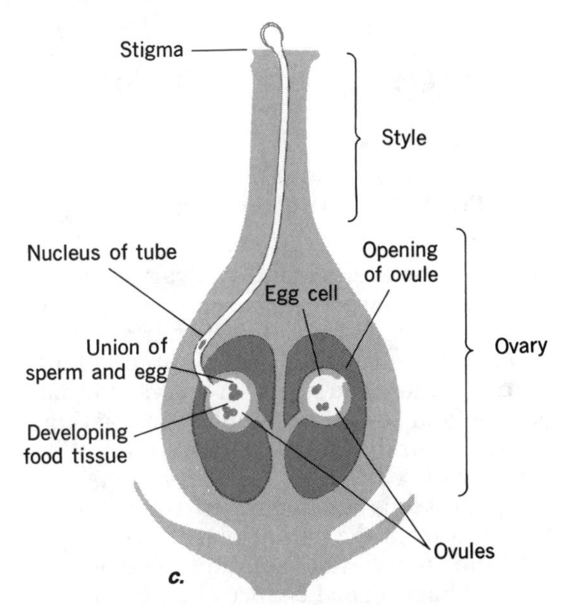

Fig. 10-2. Steps in the fertilization of a flower.

Fertilization in a Flower. As a result of pollination, several pollen grains may stick to a stigma. Then, an extension grows out of each pollen grain. This extension is called a **pollen tube** (Fig. 10-2*a*). The tube grows down through the tissues of the stigma, the style, and the ovary, and enters a tiny opening in an ovule. As the pollen tube enters the ovule, the end of the pollen tube dissolves.

By the time a pollen tube reaches an ovule, two sperm cells have formed inside the pollen tube (Fig. 10-2*b*), and one egg cell has formed inside the ovule. The sperm cells move out of the pollen tube into the tissue of the ovule (Fig. 10-2*c*). One sperm cell unites with the egg, forming a fertilized egg cell that soon develops into an *embryo plant.* The other sperm cell helps form the food tissue that develops around the tiny plant.

Embryo and Seed. As the fertilized egg in an ovule develops and grows, the ovule itself grows. As the ovule enlarges, its outer covering hardens. When ripe, the ovule and its contents are called a **seed.** A seed is composed of a hard outer wall (called the *seed coat*), an embryo, and food tissue (Fig. 10-3).

The major parts of a plant embryo are: the embryo tip, the embryo stem, the embryo root, and either one or two embryonic seed leaves, or **cotyledons.** The *embryo tip* develops into the first leaves, buds (which later give rise to other leaves), and growing point of the stem. The *embryo stem* becomes the middle part of the stem and its branches. The *embryo root* gives rise to the root and its branches. The cotyledons contain stored food, which is used by the young plant when it begins to grow. When its chloroplasts develop, the young plant can make its own food by photosynthesis.

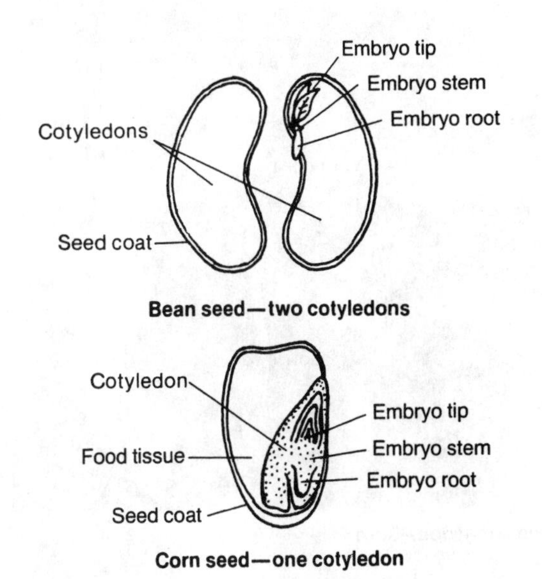

Fig. 10-3. Two types of seeds (cutaway views).

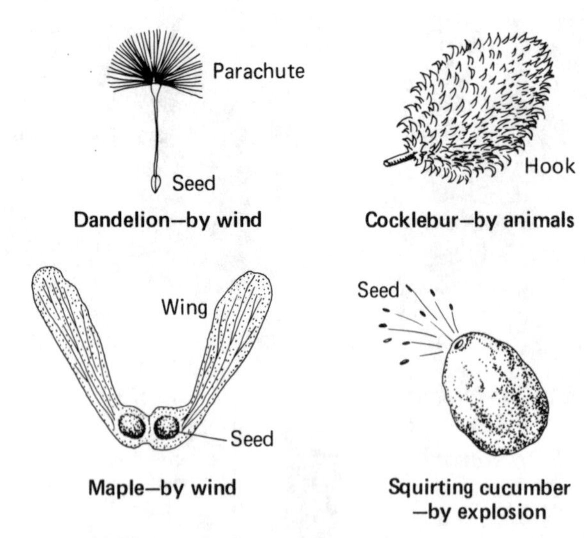

Fig. 10-4. Types of seed dispersal.

Fruit. As the ovules inside an ovary grow, the ovary itself enlarges. A ripened ovary, which encloses seeds and any flower parts still attached to it, is called a **fruit.** In some plants, the fruit is juicy and sweet; in others, it is hard and dry, and sometimes unpleasant tasting.

A fruit usually contains many seeds. When ripe, a fruit scatters its seeds. By being scattered, the seeds are separated from one another and from the parent plant. Those seeds that land in favorable soil and are not overcrowded have a good chance of sprouting and growing successfully.

The way in which seeds are scattered depends upon the kind of plant (Fig. 10-4). Fruits of some plants possess "wings" or feathery parachutes. These structures enable the wind to scatter the seeds. The fruits of other plants have hooks that catch onto fur, enabling animals to scatter the seeds. Fruits of a few plants are balloonlike sacs that are under pressure. When these sacs explode, the seeds are shot out and scattered in many directions. Other fruits are eaten by animals who then scatter the undigested seeds when they eliminate waste from their digestive systems.

The Sprouting of Seeds. After seeds leave the parent plant, they lie inactive until the next growing season. When environmental conditions are suitable, that is, when adequate moisture, oxygen, and warmth are present, the embryo within the seed begins to sprout, or **germinate.** The stages in the germination of two different types of seeds are shown in Fig. 10-5.

SEXUAL REPRODUCTION IN ANIMALS

Some simple animals, such as the hydra, planaria, and earthworm, can reproduce either asexually or sexually. All higher animals, like the frog, dog, and human, reproduce sexually.

Sex Glands. Sex cells are produced by special glands, called the **sex glands.** The sex

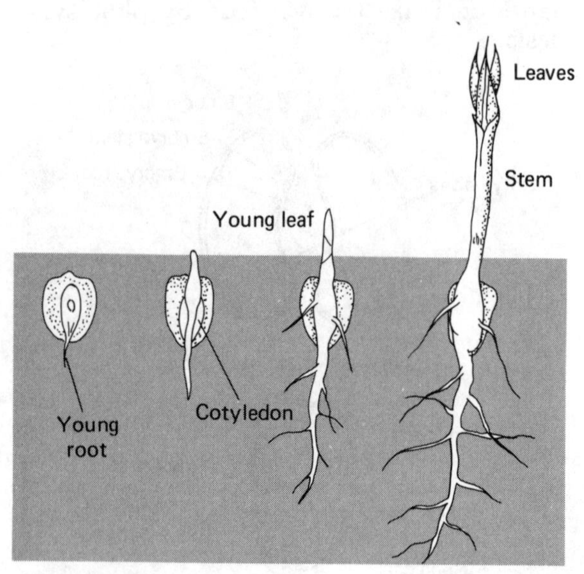

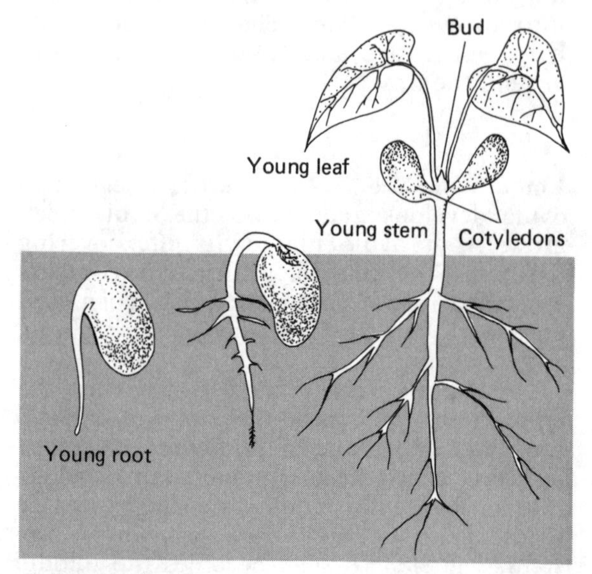

Fig. 10-5. The germination of seeds.

gland of males is called the **testis** (plural, *testes*); that of females the **ovary** (Fig. 10-6). The sex cells produced by the testis are *sperm cells;* those produced by the ovary are *egg cells.*

The sperm cells leave the testes by way of delicate tubes called **sperm ducts.** The egg cells leave the ovaries by way of thick tubes called **oviducts.**

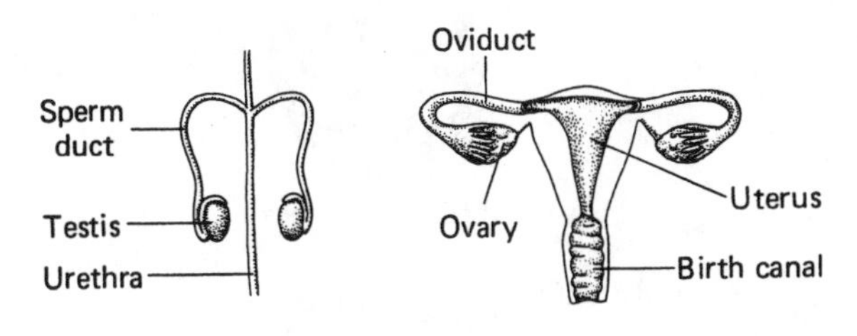

Fig. 10-6. Human male and female sex glands.

Sperm and Egg Cells. In every kind of animal, the nucleus of a sperm cell and that of an egg cell are similar. In other respects, these sex cells are markedly different (Fig. 10-7).

A sperm cell is very small. One end of the cell is composed of a nucleus and a very small amount of cytoplasm. The rest of the cell is a long, whiplike tail. By lashing this tail back and forth, the sperm cell can move ("swim") in water or body fluid. The male animal produces many more sperm cells than a female produces egg cells.

Egg cells are usually round and always much larger than sperm cells. The egg cell is composed of a nucleus and a large amount of cytoplasm. The cytoplasm contains stored food, called **yolk.** An egg cell's size depends upon the amount of yolk in it. For example, the egg cell of birds is large because it contains more yolk than that of other animals. (Note that the yolk of an animal's egg serves the same purpose as the stored food and cotyledons of seeds.) Unlike the sperm cell, the egg cell has no structure for movement. The sperm cell has to travel to the egg cell to fertilize it.

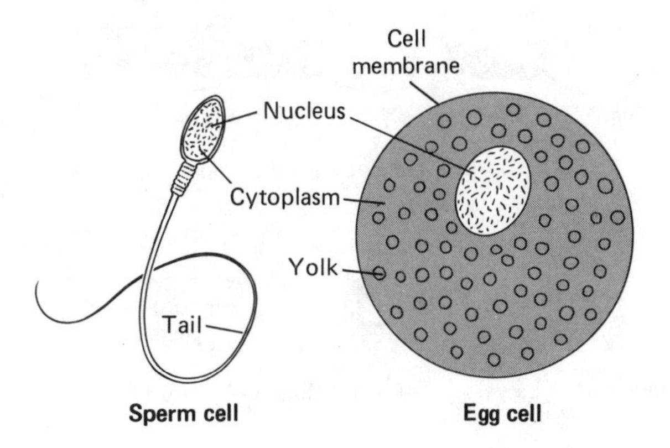

Fig. 10-7. The sex cells of animals.

Fertilization in Animals. The new generation starts when a sperm cell enters an egg cell and unites with it. As in plants, the union of a sperm cell nucleus with an egg cell nucleus is called **fertilization.** The combined cell that results from the union of sperm and egg cell nuclei is called a fertilized egg. There are two types of fertilization.

1. External Fertilization. In water-dwellers, such as fishes and amphibians, the eggs of the female usually pass out of her body through the oviducts and into the water. The sperm cells released by the male then swim to these eggs and unite with them. This type of fertilization outside the female's body is called **external fertilization.**

2. Internal Fertilization. In some water-dwellers and in most land-dwellers, the eggs of the female are kept inside her body (within the oviducts). Sperm cells are deposited by the male inside the female, swim to an oviduct in the fluid present there, and unite with the eggs. This type of fertilization inside the female's body is called **internal fertilization.** Internal fertilization occurs in insects, some fish, and all reptiles, birds, and mammals.

Development of Young Animals

1. External Development. In some kinds of animals, the young organism develops outside of the mother's body. Such *external development* is found in animals that fertilize externally, like the fishes and amphibians. It is also found in certain animals that fertilize internally, but lay eggs so that the young develop externally, such as birds and insects.

2. Internal Development. In other kinds of animals, the young develop inside the mother's body. This type of development, called *internal development,* is common among mammals, in which both fertilization and development usually take place internally. In reptiles, both forms of development occur. Most reptiles lay eggs, but some snakes give birth to live young. Some fishes that have internal fertilization and development also bear live young.

Early Stages in the Development of the Young. Whether fertilized externally or internally, the fertilized egg divides (by mitosis)

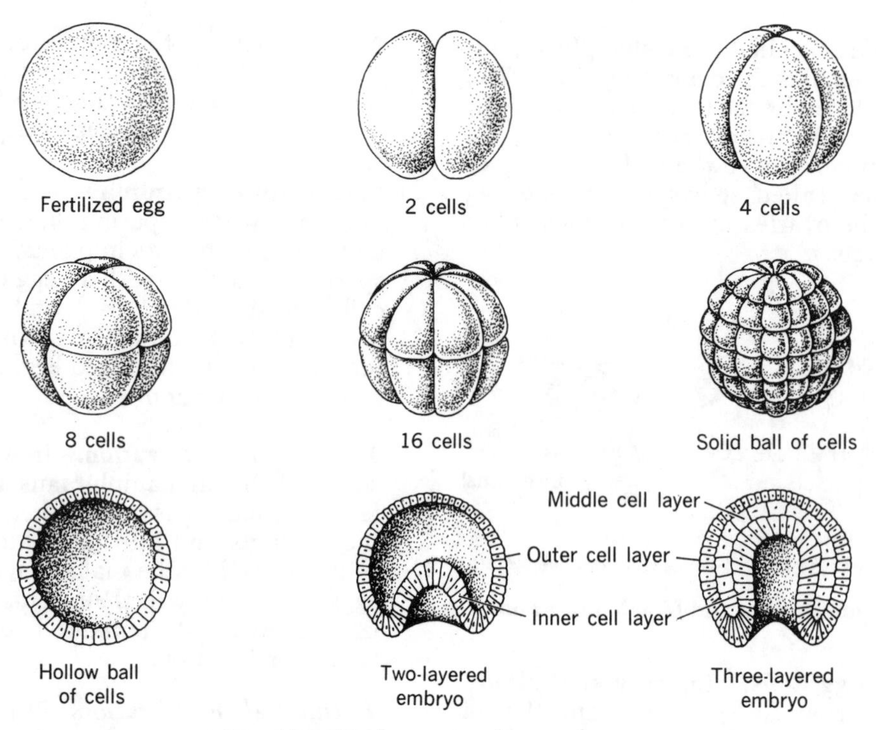

Fig. 10-8. Early stages of an embryo.

many times and forms an organism called an **embryo.** Fig. 10-8 shows the early stages in the development of an embryo. These stages are much the same in all kinds of animals.

At first, the fertilized egg divides into two cells, forming a two-celled embryo. The two-celled embryo divides into four cells; the four cells into eight; and so on, until a ball of cells is formed. Later, as still more cells form, this ball becomes hollow and one of its sides folds inward. At this stage, the embryo consists of two layers of cells. Shortly afterward, a third layer of cells forms between the outer and inner cell layers. It is from these three cell layers that all organs of the new individual grow and develop.

LIFE CYCLES OF SOME ANIMALS

The stages in the development of an organism, from the time it is a fertilized egg until it grows to adulthood, reproduces, and eventually dies, are called its **life cycle.**

Life Cycle of an Insect. In insects, the fertilization of the egg is internal but the development of the young is external. The female deposits the fertilized eggs close to a food source that will be available to the young when they hatch. Except for a few types of insects, such as ants and bees, no other care is provided for the young. Before hatching, the embryo feeds on yolk within the egg. After hatching, the young feeds on plant or animal matter. Like amphibians, insects undergo a change in body form as they develop.

Insects, like other jointed animals, are covered by a hard *exoskeleton* that limits their growth. In the molting process, the exoskeleton is shed and then formed anew, several times. Before each new exoskeleton hardens, the insect grows rapidly. After the final molting, the insect is an adult and grows no more.

Some types of insects, such as the grasshopper, undergo a gradual change during their development. In these insects, the young resem-

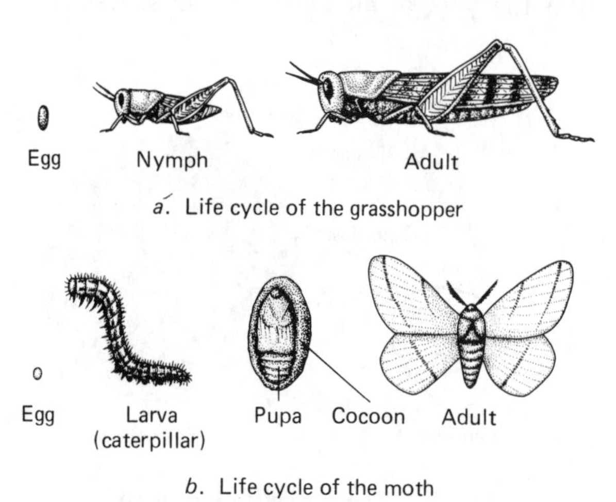

Fig. 10-9. Growth stages of insects.

bles the adult, but cannot fly because its wings are not fully developed. There are three stages in the life cycle of such insects—*egg, nymph,* and *adult* (Fig. 10-9*a*).

Many types of insects, like the moth, undergo a complete change, or **metamorphosis,** during their development. In these insects, the young is very different from the adult. There are four stages in the life cycle of such insects—*egg, larva, pupa,* and *adult* (Fig. 10-9*b*). (The *cocoon* is the pupa's cover while it is developing into the adult moth stage.)

Life Cycle of a Bony Fish. In the process called **spawning,** a female fish lays numerous eggs in the water, and the male sprays them with sperm cells. After fertilization, each developing embryo is nourished by the yolk of its egg. Before the yolk supply is exhausted, the embryo hatches into a tiny fish. At first, this fish feeds on the remaining yolk. When the yolk is used up, the fish gets its own food. The young fish grows and matures. When its sex glands (or sex organs) have developed, the fish is ready to mate and reproduce (Fig. 10-10).

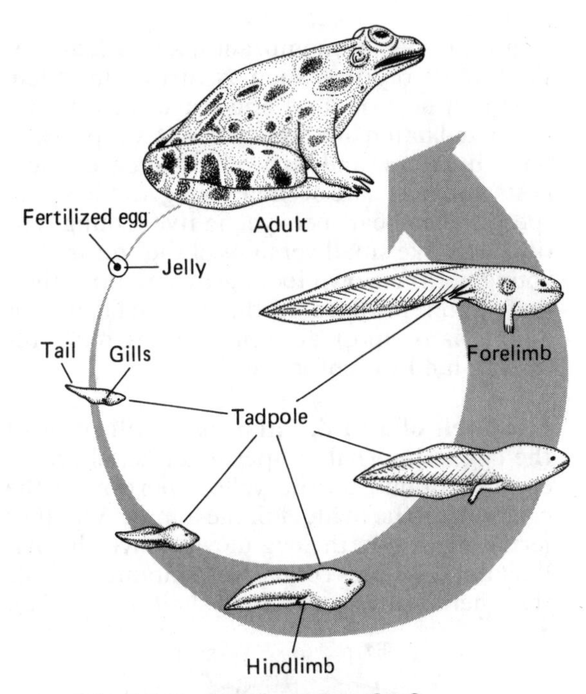

Fig. 10-11. Development of a frog.

Life Cycle of an Amphibian. In amphibians, as in bony fishes, both fertilization of the egg and development of the embryo are external and take place in or near the water. When frogs mate, layers of protective jelly form around the batch of fertilized eggs, which is then warmed by the sun. (The warmth helps the eggs develop.)

Each developing embryo is nourished by its yolk. Before long, the embryo hatches into a *tadpole,* which gets its own food, grows, and undergoes a series of body changes. After the last stage of metamorphosis, the tadpole is transformed into a frog. This series of changes from egg to tadpole to adult frog is shown in Fig. 10-11.

Like a fish, the young tadpole breathes through gills and swims by moving its flat tail. As the tadpole grows, hind limbs appear; later, forelimbs appear. Meanwhile, the body of the tadpole absorbs the tail; lungs form and replace the gills. By this time, the tadpole has become a frog and can hop out of the water and onto land. When its sex organs are fully developed, the frog can mate and reproduce. Amphibians must return to the water to reproduce.

Life Cycle of a Reptile. In reptiles, the fertilization of the egg is internal. When the egg enters the oviduct, it is fertilized by sperm. Then the egg is coated with albumen (egg

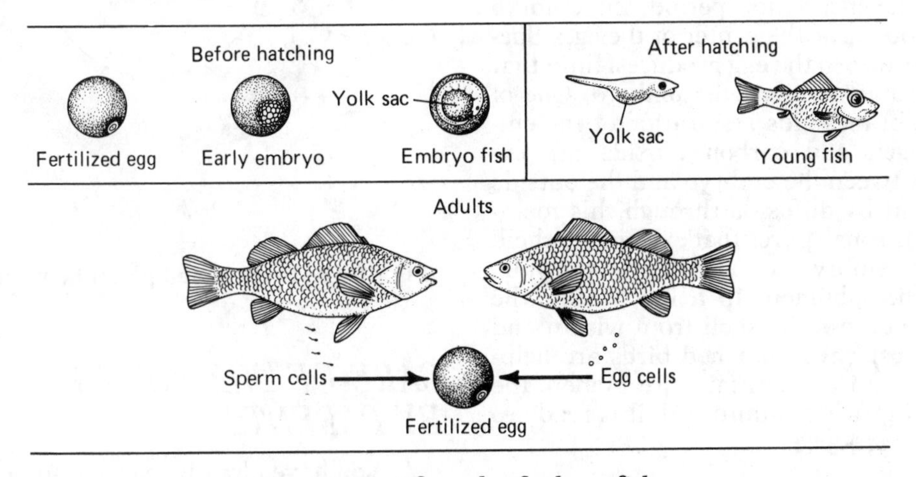

Fig. 10-10. Life cycle of a bony fish.

"white"), special membranes, and a leathery shell. Most reptiles, such as turtles, lay their eggs in a nest on land and leave them to develop and hatch on their own. A few reptiles—the alligators and crocodiles—guard their nests and take care of their young. Some snake species even bear their young live. Young reptiles look like small versions of the adults. The young get their own food, grow, mature, then mate and reproduce. In the reverse fashion of amphibians, most aquatic reptiles must return to land to reproduce.

Life Cycle of a Bird. In birds, fertilization of the egg is internal. A sperm cell fertilizes an egg cell (found on the yellow portion of the egg) within the oviduct of the female. Whether fertilized or not, the egg moves down the oviduct and becomes coated with albumen, three thin membranes, and a hard shell (Fig. 10-12).

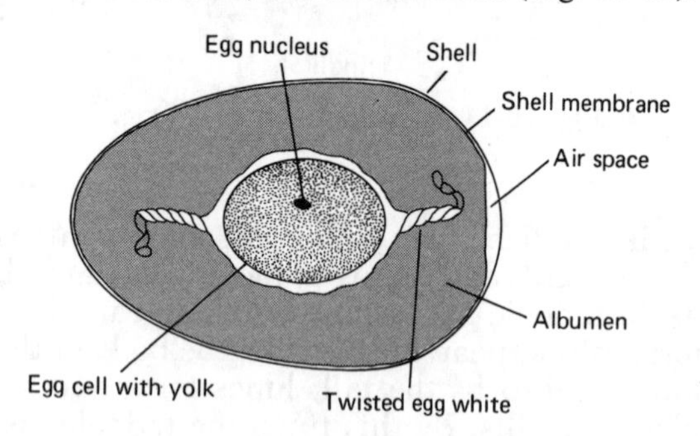

Fig. 10-12. Egg of a bird.

After being laid in a nest, the eggs are sat upon, or *incubated,* by the parent birds. In this way, the fertilized eggs are warmed by the body heat of the parents. Unless a fertilized egg is kept warm, the embryo does not develop. Birds usually guard their eggs and then take care of their young until the young can fly or can get their own food.

During the incubation period, the embryo bird develops near the center of the egg. Special membranes in the egg produce a fluid that surrounds and cushions the embryo. One of these membranes aids respiration of the embryo. Oxygen and carbon dioxide are exchanged between the embryo and the outside environment by diffusion through this membrane and through pores that exist in the shell. At first, the embryo is nourished by the yolk, later by the albumen. In a few weeks, the young bird cracks the shell from within and hatches. Most newly hatched birds are helpless and require parental care. When the young bird grows to adulthood, it is ready to mate and reproduce.

Life Cycle of a Mammal. In most mammals, both fertilization of the egg and development of the young are internal. (The exceptions to this are the few mammals that lay eggs, like the platypus, and the mammals that complete their development in a pouch, such as the kangaroo.) The very young embryo becomes attached to the inside of a strong, muscular organ, called the *uterus.* In this place, a special structure called the *placenta* develops (see Fig. 10-13). The placenta exchanges food, oxygen, carbon dioxide, and wastes between the embryo and the mother. These materials are exchanged by way of two separate sets of capillaries (tiny blood vessels). One set is connected to the circulatory system of the embryo; the other set to the circulatory system of the mother. The two sets of capillaries are not directly connected. Nutrients and oxygen from the blood of the mother's capillaries diffuse into the embryo's capillaries. Similarly, wastes from the embryo pass into the mother's bloodstream and are eventually excreted by her kidneys.

As the embryo grows, the mother's uterus, which has elastic, muscular walls, is stretched and becomes larger. At birth, the walls of the uterus automatically contract, pushing the young mammal out of the mother's body. The young then feed on milk, produced by the mother's milk glands. Mammals care for their young until the offspring can become independent. When it has fully developed, the young mammal can mate and reproduce.

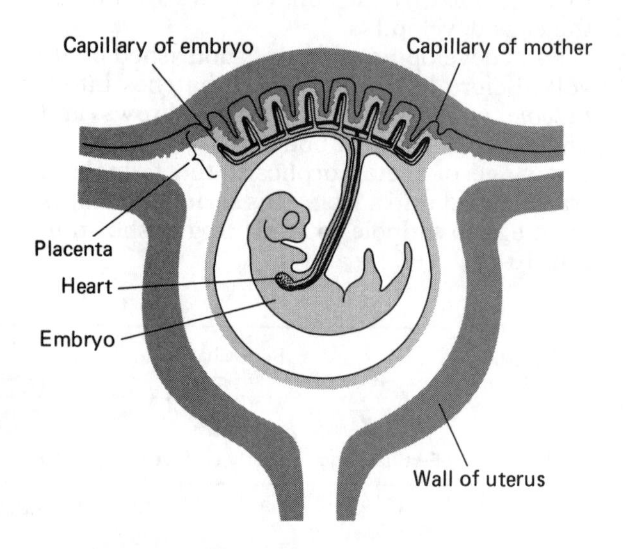

Fig. 10-13. Embryo and placenta of a mammal.

CARE OF THE YOUNG BY PARENTS

As you have already learned, some young of a species must live long enough to reproduce if the species is to continue. The survival at least of some offspring until reproductive age can be assured in one of two ways: (1) by producing so many offspring that, even without parental care, some of them are likely to live until they mate, or (2) by producing comparatively few offspring and taking very good care of them until they mature.

In those species of animals where both fertilization of eggs and development of young are external, a very large number of eggs is produced. Many of these eggs are fertilized, but they are then abandoned by the parents. Very often, other animals eat most of the fertilized eggs. This occurs in most species of aquatic invertebrates, fishes, and amphibians. A few offspring, however, do manage to reach adulthood and reproduce. In the case of insects, although fertilization is internal, development is external. Many eggs are laid, but most do not survive to hatch.

Among mammals, fertilization of eggs is internal, and development of young is also internal. These animals usually produce only a few eggs, most of which are fertilized. The young that develop are cared for by one or both parents for some time and have a high rate of survival. This type of care also occurs among birds and some reptiles, although the development of young is external. Among live-bearing reptiles (some snakes), there is no parental care and many young are born.

In humans, usually only one egg is produced at a time and is successfully fertilized. After birth, the offspring needs to be cared for by its parents for many years.

Like the animals that lay many fertilized eggs, flowering plants produce numerous seeds. And just as most animal eggs do not hatch and develop, most seeds do not take root and sprout. However, those few seeds that do land in favorable conditions have their stored food and cotyledons to help them sprout and grow.

BENEFITS OF SEXUAL REPRODUCTION

In sexual reproduction, genes from two parents contribute to the inherited traits of the offspring. When these genes join to form the fertilized egg, there is a chance for the offspring to inherit some favorable traits from each of its parents. This gives the offspring a greater variety of traits to help it survive changes in environmental conditions. People use this fact to intentionally develop and improve breeds of plants and animals that are useful to us. Sperm cells (or pollen grains, which form plant sperm cells) that have a desired characteristic are used to fertilize egg cells that have another desired characteristic. The offspring that develop often have both desired characteristics. Examples of organisms that have been improved to our benefit in this fashion are high-protein corn plants and dairy cows that produce a lot of milk. You will learn more on this subject in Chapter 11.

CHAPTER REVIEW

Science Terms

The following list contains all of the boldfaced scientific terms found in this chapter and the page on which each appears.

cotyledons (p. 109)
egg (p. 108)
embryo (p. 112)
external fertilization (p. 111)
fertilization (p. 111)
fruit (p. 110)
germinate (p. 110)
internal fertilization (p. 111)
life cycle (p. 112)
metamorphosis (p. 112)
ovary (p. 111)
oviducts (p. 111)
ovules (p. 108)
petals (p. 108)
pistil (p. 108)
pollen grains (p. 108)
pollen tube (p. 109)
pollination (p. 109)
seed (p. 109)
sepals (p. 108)
sex cell (p. 108)
sex glands (p. 110)
sexual reproduction (p. 108)
spawning (p. 113)
sperm (p. 108)
sperm ducts (p. 111)
stamens (p. 108)
testis (p. 111)
yolk (p. 111)

Matching Questions

On the blank line, write the letter of the item in column B which is most closely related to the item in column A.

	Column A	Column B
____	1. female sex organ of a plant	*a.* embryo
____	2. male sex organ of a plant	*b.* fruit
____	3. found inside a plant's ovary	*c.* placenta
____	4. embryonic seed leaves of plant	*d.* spawning
____	5. ripened plant ovary with seeds	*e.* ovules
____	6. stored food in an animal's egg	*f.* pistil
____	7. union of sperm cell and egg cell	*g.* uterus
____	8. early developing organism	*h.* cotyledons
____	9. reproductive process of fish	*i.* stamen
____	10. where young mammal develops	*j.* fertilization
		k. yolk

Multiple-Choice Questions

On the blank line, write the letter preceding the word or expression that best completes the sentence or answers the question.

1. A cell contributed by each of two parents to the next generation is called a
a. spore cell *b.* sex cell *c.* pollen grain *d.* fertilized egg — 1 ____

2. A jellyfish can reproduce either asexually or sexually, but a cat can only reproduce
a. asexually *b.* by regeneration *c.* sexually *d.* by incubation — 2 ____

3. Sperm cells are produced by the male animal's
a. testis *b.* ovary *c.* sperm duct *d.* oviduct — 3 ____

4. The tubes through which eggs leave the body of a female frog are the
a. sperm ducts *b.* yolk ducts *c.* xylem ducts *d.* oviducts — 4 ____

5. Sperm cells are able to swim in a fluid because they have
a. false feet *b.* cilia *c.* tails *d.* muscle fibers — 5 ____

6. Stored food in a fish egg is called
a. starch *b.* fat tissue *c.* glucose *d.* yolk — 6 ____

7. Which two structures unite in sexual reproduction?
a. fertilized eggs *c.* egg and sperm
b. albumen and yolk *d.* ovary and testis — 7 ____

8. During internal fertilization in a bird, sperm cells unite with egg cells in which organ of the female?
a. ovary *b.* uterus *c.* oviduct *d.* placenta — 8 ____

9. Internal fertilization and external development occur in
a. most fishes *b.* most amphibians *c.* most mammals *d.* all birds 9 ____

10. The eggs of birds have
a. no protective covering *c.* a jellylike covering
b. a hard covering *d.* a leathery covering 10 ____

11. Before hatching, the young organism is called
a. an ovule *b.* a baby *c.* a spawn *d.* an embryo 11 ____

12. As a fertilized egg develops into an offspring, the egg undergoes
a. a decrease in size *c.* three cell divisions
b. very few changes *d.* many cell divisions 12 ____

13. The young animal that hatches from a frog's egg is called a
a. cocoon *b.* pupa *c.* tadpole *d.* fertilized egg 13 ____

14. A tadpole breathes mainly through its
a. nostrils *b.* gills *c.* lungs *d.* tail 14 ____

15. The part that grows first, as a tadpole changes into a frog, is the
a. hind limbs *b.* forelimbs *c.* tail *d.* lungs 15 ____

16. A chick cannot develop inside a hen's egg unless the egg is
a. fertilized and kept warm *c.* laid in water
b. unfertilized and kept warm *d.* kept cool in a nest 16 ____

17. A chick embryo inside the egg gets oxygen through
a. the placenta *b.* its lungs *c.* pores in the shell *d.* its gills 17 ____

18. In which structure does the embryo of a rabbit develop?
a. uterus *b.* stomach *c.* ovary *d.* intestine 18 ____

19. The embryo of a cow gets rid of its waste carbon dioxide through
a. its lungs *c.* the placenta
b. the cow's large intestine *d.* the cow's stomach 19 ____

20. The embryo of an insect feeds on
a. blood *b.* yolk *c.* grass *d.* frogs 20 ____

21. Two animals that undergo a metamorphosis are the
a. frog and moth *c.* frog and snake
b. fish and frog *d.* snake and bird 21 ____

22. The sex cells of a flowering plant are located in the
a. sepals and petals *c.* stigma and nectar
b. fruit and seed *d.* stamens and pistils 22 ____

23. In order to reach an ovule, a pollen tube grows through the
a. stigma, stamen, petal *c.* sepals, stamen, stigma
b. stigma, style, ovary *d.* cotyledon, style, ovule 23 ____

24. The egg cell of a lettuce plant develops in the
a. pollen grain *b.* ovule *c.* stigma *d.* fruit 24 ____

25. The sperm cell of a potato plant develops in the
a. testis *b.* sperm tube *c.* petal *d.* pollen tube 25 ____

26. The part of a flower that becomes the seed is the
a. petal *b.* ovule *c.* sepal *d.* ovary 26 ____

27. A ripened ovary of a flower is called the
a. embryo *b.* seed *c.* pistil *d.* fruit 27 ____

28. Pollination of small, plain flowers is usually done by
a. wind *b.* insects *c.* people *d.* water 28 ____

29. In order to start growing, most seeds need
a. water, oxygen, shade
b. water, oxygen, warmth
c. yolk, water, light
d. light, pollen, nectar 29 ____

30. The greatest amount of parental care is found among organisms that
a. reproduce asexually
b. have external fertilization
c. produce many offspring
d. produce few offspring 30 ____

Modified True-False Questions

In some of the following statements, the italicized term makes the statement incorrect. For each incorrect statement, write the term that must be substituted for the italicized term to make the statement correct. For each correct statement, write the word "true."

1. Two parents are required in *vegetative* reproduction. 1 ________

2. Egg cells are produced in the *ovaries*. 2 ________

3. The smaller of the two sex cells is the *egg* cell. 3 ________

4. In general, *more* sperm cells are produced than egg cells. 4 ________

5. *Internal* fertilization is characteristic of amphibians. 5 ________

6. *External* development is characteristic of mammals. 6 ________

7. Embryo fish and very young fish feed on *yolk*. 7 ________

8. Reptiles' eggs are covered by a leathery *jelly*. 8 ________

9. The egg cell of a chicken is found on the *white* of the egg. 9 ________

10. When a grasshopper molts, it sheds the *exoskeleton*. 10 ________

Testing Your Knowledge

1. Compare a sperm cell and an egg cell with respect to:
a. the nucleus that each contains ________________________________

____

b. the amount of cytoplasm in each ________________________________

____

c. the ability of each to move ________________________________

____

2. Explain the difference between the two terms in each of the following pairs:
   a. pistil and stamen ______

   b. self-pollination and cross-pollination ______

   c. external fertilization and internal fertilization ______

   d. external development and internal development ______

3. Farmers used to spray their fields with the chemical DDT to kill off harmful insects. But then certain plants would no longer grow. Give an explanation for this occurrence. ______

4. If both cotyledons are removed from a bean seed and then the seed is covered with moist soil, a young plant fails to grow. Explain why. ______

5. Plant breeders who are trying to develop a new variety of flower sometimes cover the flower tops with plastic bags. Why do you think this is done? ______

6. The juicy part of a cherry fruit does not nourish the plant embryo in the seed, but it helps the embryo in another way. Explain how. ______

7. Explain how identical twins develop from a single fertilized egg. ______

8. What would happen to a flowering plant if its anther stalk were removed? ______

9. Explain how fruits aid in seed dispersal. ______

Gestational Surrogacy: Whose Baby Is It?

On September 19, 1990, Anna Johnson gave birth to a healthy baby boy. On October 23, 1990, Judge Richard Parslow said that the baby was not Anna's. The California judge gave the baby to Crispina and Mark Calvert who, the judge declared, were the baby's real parents. How is it possible for a judge to rule that a woman who gives birth to a baby is not the baby's real mother?

To find the answer to this strange question, let's retrace the story of Anna Johnson and the Calverts. First of all, long before Anna Johnson had become pregnant, Crispina Calvert had had an operation called a hysterectomy. In such an operation, a woman's uterus (womb) is removed. This is the organ within which a baby grows inside its mother. So Crispina Calvert could not have a baby.

Fortunately for Crispina, however, the surgeon had not removed her ovaries. These are the organs that produce a woman's eggs. Normally, when one of these eggs is fertilized by a man's sperm, a baby begins to develop. Although Crispina could produce eggs and her husband could produce sperm, Crispina had no place for a fertilized egg to grow. The Calverts' problem was finding such a place.

The "place" turned out to be Anna Johnson's uterus. Anna agreed to let the Calverts "rent" her uterus for $10,000. Here's how medical technology made it possible for this to happen.

Doctors removed eggs from Crispina's ovaries and put them in a glass laboratory dish. The doctors also got sperm from her husband Mark and put them in the same dish. While mixed together in the dish, one of Crispina's eggs became fertilized by one of Mark's sperm. The doctors then took the fertilized egg and placed it inside Anna Johnson's uterus. The egg attached to the wall of the uterus and developed into the baby that Anna gave birth to.

This procedure has a scientific name. The name is gestational surrogacy. *Gestation* means the growth and development of an egg from fertilization to birth. *Surrogacy* refers to someone who stands in for, or replaces, someone else. In this case, Anna Johnson was a surrogate for the gestation of Crispina Calvert's fertilized egg.

Gestational surrogacy means something more too. It means that a fertilized egg, containing the genes of one woman, is put inside another woman from whom it received no genes. When Anna Johnson decided that she wanted to keep the baby she had delivered, the Calverts said Anna had no right to the baby because the boy had none of her genes.

The case went to court and Judge Parslow decided that Crispina Calvert was the real mother of the baby. He said his decision was based on the fact that the Calverts were the genetic parents of the baby. This decision, however, actually contradicts California law. In California, a mother is usually defined as the woman who gives birth to a baby.

So whose baby is it anyway? What do you think?

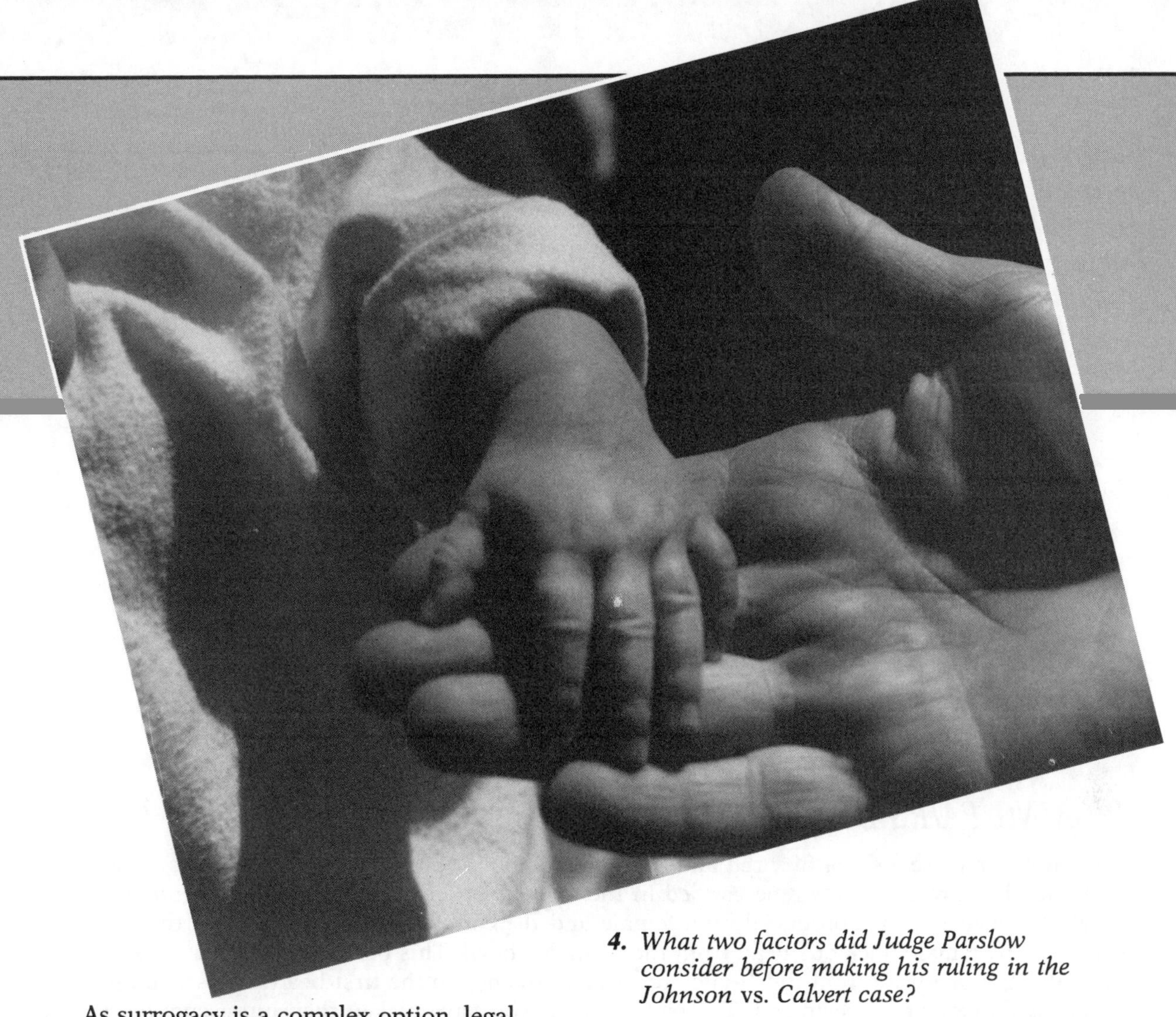

As surrogacy is a complex option, legal contracts are now drafted for both the surrogate and the intended parents. In addition, there is a series of psychological counseling and personality testing involved in the surrogacy.

1. *Judge Parslow ruled that the baby given birth to by Anna Johnson belonged to*

a. the doctors.
b. Anna Johnson.
c. Crispina and Mark Calvert.
d. the state of California.

2. *A hysterectomy involves the removal of*

a. eggs. c. the uterus.
b. sperm. d. a baby.

3. *The gestational surrogate in this story was*

a. Anna Johnson. c. Mark Calvert.
b. Crispina Calvert. d. Richard Parslow.

4. *What two factors did Judge Parslow consider before making his ruling in the Johnson* vs. *Calvert case?*

____

____

____

5. *The following are quotes that appeared in an article in* Time, *August 27, 1990. Write a short essay explaining why you agree or disagree with each quote as it relates to the Johnson* vs. *Calvert controversy.*

"Just because you donate a sperm and an egg doesn't make you a parent."

"Both women—the one who bore the child and the one who provided the egg—have some biological claim to the child."

"The gestational mother is the legal mother . . . Biologically she's taken the majority of the risks."

"That child is biologically Cris and Mark's."

11

Heredity and Genetics

LABORATORY INVESTIGATION

PICKING PAIRS BY CHANCE

A. You will receive a set of fifty red and fifty white "pop-it" beads, in separate bowls. Assume that each bead represents a gene carried in the nucleus of a sex cell. Let the contents of one bowl represent the eggs produced by a female and the contents of the other bowl the sperm produced by a male. Pick one bead from the "female" bowl. This bead represents a gene in an egg cell. Pick one bead from the "male" bowl and attach it to the first bead. The second bead represents a gene in a sperm cell. The pair of beads you have picked then represents the two genes in a fertilized egg—one from the female parent and the other from the male parent. In the same fashion, pick ten more pairs of beads.

1. What color combination(s) did you get? ____________________

2. If you continued picking pairs, what combination(s) could result? ____________________

Explain. ____________________

3. Why is it impossible to get any other color combinations? ____________________

____

4. What do the pairs of beads represent? ____________________

____

B. Mix your fifty white and fifty red beads together in one bowl. One of your classmates, who will work with you, should also mix together a set of beads. You now have a bowl containing both red and white beads. Assume that your bowl represents a female organism that produces 100 eggs. Fifty eggs carry a gene for redness, and the other fifty eggs carry a gene for whiteness. Similarly, your partner's bowl, which represents a male, contains fifty sperm cells that carry a gene for redness, and fifty sperm cells that carry a gene for whiteness.

5. If you picked one bead from the female bowl and one from the male bowl, what color combination(s) could you get? ____________________

Explain. ____________________

6. Which combination should appear most often? ______

Why? ______

______

7. What are the chances of picking a red-white combination more often then red-red or white-white? ______

______

Explain. ______

______

C. Thoroughly mix the beads in the male bowl. Do the same for the female bowl. With your eyes closed, pick a bead from the male bowl and one from the female bowl at the same time. Pop them together and hand the pair to your partner, who will tally the pair in the table below. Continue this procedure until all 200 beads have been used up.

D. Separate all the pairs of beads, again placing fifty red ones and fifty white ones in the female bowl and the other fifty red and fifty white beads in the male bowl. Repeat the procedure of picking pairs, only this time your partner will pick the 100 pairs as you keep the tally. Total the results.

	Red-Red	*Red-White*	*White-White*
First student			
Second student			
Total			

7. Are your results different from your partner's? ______

8. Why do you think the totals are not in a perfect ratio of 1:2:1 (50 red-red: 100 red-white: 50 white-white)? ______

______

Inheritance of Traits

Organisms that have survived for many generations in a particular environment possess certain traits, or **adaptations,** suitable for their way of life. When the organisms reproduce, these traits are passed on to their offspring. As a result, the offspring are adapted to the same environment and way of life as that of their parents.

The development of traits in each new generation of offspring is controlled by the genes that are inherited. All of the genes in the cells of an organism make up its *genotype.* The "expression" of the genes, or appearance of the organism, is called its *phenotype,* which is determined by the interaction of the genotype and the environment. As explained in Chapter 3, in the section on cell division and mitosis, genes are located in chromosomes.

In organisms that reproduce asexually, inheritance is relatively simple because mitosis is the major process involved in reproduction. During mitosis, the genes in the set of chromosomes in a parent cell make copies of themselves, forming two sets. Then, one set of chromosomes passes into one daughter cell, and the second set into the other daughter cell. Since the genes in the offspring cells are identical, the offspring develop the same traits as the parent.

On the other hand, in organisms that reproduce sexually, inheritance is more complicated because other processes, in addition to mitosis, are involved. As a result of these processes, the offspring are rarely exactly alike. Although they resemble one another and their parents, the offspring do differ from each other and from the parents. The differences that are present among all organisms of the same species are called **variations.**

Variations sometimes appear in organisms that reproduce asexually. However, variations occur much more often in organisms that reproduce sexually. This happens because one set of chromosomes, containing genes from the male parent, and a second set of chromosomes, containing genes from the female parent, join to form one offspring. This combination of two sets of genes makes it possible for the offspring to resemble its parents in some respects and differ from them in other respects.

In order to understand how this comes about, follow the steps explaining the formation of sperm cells and egg cells. Then you will be able to see how the chromosomes in the sperm cells and egg cells produce the next generation.

CHROMOSOME NUMBERS IN CELLS

In many-celled organisms that reproduce sexually by means of fertilization, only a few special cells are directly involved in the process. The cells that play no direct part in fertilization are called **body cells.** Some examples of these are muscle, nerve, bone, and skin cells. The nucleus of each body cell contains a full set of chromosomes. The number of chromosomes found in these cells is the same in all organisms of a particular species. However, the number of chromosomes varies in different species. For example, the body cells of all fruit flies have eight chromosomes (or four pairs) each; of pea plants, fourteen chromosomes (or seven pairs); and of humans, forty-six chromosomes (or twenty-three pairs). The chromosome number that is characteristic of the body cells of a species is called the **species number** of chromosomes. (See Fig. 11-1.)

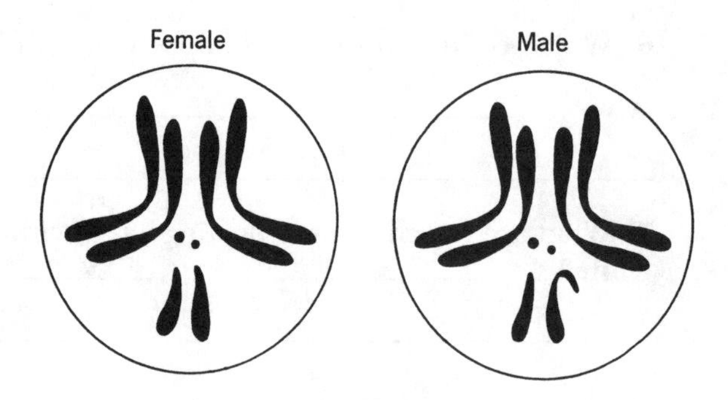

Fig. 11-1. Female and male fruit fly chromosomes.

The cells that are directly involved in fertilization are called *sex cells.* The sex cells are the egg cells and sperm cells. Nuclei of sex cells contain chromosomes, but only one-half the number found in the nuclei of body cells. For example, in the fruit fly, the sex cells contain only four chromosomes; in the pea plant, the sex cells contain only seven chromosomes; and in the human, the sex cells contain only twenty-three chromosomes. The chromosome number that is characteristic of the sex cells of a species is one-half the species number of chromosomes. This change in the chromosome number from the species number to one-half that number takes place when sex cells are formed.

The fruit fly is used as an example in the following discussion because it has a small number of chromosomes.

THE FORMATION OF SPERM CELLS

Immature sperm cells, which are located in the male reproduction glands (the testes) in animals, first contain the species number of chromosomes (Fig. 11-2*a*). As these cells develop, they go through a special kind of cell division, called **meiosis.** In this process, four new cells of equal size are formed, with one-half the species number of chromosomes in each cell. This type of division, which reduces the chromosome number to one-half the species number, is called **reduction division.** For example, a cell having a species number of eight chromosomes undergoes reduction division; each of the four new sex cells formed has just four chromosomes.

After reduction division has taken place, the shape of each developing sperm cell changes. The cytoplasm of the cell shrinks and a **flagellum**—a whiplike tail—forms. The cell that is produced is a mature (ripe) sperm cell that can move itself by the flagellum. Mature sperm cells can leave the testes.

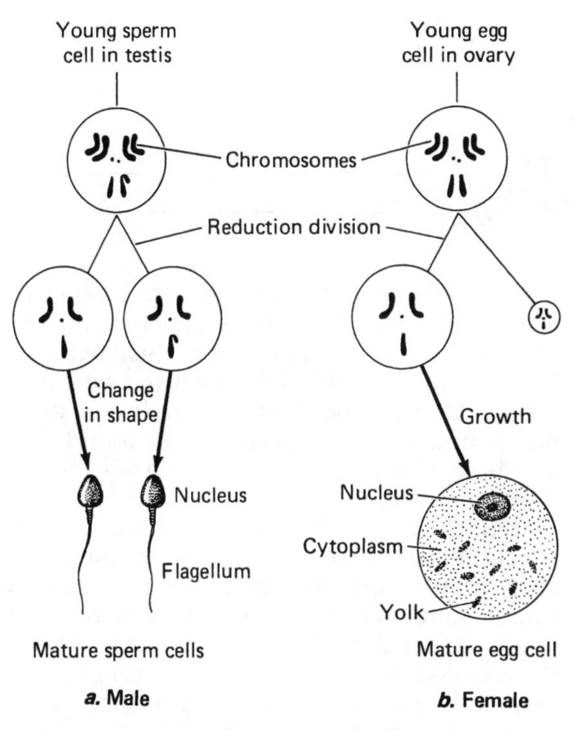

Fig. 11-2. Formation of sperm and egg cells.

THE FORMATION OF EGG CELLS

Immature egg cells, which are located in the female reproductive glands (the ovaries) in animals, also first contain the species number of chromosomes (Fig. 11-2*b*). Like the developing sperm cells, the young egg cells go through reduction division. As a result, the cells that are formed have one-half the species number of chromosomes.

Unlike the cells produced in the male, which are of equal size, the cells produced by reduction division in the female are of unequal size. One cell is large and contains a great deal of cytoplasm; the others are tiny and contain very little cytoplasm. However, all of the cells have one-half the species number of chromosomes.

The large cells are unripe egg cells, which develop further. The tiny cells eventually disappear and take no part in reproduction. Yolk (made from nutrients brought to the ovary by the blood) accumulates in the one large egg cell and, as a result, the cell grows. When growth is complete, a mature egg cell is formed. This cell contains food and, like the mature sperm cell, has one-half the species number of chromosomes. When mature, the cells are released from the ovaries. (The egg cell has no tail and cannot move itself.)

FERTILIZATION

In the process of fertilization, a sperm cell encounters and enters an egg cell. Although hundreds of sperm cells may try, only one manages to burrow into the egg. Then the chromosomes of both cells unite to form the fertilized egg. Figure 11-3 shows, in simplified form, what happens during the process of fertilization in the fruit fly.

A sperm cell of a fruit fly, containing four chromosomes (one-half the species number), and an egg cell, also containing four chromosomes, come into contact. After the sperm enters the egg, the fertilized egg contains eight chromosomes, which is the original species number of the fruit fly (found in its body cells). The fertilized egg cell then divides many times and becomes an embryo. Since all the cell divisions in this process involve only mitosis, every body cell that develops in the new individual also contains the species number (a full set) of chromosomes. Upon reaching adulthood, the new individual's ovaries or testes form the sex cells for reproducing the next generation.

You can now see that half the number of chromosomes (and thus genes) in a fertilized egg come from the male parent and half from the female parent. As each trait of the off-

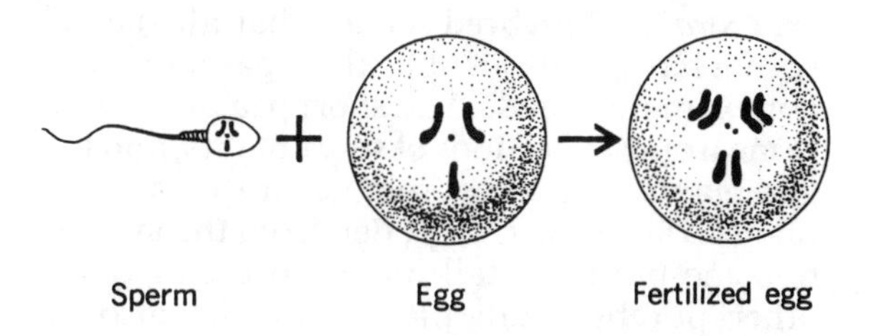

Fig. 11-3. Fertilization in the fruit fly.

spring develops, it is influenced (to varying degrees) by two genes—one from the male parent and one from the female.

In the next section, you will learn about the interactions between the two genes of a trait, and the development of traits that may resemble, or differ from, those of one or both parents.

Heredity

Heredity is the passing on of traits, or characteristics, from parents to offspring. **Genetics** is the science that studies heredity. An Austrian monk, *Gregor Mendel* (1822–1884), who conducted experiments using garden pea plants, was the first to succeed in explaining heredity. He did this even before chromosomes and genes were discovered. Mendel found that there were three fundamental laws that determine how heredity operates. It has since been found that Mendel's Laws apply generally to all sexually reproducing organisms.

MENDEL'S EXPERIMENTS

Among the inherited traits of the pea plant that Mendel studied were the length of the stem, the color of the flower, the form of the seed, and the color of the seed. He observed that there were opposite (or contrasting) characteristics for each of these traits. For example, some stems were tall and others were short; some flowers were purple and others were white.

Mendel started his experiments with *purebred strains.* **Purebred** means that all the offspring have traits just like their parent plants. For example, Mendel had short pea plants that came from generations of short plants, and tall pea plants that came from generations of tall plants. He transferred pollen from the anthers of some purebred tall plants to the pistils of other purebred tall plants. Mendel also allowed each purebred tall plant to self-pollinate.

After collecting all the seeds that developed, Mendel planted them. He found that all the seeds grew into tall plants like their parents. In other words, 100% of the offspring were tall. When Mendel pollinated purebred short plants in a similar experiment, he found that purebred short parents produced all (100%) short offspring.

Mendel then artificially placed pollen from purebred tall plants on the pistils of purebred short plants and, also, pollen from purebred short plants on the pistils of purebred tall plants. All of the offspring that resulted from these crosses were as tall as their tall parents. Only tallness, the contrasting character to shortness, appeared in the first generation offspring, or F_1. Mendel called the characteristic that appeared in the F_1 the **dominant character.** The characteristic that did not appear in the F_1 Mendel called the **recessive character.** Such offspring of contrasting parents, which exhibited only the dominant character, he called **hybrids.** The traits that Mendel referred to as "characters" we now know to be controlled by **dominant genes** and **recessive genes.**

Law of Dominance. From the results of numerous experiments of this type, Mendel was able to state his first law of heredity, the Law of Dominance: *When two parents that are pure for contrasting characters are crossed, the dominant character appears in the hybrid and the recessive character is hidden.*

Genetics of Dominance. Each body cell and each immature sex cell of purebred parents, such as tall pea plants, has two sets of chromosomes—the species number of chromosomes. Therefore, each of these cells contains two genes for tallness. As you learned in the preceding section, when a parent produces sex cells, the number of genes (or chromosomes) is reduced to one-half the species number; so, each sex cell would contain only one gene for tallness. As the egg and sperm cells unite, they form a fertilized egg that contains a gene for tallness from one parent and another gene for tallness from the other parent. Later, when the fertilized egg develops into a young plant—having a combination of two genes for tallness—it grows up to be tall like its parents. The chemical reactions controlled by the DNA of genes (the genotype), along with environmental factors, determine the final appearance (the phenotype) of the offspring.

By using a few symbols and a diagram called the *Punnett square,* you will be able to see the relationship between the combination of genes that an organism inherits and its final appearance. The Punnett square method is also useful in working out problems in heredity. In this method, letters are used to represent genes: a capital letter (such as T) for the dominant gene and a small letter (such as t) for the recessive gene. These symbols are explained in a key at the beginning of the problem.

Example 1. Cross two pure tall pea plants, showing the gene combinations and the appearance of their offspring.

Key: Let T = gene for tallness (dominant)
t = gene for shortness (recessive)

	Female		*Male*
Parents:	pure tall	×	pure tall
Genes of Parents:	TT	×	TT
Genes in Sex Cells:	T		T

Possible Fertilizations:

	Sperm Cell T
Egg Cell T	TT

Offspring (F_1):

Gene Combinations	*Appearance*
all TT	100% tall (pure)

You can readily see that Mendel obtained all tall pea plants from parents that were purebred tall, because the sex cells of the parents contained genes for tallness only.

Example 2. Cross two pure short pea plants, showing the gene combinations and the appearance of their offspring.

Key: Same as in Example 1.

	Female		*Male*
Parents:	pure short	×	pure short
Genes of Parents:	tt	×	tt
Genes in Sex Cells:	t		t

Possible Fertilizations:

	Sperm Cell t
Egg Cell t	tt

Offspring (F_1):

Gene Combinations	*Appearance*
all tt	100% short (pure)

You can also see that Mendel obtained all short pea plants from parents that were purebred short, because the sex cells of the parents contained genes for shortness only.

Example 3. Cross a pure tall pea plant with a pure short pea plant, showing the gene combinations and the appearance of their offspring.

Key: Same as in Example 1.

	Female		*Male*
Parents:	pure tall	×	pure short
Genes of Parents:	TT	×	tt
Genes in Sex Cells:	T		t

Possible Fertilizations:

	Sperm Cell t
Egg Cell T	Tt

Offspring (F_1):

Gene Combinations	*Appearance*
all Tt	100% tall (hybrid)

Here you can see that Mendel obtained all tall offspring when crossing parents with contrasting characters, because the only possible combination of genes at fertilization was Tt—one dominant gene from the pure tall parent and one recessive gene from the pure short parent. (Part A of your laboratory investigation demonstrated why this was the only possible combination.) As a result of a chemical reaction between the dominant and recessive genes, only one of the genes—the dominant one—was able to show its effect. In this case, the dominant gene controlled the development of a tall stem in each hybrid individual. However, the recessive gene is still present in every hybrid, but it is inactive and is overshadowed by the dominant gene. Hybrids have two unlike genes for a trait, in contrast to purebreds, which have two genes that are alike for a trait.

Law of Segregation and Recombination. Mendel crossed many first-generation hybrid tall pea plants and obtained their seeds. After planting these seeds, Mendel found that some plants of this second generation (or F_2) were tall, whereas others were short. When he counted the numbers of tall and short offspring, he found that the ratio of plants showing the dominant character to those showing the recessive character was 3:1. From the results of many experiments of this type, Mendel was able to state his second law of heredity, the Law of Segregation and Recombination: *When two hybrids are crossed, the recessive character, which had been hidden in the F_1 generation, is segregated (separated) from the dominant character. At fertilization, there is a chance for the recessive characters to recombine (come together again) and appear in some members of the F_2 generation.*

Genetics of the 3:1 Ratio. As sex cells mature and reduction division occurs, the two sets of chromosomes in the immature sex cells separate, and one set of chromosomes moves into each new cell. As a result, every egg and sperm cell contains only one of any pair of genes. When fertilization takes place, there are chances for like or unlike genes to meet. Parts C and D of your laboratory investigation demonstrated these chances:

1. If a sperm having a dominant gene unites with an egg having a dominant gene, the offspring has two dominant genes and shows the dominant character (pure).
2. If a sperm having a dominant gene unites with an egg having a recessive gene, the offspring has one dominant gene and one recessive gene and shows the dominant character (hybrid).
3. If a sperm having a recessive gene unites with an egg having a dominant gene, the offspring has one dominant gene and one recessive gene and shows the dominant character (hybrid).
4. If a sperm having a recessive gene unites with an egg having a recessive gene, the offspring has two recessive genes and shows the recessive character (pure).

As a result, in the F_2 generation there are three chances that an offspring will show the dominant character to one chance that it will show the recessive character (Fig. 11-4).

Note that what Mendel called *segregation* takes place during what is now called *reduc-*

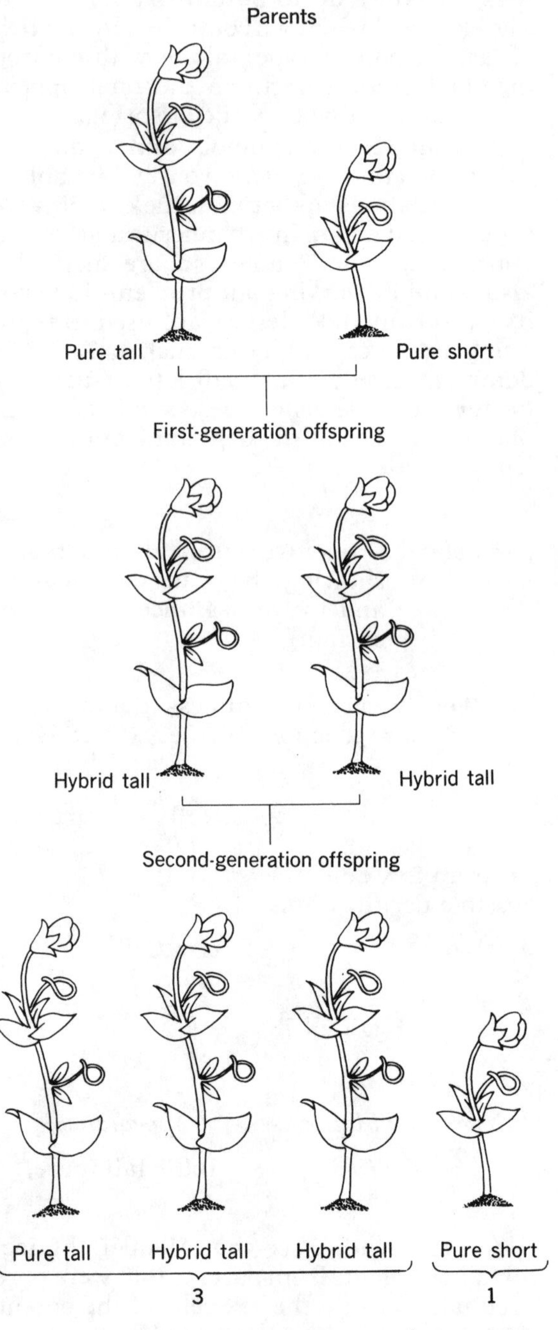

Fig. 11-4. The 3:1 ratio in the second generation.

tion division; and what he called *recombination* takes place at *fertilization.*

The reason for the 3:1 ratio becomes clearer when we work out a genetic cross of two hybrids with a Punnett square.

Key: Let T = gene for tallness (dominant)
t = gene for shortness (recessive)

	Female		*Male*
Parents (F_1):	hybrid tall	×	hybrid tall
Genes of Parents:	Tt	×	Tt
Genes in Sex Cells:	T or t		T or t

Possible Fertilizations:

	Sperm Cells T	Sperm Cells t
Egg Cells T	TT	Tt
Egg Cells t	Tt	tt

Offspring (F_2):

Gene Combinations	*Appearance*
¼ TT	25% tall (pure)
²⁄₄ Tt	50% tall (hybrid)
¼ tt	25% short (pure)

Note that the ratio of the different gene combinations (1/4 TT:2/4 Tt:1/4 tt) is 1:2:1. Note further that since all tall plants look alike, regardless of the genes within them, there are really 75% tall plants and 25% short plants. This ratio, 75%:25% (or 3:1), is the ratio of the appearance (phenotype) of tall to short plants, whereas 1:2:1 is the ratio of the gene combinations (genotype) in the same F_2 generation.

Law of Independent Assortment. In some of his experiments, Mendel used plants that were hybrid for two traits. For example, he crossed plants that were hybrid tall and had hybrid yellow seeds with others of the same type. (In pea plant seeds, yellow is dominant and green is recessive.) He found that the ratio relating to the appearance of the offspring was 9:3:3:1. That is, out of every sixteen offspring, nine were tall and yellow-seeded; three were tall and green-seeded; three were short and yellow-seeded; and one was short and green-seeded. From the results of many such experiments, Mendel was able to state his third law of heredity, the Law of Independent Assortment (or Unit Characters): *Each character behaves as a unit and is inherited independently of any other character.*

Genetics of the 9:3:3:1 Ratio. It is now known that the genes for height (tall or short) and the genes for seed color (yellow or green) lie in different pairs of chromosomes of the two sets of chromosomes present in pea plant body cells and young sex cells. When the immature sex cells undergo reduction division, their two sets of chromosomes and genes move apart and enter separate egg cells and sperm cells. At fertilization, several different gene combinations can occur. The ratio of these combinations that appears is 9:3:3:1. This ratio represents the four combinations that can result. Fully written out, this means that of the sixteen offspring, there are

9/16 tall plants with yellow seeds
3/16 tall plants with green seeds
3/16 short plants with yellow seeds
1/16 short plants with green seeds

By separately adding up all the tall and all the short plants, we find that there are twelve tall plants to four short plants—a ratio of 3:1.

By separately adding up all the yellow-seeded plants and all the green-seeded plants, we find that there are twelve yellow to four green—another ratio of 3:1.

These ratios show that the ratio for height has no effect on the ratio for seed color and that the genes for height have no effect on the genes for seed color. In other words, although the genes for height and the genes for color are inherited at the same time, they act (are *sorted*) independently of each other. The two ratios of 3:1 also show that the 9:3:3:1 ratio is a combination of two separate 3:1 ratios.

INCOMPLETE DOMINANCE

Since Mendel's time, it has been discovered that, in some cases, there are exceptions to the Law of Dominance. The appearance of an offspring's trait can be a mixture of the contrasting characters of its purebred parents. In these cases, the reaction between the contrasting genes results in a blended trait that does not resemble the pure trait of either parent. This type of inheritance is called *blending inheritance,* or **incomplete dominance.**

Examples of Incomplete Dominance. When purebred red-flowered four-o'clock plants are crossed with purebred white-flowered four-o'clock plants, their offspring have pink flowers. Although these offspring are pink, the genes for redness and whiteness do not change. This can be seen by working out the crosses for the F_1 and F_2 generations of offspring.

Example 1. Cross a red-flowered four-o'clock plant with a white-flowered four-o'clock plant, showing the gene combinations and the appearance of their offspring.

Key: Let R = gene for red flower
W = gene for white flower

Neither gene is dominant.

	Female		*Male*
Parents:	pure red	×	pure white
Genes of Parents:	*RR*	×	*WW*
Genes in Sex Cells:	*R*		*W*

Possible Fertilizations:

	Sperm Cell *W*
Egg Cell *R*	*RW*

Offspring (F_1):

Gene Combinations	*Appearance*
all *RW*	100% pink (hybrid)

Example 2. Cross two pink four-o'clock plants, showing the gene combinations and the appearance of their offspring.

	Female		*Male*
Parents (F_1):	hybrid pink	×	hybrid pink
Genes of Parents:	*RW*	×	*RW*
Genes in Sex Cells:	*R* or *W*		*R* or *W*

Possible Fertilizations:

Egg Cells \ Sperm Cells	*R*	*W*
R	*RR*	*RW*
W	*RW*	*WW*

Offspring (F_2):

Gene Combinations	*Appearance*
¼ *RR*	25% red (pure)
²⁄₄ *RW*	50% pink (hybrid)
¼ *WW*	25% white (pure)

Note that in the F_2 offspring, the ratios of both the gene combinations (genotype) and the appearance (phenotype) of the offspring are the same, that is, 1 : 2 : 1. Therefore, this type of cross still follows Mendel's Law of Segregation and Recombination.

CHAPTER REVIEW

Science Terms

The following list contains all of the boldfaced scientific terms found in this chapter and the page on which each appears.

adaptations (p. 123)
body cells (p. 124)
dominant character (p. 126)
dominant genes (p. 126)
flagellum (p. 125)
genetics (p. 126)
heredity (p. 126)
hybrids (p. 126)
incomplete dominance (p. 129)
meiosis (p. 124)
purebred (p. 126)
recessive character (p. 126)
recessive genes (p. 126)
reduction division (p. 124)
species number (p. 124)
variations (p. 124)

Matching Questions

On the blank line, write the letter of the item in column B which is most closely related to the item in column A.

	Column A	Column B
____	1. all the genes of an organism	*a.* variations
____	2. muscle, nerve, and bone cells	*b.* phenotype
____	3. a whiplike tail	*c.* sex cells
____	4. appearance of an organism	*d.* reduction division
____	5. passing on of parents' traits	*e.* genetics
____	6. cells involved in fertilization	*f.* genotype
____	7. different traits within a species	*g.* dominant genes
____	8. study of inheritance of traits	*h.* flagellum
____	9. chromosome number is halved	*i.* recessive genes
____	10. produce purebred short pea plants	*j.* body cells
		k. heredity

Multiple-Choice Questions

On the blank line, write the letter preceding the word or expression that best completes the sentence or answers the question.

1. The group that includes only body cells is
a. oviduct cells and egg cells
b. sperm cells and egg cells
c. sperm duct cells and sperm cells
d. oviduct cells and sperm duct cells — 1 ____

2. The normal, or species, number of chromosomes in the human is
a. 23 *b.* 12 *c.* 46 *d.* 50 — 2 ____

3. If the species number of chromosomes is 4, how many chromosomes does a sperm cell of the species have?
a. 2 *b.* 4 *c.* 8 *d.* 1 — 3 ____

4. When a cell of the testis ripens into a sperm, the cell
a. undergoes division and then enlarges
b. changes shape and then enlarges
c. undergoes division and then changes shape
d. undergoes division and then enters the ovary — 4 ____

5. If the number of chromosomes in an egg cell is 4, the species number of chromosomes is
a. 2 *b.* 4 *c.* 16 *d.* 8 — 5 ____

6. In animals, when an egg cell is formed in the ovary, another type of cell produced at the same time is
a. a sperm that disappears
b. a sperm that fertilizes it
c. a yolk cell that gathers food
d. a tiny cell that disappears — 6 ____

7. In a fertilized egg that has 4 chromosomes,
a. 2 came from the mother and 2 from the father
b. 3 came from the mother and 1 from the father
c. 4 came from the mother and 4 from the father
d. 4 came from the mother and 0 from the father
7 ____

8. The laws of heredity were first stated by
a. Francisco Redi
b. Charles Darwin
c. Gregor Mendel
d. Robert Hooke
8 ____

9. When purebred yellow-seeded pea plants are crossed with other purebred yellow-seeded pea plants, the plants of the next generation have seeds
a. all of which are yellow
b. all of which are green
c. all of which are white
d. some of which are yellow
9 ____

10. The offspring of parents that are pure for opposite traits are called
a. hybrids *b.* dominants *c.* recessives *d.* mixtures
10 ____

11. The crossing of pure tall pea plants with pure short pea plants produces offspring that are
a. 50% tall and 50% short
b. 100% short
c. 100% medium
d. 100% tall
11 ____

12. When pure white guinea pigs are crossed with other pure white guinea pigs, we can expect their offspring to be
a. 100% gray
b. 100% black
c. 100% white
d. 50% black and 50% white
12 ____

13. When does the number of chromosomes in a cell halve?
a. during all cell divisions
b. only during mitosis
c. only during reduction division
d. during fertilization
13 ____

14. The parts on the chromosomes that control the development of the inherited traits of an organism are the
a. seeds *b.* genes *c.* beads *d.* flagella
14 ____

15. White mice mated with white mice normally produce only white mice because the genes for color in the parents are
a. hybrid *b.* the same *c.* weak *d.* different
15 ____

16. If one parent has the gene combination *AA* and the other has *aa*, the gene combination in the offspring will be
a. *AA* *b.* *Aa* *c.* *aa* *d.* none of these
16 ____

17. A 3:1 ratio in offspring is the result of processes stated in the law of
a. dominance
b. recessiveness
c. segregation and recombination
d. independent assortment
17 ____

18. Although genes separate before reproduction, they have a chance to come together again in different combinations during the process of
a. fertilization *b.* mitosis *c.* segregation *d.* spore formation
18 ____

19. A student tossed two pennies at a time for 100 times. The results were: both heads—25; one head and one tail—46; both tails—29. Which cross could result in a similar ratio?
a. *Aa* × *Aa* *b.* *Aa* × *aa* *c.* *Aa* × *AA* *d.* *AA* × *aa*
19 ____

20. If a plant having red flowers is cross-pollinated with one having yellow flowers and their offspring produce orange flowers, the color change was probably the result of
a. dominance *b.* recessiveness *c.* unit characters *d.* incomplete dominance
20 ____

21. The crossing of parents that are hybrids for two traits would most likely result in a ratio of
a. 12:1 *b.* 9:3:3:1 *c.* 9:1 *d.* 1:1 21 ____

22. In guinea pigs, rough fur is dominant over smooth fur. If two hybrid rough guinea pigs are mated and have six offspring,
a. all of them should have rough fur
b. most of them should have rough fur
c. most of them should have smooth fur
d. all of them should have smooth fur 22 ____

23. Two pink four-o'clock hybrids are cross-pollinated. Of their offspring,
a. all should be pink *c.* all should be white
b. all should be red *d.* some should be pink 23 ____

24. When one of two traits can be inherited without the other, the gene for these two traits are said to be
a. dominant *b.* recessive *c.* blended *d.* independent 24 ____

25. Which of the crosses below best illustrates Mendel's principle of segregation?
a. *BB x bb* *b.* *Bb x Bb* *c.* *bb x bb* *d.* *BB x BB* 25 ____

Modified True-False Questions

In some of the following statements, the italicized term makes the statement incorrect. For each incorrect statement, write the term that must be substituted for the italicized term to make the statement correct. For each correct statement, write the word "true."

1. The traits of organisms that reproduce either sexually or asexually are controlled by *chloroplasts.* 1 ____________

2. Offspring of one parent are more likely to be *the same* than are the offspring of two parents. 2 ____________

3. Cells that are not directly involved in sexual reproduction are called *body* cells. 3 ____________

4. A gene that is overshadowed by a dominant gene is called a(an) *incomplete* gene. 4 ____________

5. If *BB* represents a purebred, and *bb* the opposite purebred, then *Bb* represents a *variation.* 5 ____________

6. That every trait is inherited separately from every other trait is stated by the Law of *Dominance.* 6 ____________

7. Reduction division occurs when a cell undergoes *mitosis.* 7 ____________

8. Body cells contain the *species number* of chromosomes. 8 ____________

9. A sperm cell leaves the testis by moving its *cytoplasm.* 9 ____________

10. Blending inheritance is also known as *incomplete dominance.* 10 ____________

Testing Your Knowledge

1. In each of the following write the number of the proper choice in the space provided. Check your answer with a Punnett square.

a. If all of a large number of offspring of a garden pea cross are *Tt*, the parents were most likely
(1) *Tt* × *Tt* (2) *TT* × *tt* (3) *TT* × *Tt* (4) *Tt* × *tt* ___

b. When red and pink four-o'clocks are crossed, the color of the offspring will be
(1) some pink, some white
(2) all pink
(3) some red, some pink
(4) all red ___

c. If many tall pea plants are crossed, which result would be the *least* likely?
(1) 100% pure tall
(2) 50% pure tall, 50% hybrid tall
(3) 50% short, 50% tall
(4) 25% short, 75% tall ___

d. A cattleman buys a black bull—supposedly a purebred (pure black). Knowing that black color in cattle is dominant over red color, he decides to make sure that the black bull is purebred by mating it with several red cows. If the bull is pure, the offspring should be
(1) all black
(2) all red
(3) 3 black:1 red
(4) 1 black:1 red ___

2. Label each of the blanks indicated.

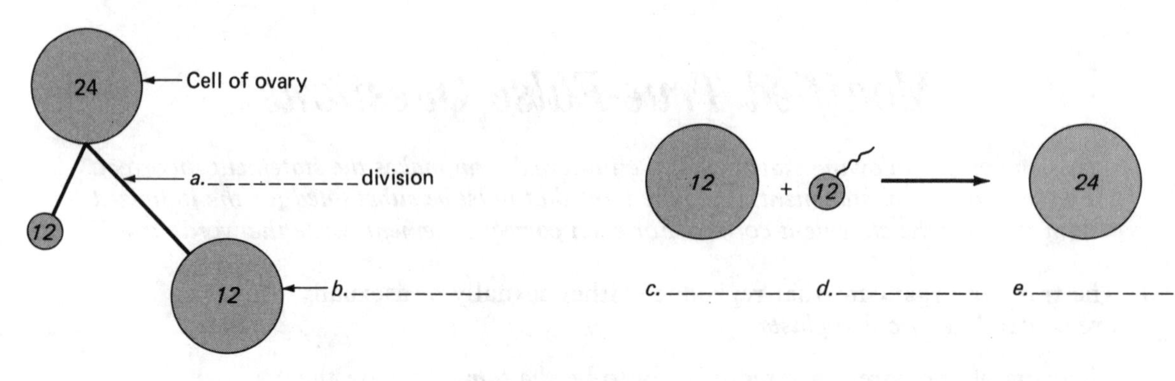

3. Show a labeled Punnett Square for the following genetic cross:

A pair of roan cattle was mated. Roan, white, and red offspring were produced.

a. show the F_1 cross that produced the roan parents. ___
b. show the F_2 cross that produced the three different types of offspring. ___

12

Heredity and Evolution

LABORATORY INVESTIGATION

STUDYING HEREDITY IN A FAMILY

You are going to survey the earlobe form, or shape, in either your family or a family that you know. Earlobes are considered to be either attached or free (unattached) in form. (See Fig. 12-1.)

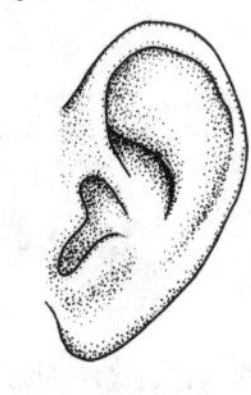

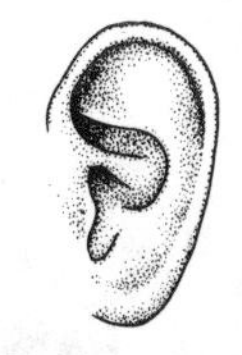

Fig. 12-1. Free and attached earlobes.

A. Observe the earlobe form of the brother(s), sister(s), father, mother, uncle(s), aunt(s), and, if possible, the grandparents on the mother's side and on the father's side. Record your observations in the following table:

Relative	*Ear Form*	*Relative*	*Ear Form*
Self (if applicable)		Uncle(s) (mother's side)	
Brother(s)		Aunt(s) (mother's side)	
Sister(s)		Grandfather (father's side)	
Father		Grandmother (father's side)	
Mother		Grandfather (mother's side)	
Uncle(s) (father's side)		Grandmother (mother's side)	
Aunt(s) (father's side)		Other relatives:	

B. Examine the family history (pedigree) chart on the next page (Fig. 12-2). Inside each symbol record "F" for earlobes that are free (unattached) and "f" for earlobes that are attached.

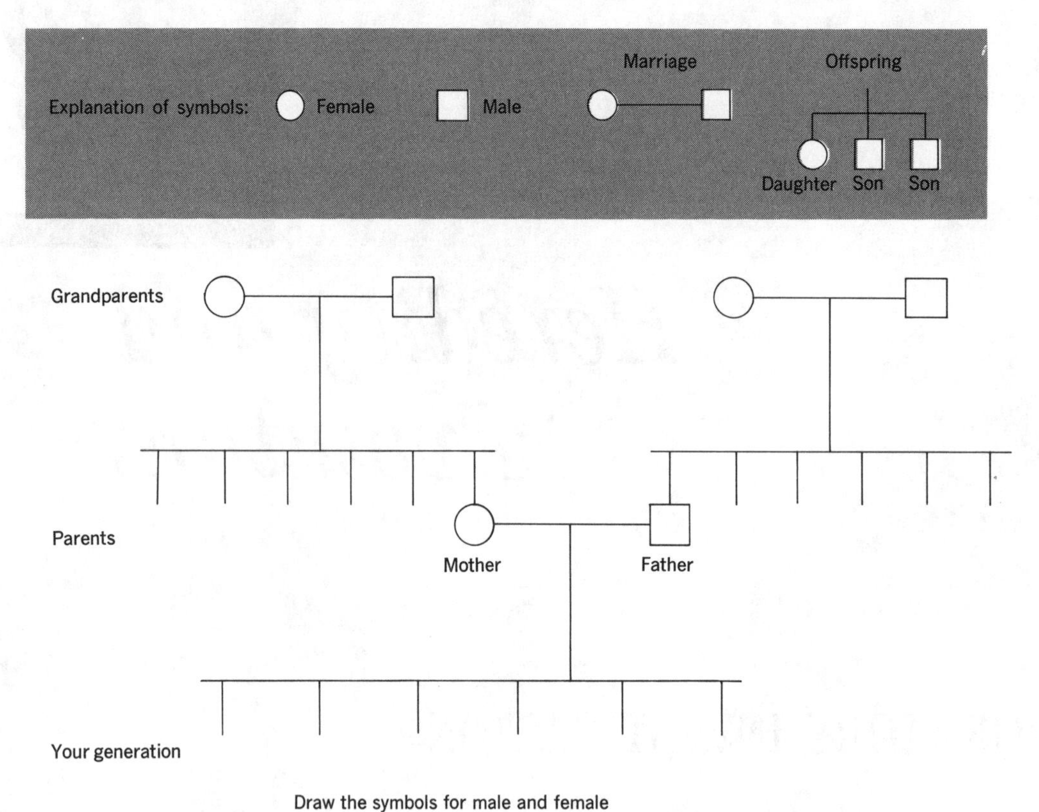

Fig. 12-2. Family history chart.

C. Using the pedigree chart as you have filled it in for the surveyed family, answer the following questions:

1. What evidence is there in this family that earlobe form is inherited? ______________

2. Which earlobe form appears to be dominant? ______________

Explain the reason for your answer. ______________

3. What gene combination for earlobe form do you probably have (if this is your family)?

____

Explain your reasoning. ______________

____

4. What gene combination for earlobe form does the mother probably have? ______________

What gene combination for earlobe form does the father probably have? ______________

Explain your reasoning in both cases. ______________

____

5. What gene combinations for earlobe form do/did each of the grandparents probably have?

____

Human Inherited Traits

The study of human inheritance is very complicated, but scientists have discovered many facts about human hereditary traits. What is known about a few of these traits is shown in Table 12-1 below. The information given was gathered after researchers studied numerous family histories and large numbers of hospital and medical records. Scientists are now involved in many long-term research projects investigating the chromosomes and genes of people to learn more about inheritance of various traits and conditions.

INHERITANCE AND SEX

Sex Determination. The sex of an offspring is determined by genes that lie in a particular pair of chromosomes, called the **sex chromosomes.** In one sex, the two sex chromosomes are alike in appearance; in the other, they are not (see Fig. 12-3). In the human female, the two sex chromosomes are alike. Each of these sex chromosomes is called an *X* chromosome.

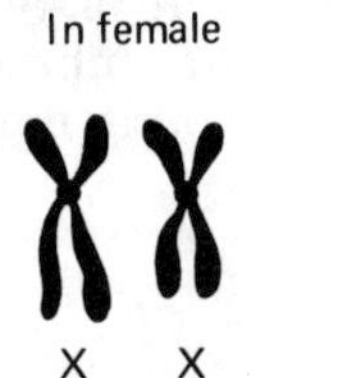

Fig. 12-3. Human female and male sex chromosomes.

Thus, a female has two *X* chromosomes, or *XX*. In the human male, one sex chromosome looks like an *X* chromosome of the female and the other is noticeably different. This different chromosome is called the *Y* chromosome. Thus, a male has both an *X* and a *Y* chromosome, or *XY*. The *Y* chromosome is much shorter than the *X* chromosome.

In the preceding chapter you learned that every body cell and every immature sex cell has the species number (or two sets) of chromosomes. After reduction division, each sex cell that is formed has one set of chromosomes. In an adult female, every young egg cell has the *XX* combination of chromosomes. As a result of reduction division, every mature egg cell has only one *X* chromosome. In an adult male, every young sperm cell has the *XY* combination of chromosomes. As a result of reduction division, some mature sperm cells have an *X* chromosome, and some have a *Y* chromosome.

When a sperm cell and an egg cell unite during fertilization, the combination of chromosomes that results determines whether the new individual will be a male or female. The egg cell, having an *X* chromosome from the mother, may unite with a sperm cell that carries either an *X* or a *Y* chromosome from the father. If the chromosome combination happens to be *XX*, the offspring will be a female. If the combination happens to be *XY*, the offspring will be a male. The combination that occurs is a matter of chance. This can be seen by working out the Punnett square below.

Key: Let *XX* = female
XY = male

Parents:	female ×	male
Sex Chromosomes:	*XX* ×	*XY*
Chromosomes in Sex Cells:	*X*	*X* or *Y*

Possible Fertilizations:

	Sperm Cells *X*	*Y*
Egg Cells *X*	*XX*	*XY*

Offspring:

Chromosome Combinations	*Sex*
½ *XX*	50% females
½ *XY*	50% males

The ratio of 50% females to 50% males is a 1:1 ratio. This ratio can only result when there are very many fertilizations. Where there is only one offspring, there is a fifty-fifty chance for it to be a male or a female. Where there are four girls in a family, the *X* and *X* combination at fertilization occurred four times, whereas the *X* and *Y* combination did not, by chance, occur at all. In a family that has all boys, the opposite would have occurred. However, in the world population as a whole, the ratio of males to females is very close to a 1:1 ratio. That is, the numbers of males and females are approximately equal.

Sex-Linked Traits. Like all other chromosomes, the sex chromosomes carry specific genes. In humans, the sex chromosomes carry such genes as those that determine color vision and blood clotting. Since the genes for

TABLE 12-1. SOME INHERITED FEATURES IN HUMANS

Color of eyes: dark dominant over light
Color of hair: dark dominant over light
Distribution of hair color: white forelock dominant over normal
Amount of hair: complete hairlessness dominant over normal
Speed of blood clotting: normal dominant over slow (hemophilia: sex-linked)
Blood type: Rh positive dominant over Rh negative
A dominant over O
B dominant over O
AB, blend of A and B
Ability to see colors: normal dominant over color blindness (both red and green appear gray: sex-linked)
Color of skin: normal color dominant over no color (albino); dark dominant over light
Number of digits: extra finger and toe (a mutation) dominant over five fingers and toes
Chin cleft: cleft (dimple) dominant over no cleft in middle of chin.
Tasting the chemical PTC: bitter dominant over no taste

these traits are carried on the sex chromosomes, the traits controlled by these genes are called **sex-linked traits.** Normal color vision is dominant over *color blindness,* which is the inability to distinguish between red and green. Normal blood clotting is dominant over *hemophilia* (bleeder's disease), or delayed clotting time. In general, color blindness and hemophilia occur more often in males than in females. This is because the shorter *Y* chromosome does not have a gene for the trait and the *X* chromosome's gene is for the recessive trait. As a result, only the recessive gene has an effect, since there is no dominant gene present to mask it.

THE HEREDITY OF TWINS

In most kinds of animals, the female produces many eggs at one time. After fertilization, the eggs develop at the same time into many offspring. In humans, one offspring is usually produced at a time. Sometimes, two or more offspring may be born at the same time. A tendency to produce more than one offspring at a time seems to be inherited. The production of more than one offspring can result either from the development of two or more fertilized eggs into separate embryos or from the splitting of a single young embryo into two or more embryos.

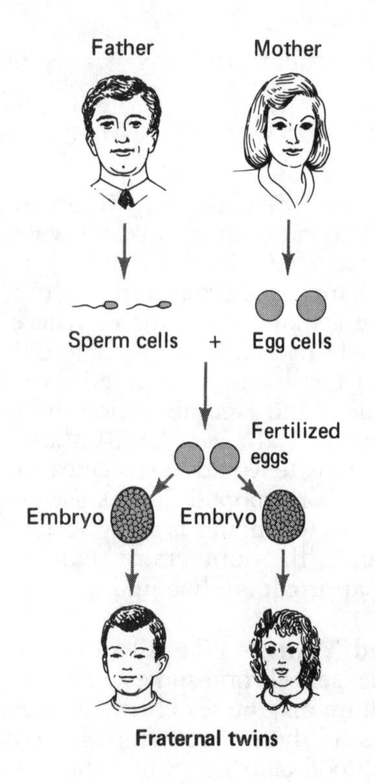

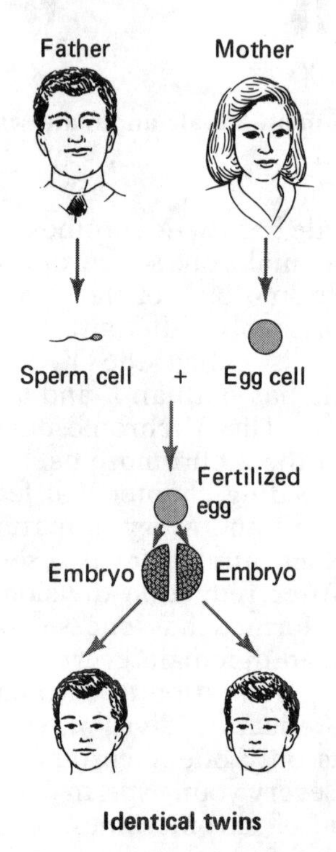

Fig. 12-4. Human twins.

Identical Twins. Sometimes, after an egg has been fertilized, it develops into a two-celled embryo and the two daughter cells separate instead of staying together. Then, each daughter cell undergoes many cell divisions, forming two separate embryos. In this manner, two individuals, called **identical twins,** are produced (Fig. 12-4*a*). Such twins are identical because they arise from the same egg and sperm. As a result, they have the same genes, look alike, and are of the same sex.

There have been many cases in which identical twins have been separated in infancy and brought up in different environments by different sets of parents. Recent studies have revealed that, when these twins grew up, they not only looked alike but they showed many similarities in character, ability, and personality. Identical twins who have been brought up together by the same (that is, natural) parents are usually not any more alike than twins who have been raised separately.

Fraternal Twins. Occasionally, a human ovary releases two eggs at a time. Each egg is then fertilized separately, and two separate embryos that become twin individuals develop (Fig. 12-4*b*). Since such twin offspring arise from two different fertilized eggs, they have different combinations of genes, do not look any more alike than any other brothers and sisters, and may be of different sexes. These twins are called **fraternal twins.**

Fraternal twins that have been brought up together in the same environment by their own parents are as different as any other two children in a family.

ENVIRONMENT AND HEREDITY

Physical traits, such as height, eye color, and skin color, are controlled by genes and are therefore inherited. Studies of the heredity of human twins and of the offspring of many kinds of organisms show that, in addition to physical traits, some types of behavior are controlled by genes. Examples of such traits are reflexes and instincts.

Studies also have shown that the development of some characteristics depends upon the environment as well as heredity. For example, an organism may inherit the ability to develop strong bones and to grow to a large size (genotype), but a poor diet can lead to disease and stunted growth (phenotype). Similarly, basic mental ability and special talents appear to be inherited, but both kinds of traits require a suitable environment for their fullest development. Thus, a talented child who is exposed to music has a better chance of becoming a good musician.

A favorable environment aids the development of desirable traits not only in humans, but also in the plants and animals that are important to people. For example, when wheat plants of the same pure breed are grown in fertile soil, they produce more grain then when they are grown in poor soil. Cows that are well-fed and kept free of disease and pests produce more milk than do cows of the same breed that are kept in less favorable conditions.

Evolution

You have learned that organisms inherit both physical and behavioral traits, and that the development of these traits can be affected by the environment. In the previous chapter you also learned that organisms have various adaptations, or traits, that make them well-suited to their environment. But environments do not always remain the same. Organisms having traits that enable them to adapt to changes in their environment are often more likely to survive and reproduce than organisms that do not. **Evolution** is the process of change that occurs in living things over time.

Some species of organisms have changed very little over millions of years. Others have become **extinct,** that is, none of their kind exist today. However, many species have *evolved,* or changed, into different types of life forms. (See Fig. 12-5*a*, *b*, and *c*.) Genetic changes that occur in organisms are passed on to their offspring. These genetic changes provide the basis for evolution.

For centuries, various scientists had come up with the idea that organisms change over time. Many people refused to accept this idea. But it was the English naturalist *Charles Darwin* (1809–1882) who came up with the most

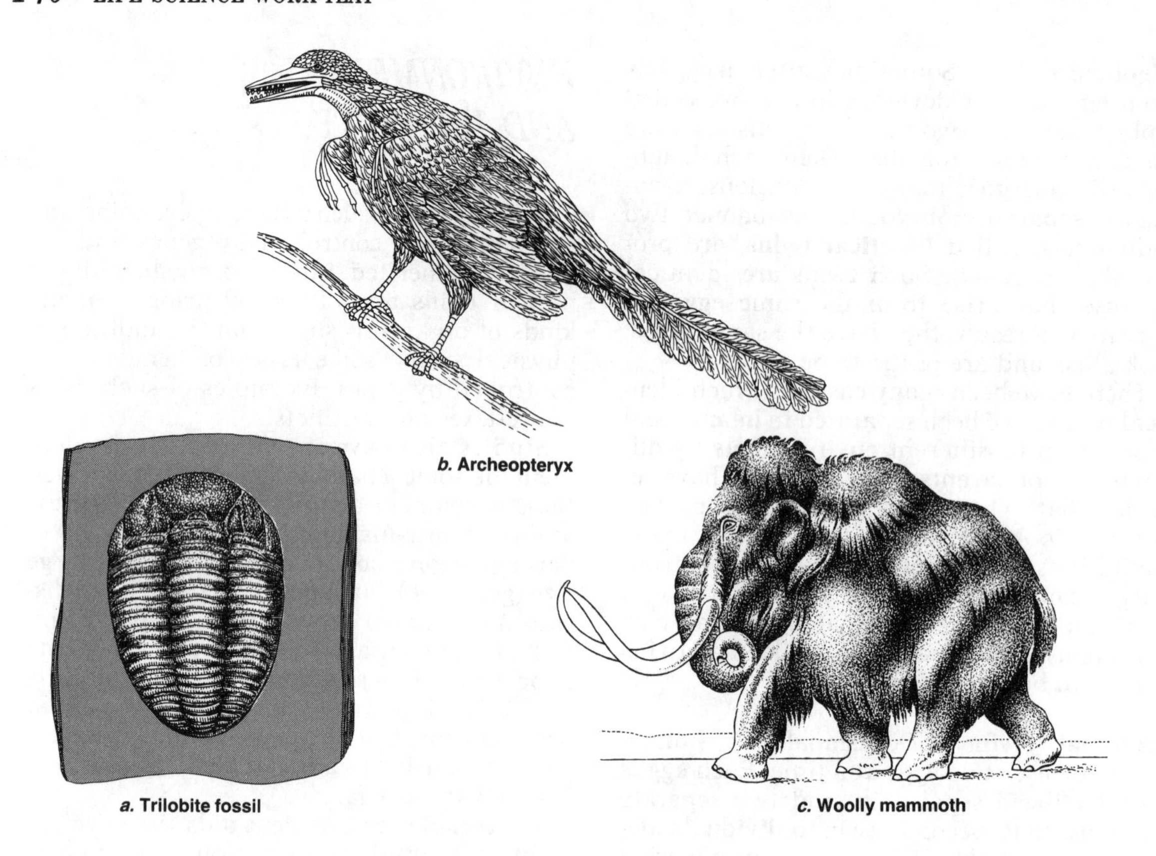

Fig. 12-5. Examples of extinct organisms.

convincing theory to explain how organisms evolve. Although Darwin devised his theory at about the same time that Mendel developed his laws of inheritance, the two men were not aware of each other's work. However, Darwin's *theory of evolution by natural selection* is well supported by what is now known about genetics. In addition, studies of the abundant fossil record, biochemistry, anatomy, and embryology all provide further support for the theory of evolution. (See Fig. 12-6 for an example of similarities in anatomy, and Fig. 12-7 for similarities in embryology—both of which show evolution from the same ancestor.)

ACQUIRED AND INHERITED TRAITS

The study of heredity shows that differences among organisms of the same species, and even within a family, can result from the normal operation of the laws of heredity (see Chapter 11). In addition, differences unrelated to heredity can occur. Some traits, which are acquired during an individual's lifetime, are due to ordinary differences in the organism's environment. These different traits are called **acquired traits.** For example, if one member of a set of identical male twins trains to be a weight lifter and the other does not, the weight lifter will develop stronger muscles than his brother. The development of strong muscles only affects certain body cells. Such acquired traits are not inherited because they do not affect the genes in sex cells. Therefore, the acquired trait for large muscles cannot reach the next generation.

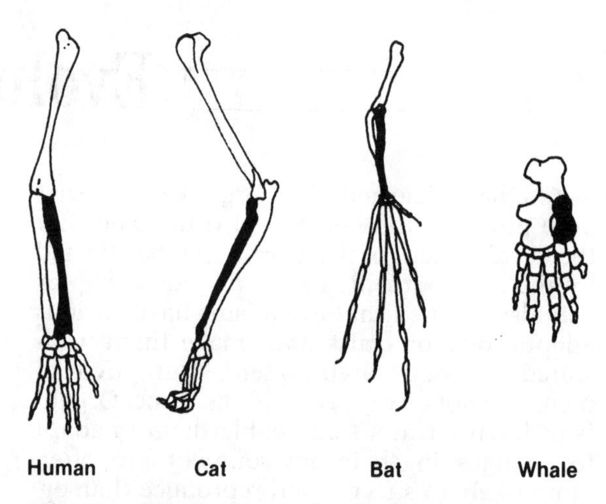

Fig. 12-6. Similarities in forelimb anatomy of mammals.

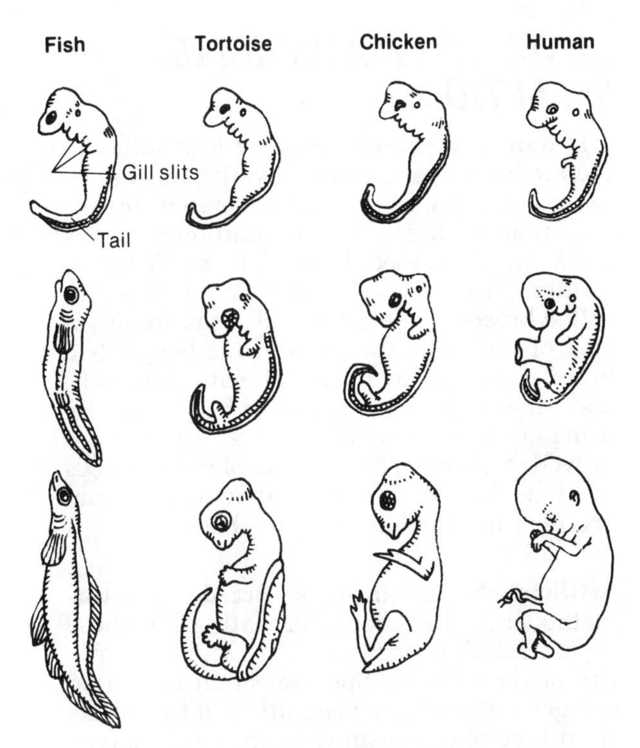

Fig. 12-7. Similarities in animal embryos.

The French scientist *Jean Baptiste de Lamarck* (1774–1829) proposed the idea that organisms can evolve by the inheritance of acquired traits. Lamarck believed that organisms responded to changes in their environment by developing new traits or discarding old traits, and that these new traits could then be passed on to offspring. For example, Lamarck would have thought that strong muscles from weight lifting could be inherited, but we know that this is not true. The work of Darwin, Mendel, and later scientists proved Lamarck's theory to be wrong.

Sometimes, new traits that can be passed on to offspring do develop. This type of trait is called an **inherited trait.** New inherited traits result from a change in one or more genes in a sex cell of at least one of the parents. When such a change in genes occurs, it is permanent (unless it kills the organism) and it may be passed on to succeeding generations.

MUTATIONS

Permanent changes in genes (or chromosomes) may occur naturally and suddenly. These changes are called **mutations.** The individual that bears such a new trait is called a **mutant.** For example, in a certain species of green frog, white offspring are sometimes produced. When such frogs are bred, the white color is found to be inherited. In this case, the new color is the mutation, and the white frog (an *albino*) is a mutant.

Hugo De Vries (1848–1935) was the first scientist to study mutations experimentally. He did this research on mutant evening primrose plants that had flowers much larger than normal. De Vries found that their descendants continued to bear large flowers.

Thomas Hunt Morgan (1866–1945) discovered such mutations as white eyes in the fruit fly. In addition, Morgan and his co-workers proved that mutations could result not only from changes in genes, but also from breaks in chromosomes.

Hermann J. Muller (1890–1967) discovered that penetrating forms of radiation such as X rays could cause genes to change permanently. Muller's work also has shown that rays from radioactive materials can reach reproductive (sex) cells and cause their genes to mutate more frequently than usual. Additional studies, based on his observations, indicate that radioactive fallout from nuclear explosions can present a hazard to future generations because of such genetic mutations.

Natural ultraviolet radiation from sunlight and chemicals in the environment can also cause mutations. Some mutations are deadly, or lethal, to organisms. Others may be neutral or beneficial. The variations in traits that arise from genetic changes enable the evolution of different species to occur.

ADAPTATIONS AND NATURAL SELECTION

Mutations that better enable an organism to survive and reproduce become adaptations for individuals in a population. Darwin realized that more organisms are born than can survive to reproduce. This is because living things have to struggle to survive, and relatively few make it to maturity. Those organisms having adaptations that make them better suited, or more "fit," for their environment are more likely to live to reproduce and pass on their traits. Darwin described the process by which more "fit" individuals survive to reproduce as **natural selection.** He meant that only those individuals with better adaptations were "selected by nature" to survive. Over the years, as environments change, natural selection continues to act on all individuals in a population. As a result, populations of species undergo evolution.

SPECIATION

The process by which new species evolve over time is called **speciation.** Darwin thought of speciation as a slow process that occurred as natural selection worked on populations for many thousands of years.

Some modern scientists propose that speciation may in some cases occur very rapidly. Genetic mutations provide the variations in traits that can give rise to new species. But it is usually physical separation of populations, known as **geographic isolation,** that causes new species to evolve. For example, geographic isolation can occur when organisms migrate and end up living on opposite sides of a mountain range, body of water, or on separate islands. Over time, as the organisms adapt to these different areas, they are more likely to evolve into different species. Darwin first noticed the effects of geographic isolation when he observed several closely related species of birds, called finches, that lived on a group of islands in the Pacific Ocean. He reasoned that the birds had all evolved from one ancestral finch species from South America that had come to the islands many years before. As these birds adapted to the variety of conditions on the different islands, speciation eventually occurred. Each species of bird had a different beak shape and size, suited to the food sources on its particular island (Fig. 12-8).

Fig. 12-8. Examples of Darwin's finches.

BENEFITS OF ARTIFICIAL SELECTION

In nature, organisms that are best suited to their environment evolve by the process of natural selection. In human societies, organisms that are best suited to fulfilling people's needs are developed, or bred, by **artificial selection.** Groups of similar organisms are called **breeds.** For example, all dogs are members of the same species, but the beagle is a breed of dog. Many breeds of plants, animals, and other organisms are used by people. By using our knowledge of heredity, we have improved or changed the features of many breeds so that they benefit us. Plant and animal breeders use the following methods:

Artificial Selection of a Breed. In this method, breeders choose for mating only those organisms that possess the best characteristics of their breed. Such selection is carried out generation after generation until the desired type of organism is produced. This type of artificial selection is also referred to as **controlled breeding.**

A breed of wheat that resists the disease called wheat rust originated from a few rust-resistant seeds. At first, the plants that grew from these seeds did not produce much grain. So, for many generations thereafter, only those rust-resistant plants that gave the largest crops were selected. Eventually, an improved breed called Kanred wheat was established, which resists the rust disease and yields a large quantity of grain.

In a similar fashion, a breed of speedier horses was developed. Generation after generation, breeders selected for mating only the swiftest horses, until thoroughbred racehorses were produced. In controlled breeding, the selected animals either mate or are **artificially inseminated,** a procedure in which the male's sperm is placed in the female's oviduct or uterus.

Hybridization. This is the method by which organisms of different breeds, each having a desirable trait, are crossed. Thus, **hybridization** is also called **crossbreeding.** Very often, hybrid organisms grow to larger size and are more vigorous than either of the parental breeds.

Hybrid corn, which is the major type of corn grown in the United States, is developed from four pure types of corn plants. One type produces large ears; another, straight rows of kernels; a third, many kernels per ear; and a

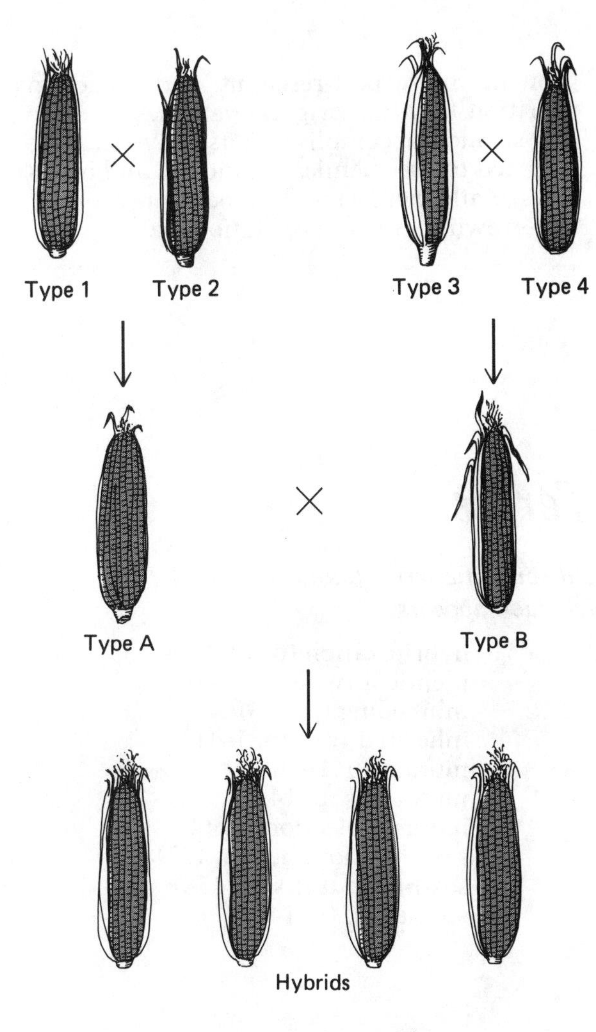

Fig. 12-9. Hybrid corn plants.

fourth, strong stalks. To produce hybrid corn, two of these types are cross-pollinated and, at the same time, the other two types are also cross-pollinated. Then, the offspring having the desirable traits from the first and second crossings are cross-pollinated to produce the hybrid corn. The hybrid corn that is produced has all four desirable traits from the pure types. (See Fig. 12-9.) Hybrid animals also have been developed that have a combination of beneficial traits.

Inbreeding. This is the method by which breeders mate only closely related organisms within a breed. The **inbreeding** of many successive generations of such related organisms often produces a breed in which all the offspring are very similar. The breed becomes pure because the genes of every organism are very much alike. Inbreeding is usually used after a breed has been improved by some other method. Inbreeding has been used to further improve such breeds as the garden pea and Holstein cattle.

After breeders produced the sweet, green garden pea, they allowed the plants to inbreed. Since peas are normally self-pollinated, the process of inbreeding continues to produce plants of this same variety.

Holstein cows, which are noted for the large volume of milk they produce, are bred only with Holstein bulls in order to produce more offspring with the same desirable characteristics.

Utilizing Mutations. Some mutations that people find useful are given special care by breeders. Seedless oranges and hornless cattle are examples of valuable breeds that arose by natural mutation.

Seedlessness in oranges arose as a mutation on a tree that before produced only oranges with seeds. This mutation caused the bud to grow into a branch bearing only seedless oranges. By grafting twigs from this branch to stems of young seeded orange trees, breeders have produced many trees that now bear seedless oranges.

Hornlessness in cattle originated as the result of a mutation in a breed that always had long horns. (Long horns are sawed off by cattlemen to prevent injury to both cattle and cattle workers.) Breeding experiments showed that the hornless condition was dominant and that the development of horns was recessive. After a few hornless individuals were obtained, they were inbred. By continued inbreeding, a new breed of hornless cattle was established.

At the Brookhaven (New York) laboratories of the Nuclear Regulatory Commission, many experiments are carried out to increase the rate at which mutations arise in plants that are of use to people. As mutants appear, they are examined for their possible value in improving existing kinds of plants. If the mutants seem valuable, they are crossed with an existing breed. An example of a plant improved by this method is the rust-resistant oat plant.

Genetic Engineering. Scientists develop new breeds of animals and new varieties of plants by inserting genes for desirable traits from one species into the cells of another. By this process of **genetic engineering,** researchers have been able to transfer genes from people and other animals into a variety of species to affect their rates of growth, among other traits. For example, scientists have been able to transfer a human growth hormone gene into the fertilized eggs of goldfish, to make the fish grow larger faster. Genes from other animals have been transferred into a variety of fish, chickens, hogs, cattle, and sheep to produce breeds that are larger and faster-growing, more disease-resistant, and more productive (in terms of eggs or milk, for example). Scientists have also genetically engineered varieties of grain, fruits, and vegetables that are tastier, faster-growing, more pest-resistant, and higher in nutrition than the original varieties of those crops. Such accomplishments may decrease the need to use chemical pesticides and insecticides, while increasing the food supply for our fast-growing human population.

CHAPTER REVIEW

Science Terms

The following list contains all of the boldfaced scientific terms found in this chapter and the page on which each appears.

acquired traits (p. 140)
artificial selection (p. 142)
artificially inseminated (p. 142)
breeds (p. 142)
controlled breeding (p. 142)
crossbreeding (p. 142)
evolution (p. 139)
extinct (p. 139)
fraternal twins (p. 139)
genetic engineering (p. 143)
geographic isolation (p. 142)
hybridization (p. 142)
identical twins (p. 139)
inbreeding (p. 143)
inherited trait (p. 141)
mutant (p. 141)
mutations (p. 141)
natural selection (p. 141)
sex chromosomes (p. 137)
sex-linked traits (p. 138)
speciation (p. 142)

Matching Questions

On the blank line, write the letter of the item in column B which is most closely related to the item in column A.

Column A

____ 1. arise from same fertilized egg
____ 2. arise from two different fertilized eggs
____ 3. process of change in life forms over time
____ 4. describes species that no longer exist
____ 5. traits that develop due to environment
____ 6. traits that develop due to heredity
____ 7. physical separation of populations
____ 8. sudden, permanent changes in genes
____ 9. term for artificial selection of breeds
____ 10. more "fit" organisms survive

Column B

a. acquired traits
b. mutations
c. geographic isolation
d. fraternal twins
e. natural selection
f. evolution
g. identical twins
h. inherited traits
i. extinct
j. genetic engineering
k. controlled breeding

Multiple-Choice Questions

On the blank line, write the letter preceding the word or expression that best completes the sentence or answers the question.

1. A human condition that is not inherited is
a. hemophilia *b.* hunger *c.* color blindness *d.* albinism 1 ____

2. A human disease that is inherited is
a. albinism *b.* Rh positive blood *c.* hemophilia *d.* typhus fever 2 ____

3. The sex of an animal is determined by the
a. chromosomes contributed by the parents
b. diet of the mother
c. chromosomes of the grandparents
d. number of male offspring already born 3 ____

4. Every egg cell of a human has
a. one *Y* chromosome
b. two *X* chromosomes
c. one *X* and one *Y* chromosome
d. one *X* chromosome 4 ____

5. Whether an offspring will be male or female is determined by chance at the moment of
a. cell division
b. reduction division
c. fertilization
d. birth 5 ____

6. Which statement about sex determination is true?
a. *XY* chromosomes are found only in males.
b. *XY* chromosomes are found only in females.
c. More females are born than are males.
d. The mother's diet influences the sex of a child before birth. 6 ____

7. Red-green color blindness is most common in
a. females *b.* males *c.* mutants *d.* albinos 7 ____

8. Whether identical twins will be born depends on what happens to
a. the fertilized egg cell
b. the *X* and *Y* chromosomes
c. the mother's diet
d. the sperm cells 8 ____

9. A person's athletic ability is determined by his or her
a. heredity only
b. education
c. environment and heredity
d. environment only 9 ____

10. The chromosome number in the body cells of human fraternal twins is
a. 43 in one and 46 in the other
b. 46 in each
c. 92 in each
d. 23 in each 10 ____

11. Twins born from two eggs fertilized at the same time are known as
a. identical twins
b. hybrid twins
c. Siamese twins
d. fraternal twins 11 ____

12. When a breeder chooses only the best animals or plants for mating, he or she uses a method of artificial
a. hybridizing *b.* sex linkage *c.* mutation *d.* selection 12 ____

13. Inbreeding is used to improve and
a. produce new types
b. maintain existing types
c. cause mutations
d. increase variation 13 ____

14. The desirable characteristics of two different breeds of wheat plants may be combined in the offspring by
a. selection *b.* grafting *c.* cross-pollination *d.* self-pollination 14 ____

15. A cat breeder can determine that a tailless cat is a mutant, if the cat
a. does not grow a tail for 5 years
b. remains tailless after its diet is changed
c. is mated to a tailed cat and the offspring are tailless
d. grows more fur than its parents 15 ____

16. Harmful changes in the genes of sex cells can be caused by
a. crossing hybrids
b. incomplete dominance
c. fertilization by three sperms
d. radioactive fallout 16 ____

17. Who first discovered that X rays could reach the genes of sex cells and cause mutations?
a. Morgan *b.* Mendel *c.* Muller *d.* De Vries 17 ____

18. An albino deer is rare. This color probably occurs as a result of
a. a natural mutation
b. an artificial breeding
c. an artificial mutation
d. a dominant trait 18 ____

19. A male dog is trained to sit up and beg. It is then mated with a female dog that has been trained the same way. Puppies born to these dogs should be able to do the same tricks
a. immediately after birth
b. without any training
c. with the same training
d. better than the parents 19 ____

20. A pure breed of roses always produces red flowers. The sudden appearance of a yellow rose on this bush would be evidence of
a. grafting *b.* mutation *c.* dominance *d.* segregation 20 ____

21. The process of change that occurs in organisms over time is
a. extinction
b. mutations
c. evolution
d. crossbreeding 21 ____

22. New traits in genes that can be passed on to offspring are
a. acquired traits
b. inherited traits
c. artificial traits
d. hybrid traits 22 ____

23. Geographic isolation of populations can lead to the process of
a. speciation
b. artificial selection
c. controlled breeding
d. hybridization 23 ____

24. Species that no longer exist are said to be
a. inbred *b.* hybridized *c.* mutated *d.* extinct 24 ____

Modified True-False Questions

In some of the following statements, the italicized term makes the statement incorrect. For each incorrect statement, write the term that must be substituted for the italicized term to make the statement correct. For each correct statement, write the word "true."

1. The chromosome combination of *X* and *Y* is found in the *males* of humans. 1 ______________

2. In a large population, the number of males to females is in a ratio of approximately *1:2.* 2 ______________

3. *Fraternal* twins are always of the same sex. 3 ____________

4. *Eye color* is an example of a trait that is inherited. 4 ____________

5. Kanred wheat is an example of a plant that has been improved by *selection* and controlled breeding. 5 ____________

6. Hybrid corn was developed through *inbreeding.* 6 ____________

7. After eating wild onions, cows give milk that has the odor of onions. This is an example of an *inherited* trait. 7 ____________

8. A change in a gene produces a new *acquired* trait. 8 ____________

9. Scientists can transfer genes for desirable traits between different species by *genetic engineering* techniques. 9 ____________

10. Hornless animals have been known to arise naturally from horned animals as a result of *reduction division.* 10 ____________

11. A mutation is a change in the *oxygen* base pairs of DNA 11 ____________

12. The cloning of cells involves the process of *meiotic* cell division. 12 ____________

13. A cross of a red cow with a white bull produces all roan offspring. This inheritance is called *dominance.* 13 ____________

Testing Your Knowledge

1. Use the Punnett square method to answer each of the following questions:
   a. What is the possibility that a child will be born a boy or a girl?

   ____

   b. If a woman who is recessive for hemophilia (that is, carries just one gene for the condition) marries a man who has hemophilia (that is, carries the gene on his *X* chromosome), what are the possible results for the gene combinations of their offspring?

   ____

   ____

2. Explain why speciation can occur in a population of birds that becomes physically separated on two different islands.

   ____

   ____

3. A potato breeder developed a breed that produced large and tasty potatoes. He found that these potatoes often rotted before he could harvest all of them. Another breeder had small potatoes that never rotted before harvesting. Using your knowledge of heredity, explain what technique could be used to develop large potatoes that don't rot early.

   ____

   ____

4. Explain the difference between the two terms in each of the following pairs:

a. fraternal twins and identical twins ______________________________

____

b. inbreeding and crossbreeding ______________________________

____

c. inherited trait and acquired trait ______________________________

____

d. natural selection and artificial selection ______________________________

____

5. Compare and contrast the events in mitotic and meiotic cell division.

____

____

6. Why must sperm and egg cells contain one-half the species number of chromosomes rather than the full number? ______________________________

____

____

7. Explain the differences between the production of sperm cells and egg cells. ______________

____

A Mysterious Extinction: What Killed Off the Dinosaurs?

If you were to search our entire planet, you would not find a single living dinosaur. Yet at one time, millions of these fascinating creatures roamed the Earth.

Although no one has ever seen a living dinosaur—dinosaurs vanished long before human beings evolved—people have long been curious about the fate of these mysterious reptiles. What, scientists ask, caused the dinosaurs to become extinct some 65 million years ago?

You might wonder, what makes us so sure about the date of the dinosaurs' extinction? First of all, scientists are able to determine the age of rocks. They also know that younger rocks usually lie on top of older rocks. So when scientists dig deeper and deeper into the earth, they are really digging farther and farther back in time. They are, in effect, traveling into the past.

As it turns out, scientists digging through many layers of rock can find no dinosaur fossils—such as bones—in rocks younger than 65 million years old. However, there are plenty of dinosaur fossils in older rocks. These findings support the idea that dinosaurs became extinct about 65 million years ago. But what could have caused them to become extinct?

Scientists considered a number of possibilities. These included rapid changes in climate. Maybe the climate became too cold, too hot, too wet, or too dry for the survival of dinosaurs or the plants and animals they fed on. But what could cause such a drastic change? Lots of dust and smoke in the air, suggested some researchers. For one thing, this would block out sunlight and, as a result, temperatures would decrease.

But this suggestion only led to another question. What could cause a cloud of dust and smoke so large that it would affect the climate of the entire Earth? The eruption of thousands of volcanoes seemed one reasonable answer, which some scientists proposed. Then a strange clue was literally dug up from beneath the Earth's surface.

A group of scientists discovered a layer of rock rich in the element iridium. What's more, the layer of rock was about 65 million years old. Was this a meaningless coincidence? No, declared the researchers. They pointed out that rocks on Earth generally contain small amounts of iridium. Rocks from space, however, often contain lots of it!

Based partly on this evidence, these scientists proposed that a huge comet or asteroid crashed into the Earth about 65 million years ago. This catastrophic impact would have sent millions of tons of dirt, dust, and smoke into the air, blocking much of the sun's light from reaching the Earth's surface.

Maybe so, said other scientists. But such a collision would have blasted out a huge crater. So where's the crater? they asked. Tiny bits of glass found around the Caribbean Sea provided a clue. These bits of glass, called tektites, are produced when large objects from space crash into the Earth. This led some researchers to believe that if they looked hard enough, they might find a crater in or near the Caribbean Sea.

This theory was proposed by the physicist Luis Alvarez and his son Walter, a geologist. They believed that an asteroid approximately 10 miles in diameter hit the Earth about 65 million years ago. The impact scattered dust and debris into the atmosphere, caused huge fires from the hot debris, severe storms with high winds, highly acidic rain, and perhaps volcanic activity. The impact would have caused changes in the Earth's atmosphere, blocking of the sunlight to the planet and the lowering of temperatures globally. The animals that could not adapt (dinosaurs) would die and become extinct. The theory was supported by the discovery of an *impact crater* at the tip of the Yucatan Peninsula in the Gulf of Mexico.

1. *The dinosaurs became extinct about*

a. 6,500 years ago.
b. 65,000 years ago.
c. 650 thousand years ago.
d. 65 million years ago.

2. *An example of a fossil is*

a. a comet.
b. a bone.
c. iridium.
d. a crater.

3. *Rock layer A lies on top of rock layer B. Rock layer A is probably*

a. older than B.
b. younger than B.
c. the same age as B.
d. older or younger than B.

4. *Describe the evidence that supports the hypothesis that an object from outer space caused the extinction of the dinosaurs.*

____

____

5. *As mentioned above, scientists think a cooling climate may have caused the dinosaurs' extinction. Now, scientists think our global climate may be warming. On a separate sheet of paper, write a brief essay describing how such a change in the world's climate may affect living things today. Discuss which regions you think might be most affected by a warming climate and why. NOTE: To write your essay, you may have to do research.*

13

How Plants Make Their Food and Live

LABORATORY INVESTIGATION

MOVEMENT OF WATER IN A PLANT

A. Gently break off a leaf from a geranium plant. Be sure to include the leaf stalk, or *petiole,* along with the leaf *blade.*

B. Take a 10-cm-square piece of aluminum foil and punch a hole in its center with a pencil. Make the hole just wide enough to fit the petiole of the leaf. Insert the petiole into the hole and pull it down as far as it will go. Be careful not to break the petiole. Make a waterproof seal by dripping candle wax around the top of the hole through which the petiole was inserted.

C. Set the petiole into a clear glass or plastic cup three-quarters full of water. The end of the petiole should reach into the water. Bend the foil down around the cup's edge, making a tight seal around the glass. Place an inverted cup that is clean and dry over the leaf (see Fig. 13-1). Put the setup in a sunny place. Inspect the cups and the leaf at intervals for 2 days.

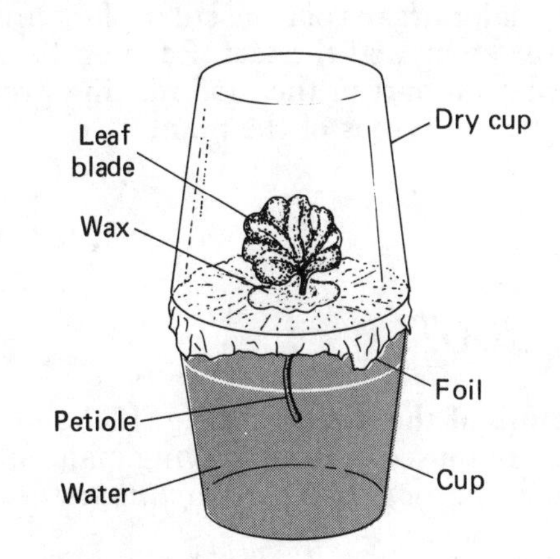

Fig. 13-1. Setup for geranium leaf in cup.

1. Describe any changes that have occurred. ______________________________

____

2. Where did the water in the upper cup come from? ______________________________

____

3. Explain your reasoning. ____________________

____

4. How do leaf blades get their water? ____________________

____

5. Why is it important for water to reach the leaves? ____________________

____

The Food of Plants

In general, plants are stationary organisms that are unable to search for food as animals do. Like animals, however, plants require energy from food to live. Plants can manufacture their own food from simple raw materials that they absorb from their environment. In this process, some of the energy of sunlight is used in chemical reactions that produce food. Water is one of the other raw materials a plant uses in producing food. Because plants (and algae) make their own food, they are called *producers*. This chapter studies the food-making system of the vascular, or higher, plants.

As you observed in your laboratory investigation, water can pass up through a petiole, into the leaf blade, and out to the atmosphere. A study of a complete plant reveals not only how excess water gets out of the plant, but how water and other raw materials enter the plant. Such a study also reveals how the raw materials are made into food and how the food is used to provide energy for the life processes of the plant.

Vascular Plants

The major organs of a vascular plant are the **root,** the **stem,** and the **leaf.** Each of these organs plays a part in the food-making process and other activities of the plant.

THE ROOT

Structure of the Root. As Fig. 13-2 shows, a root is composed of the following major parts: *growing tip, root hairs, conducting tissue,* and *storage tissue.*

The growing tip contains the *growing tissue.* Cells of this tissue continually divide and add new cells to the root. These new cells then enlarge. As a result of the division of cells and their enlargement, the root increases in size. The very tip of the growing tissue is covered by a protective layer of cells called the *root cap.*

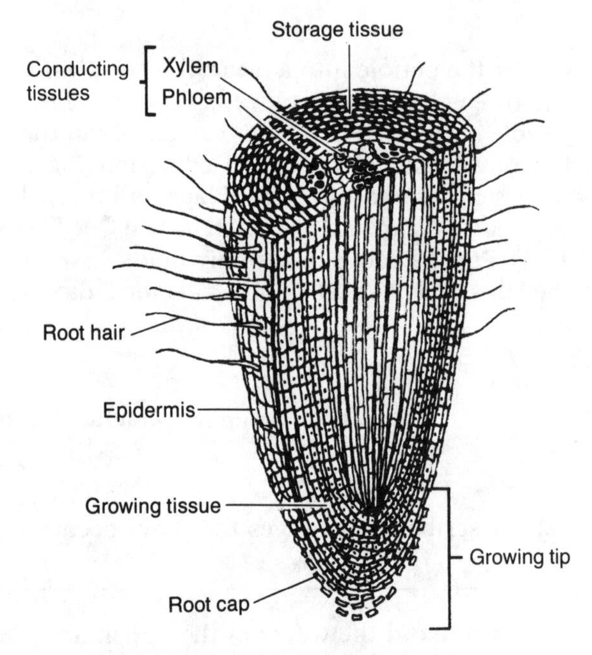

Fig. 13-2. Portion of a root.

Above the growing tissue is a region where small projections grow out from many of the outermost cells of the root. These projections, called **root hairs,** push their way between particles of soil. Each cell that has a root hair projecting from it has a large vacuole full of water and many dissolved substances.

Conducting tissue is located near the center of the root. The cells of conducting tissue are long and tubelike. There are two types of conducting-tissue cells—*xylem* and *phloem.* **Xylem** cells conduct liquids upward. **Phloem** cells conduct liquids downward. Mature xylem cells are hollow, because they contain no cytoplasm. As a result, water passes easily through xylem tissue.

Storage tissue is located between the *epidermis* (the outermost cell layer) and the central conducting tissue. The cells of storage tissue possess large vacuoles and many *granules* in their cytoplasm. The granules are frequently composed of carbohydrates and proteins.

Functions of the Root. The most important activities carried on by the root are *absorption, anchorage, conduction,* and *storage.*

Absorption. Root hairs (Fig. 13-3) absorb water and dissolved mineral salts from the soil. Water enters the root hairs by osmosis; minerals enter the root hairs mainly by diffusion.

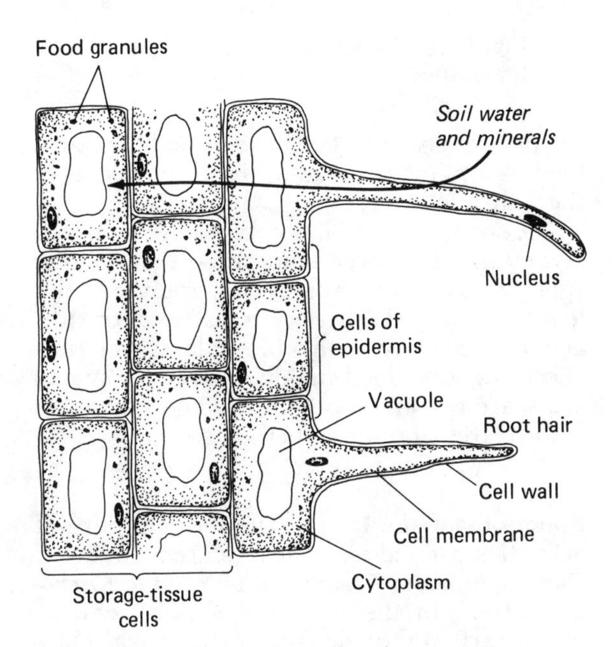

Fig. 13-3. Absorption by root hairs.

Anchorage. Roots, being underground, anchor (firmly hold) the base of the plant securely in the soil.

Conduction. Substances that have entered root hairs diffuse from cell to cell within the root tip until these substances reach the xylem. The xylem conducts the substances upward. The phloem conducts food downward from the upper parts of the plant to the cells of the root tip.

Storage. When plants make more food than they can use during a particular period of time, most of the excess food usually passes into the storage-tissue cells of the root. At times when the plant lacks food, the storage-tissue cells release the stored material for the plant's use.

Value of Roots. Because roots store plant food, they are a source of food for people and other animals. The beet, carrot, sweet potato, and turnip are roots that serve as food. Other useful root products include flavorings, such as licorice.

THE STEM

Structure of the Stem. As Fig. 13-4 shows, a stem is composed of the following major layers: *conducting tissue, growing tissue,* and *storage tissue.*

Conducting tissue consists mainly of cells of xylem and phloem, which are connected to the xylem and phloem cells of the root.

Growing tissue consists of cells that divide and add new cells to the length and width of the stem. A stem's length is increased by growth at the *terminal buds*—special areas at the tips of the branches. Other cells that continually divide add thickness to the stem. Some plants, such as the rose bush, possess a growing layer and live for many years. Other plants, like the dandelion, possess a growing layer, but live only one year. Still other plants, such as grass, lack a growing layer in the stem. These stems grow in thickness only as a result of the enlargement of their existing cells, and they usually die down to ground level each year. Woody plants, like trees, usually show annual growth rings in their stems, or *trunks.*

The storage tissue layer consists of cells in the center of the stem and cells that surround the conducting tissue.

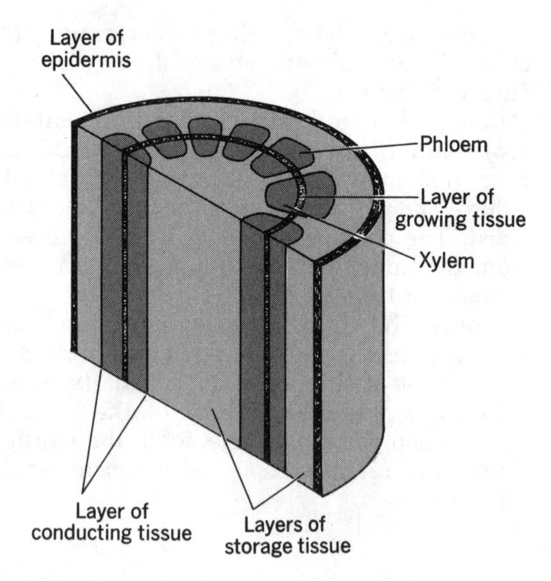

Fig. 13-4. Cutaway view of a stem.

Functions of the Stem. The most important activities carried on by the stem are *support, conduction,* and *storage.*

Support. The stem supports (holds up) the leaves and flowers of the plant, helping the leaves to get sunlight.

Conduction. Xylem and phloem cells make up the conducting tubes in the stem that transport materials between the roots and leaves.

Storage. The stems of many plants serve as storage areas for food, water, and other substances.

Value of Stems. People use the stems of the sugar cane, asparagus, and white-potato plant as food. We also use the trunks of trees such as the maple, oak, and pine for lumber (wood is mainly xylem tissue). Other stem products include paper, rubber, gum, maple syrup, and turpentine. Animals use stems as a source of food and shelter, too.

THE LEAF

Structure of the Leaf. A microscopic examination of a leaf cut crosswise (Fig. 13-5) reveals five major regions: *upper epidermis, lower epidermis, veins, palisade layer,* and *spongy layer.*

The upper epidermis of the leaf consists of a single layer of cells. These cells secrete a waxy protective coating, which helps prevent water loss from the leaf.

Except for the many pairs of **guard cells,** the cells of the lower epidermis of the leaf are much like those of the upper epidermis. A **stoma,** or pore, lies between each pair of guard cells. These cells regulate the size of the *stomata* (pores), letting gases and water vapor in and out.

Veins, or *vascular bundles,* are composed mainly of xylem and phloem tubes, which are connected to the xylem and phloem tissues of the stem and root.

The palisade layer lies just below the upper epidermis. This region may consist of one or more layers of elongated, boxlike cells. Each cell contains numerous **chloroplasts,** the bodies that store chlorophyll. **Chlorophyll** is the green pigment (colored substance) that traps the light energy needed for food-making. Most food of a plant is made in the chloroplasts of the palisade layer.

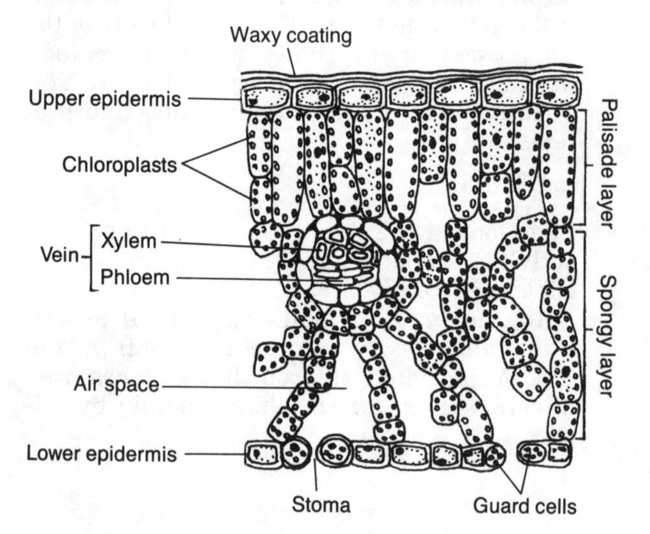

Fig. 13-5. Portion of a leaf, cut crosswise (magnified).

The spongy region lies between the palisade layer and the lower epidermis. Cells of the spongy layer are somewhat rounded and loosely arranged. These cells are used for storage and also contain chloroplasts, so they carry out some food-making. Large air spaces are scattered among the cells. The air spaces lead to the stomata, where gas exchange occurs.

Functions of the Leaf. The most important activities carried on by leaves are as follows:

Photosynthesis: Formation of Simple Carbohydrates. In the presence of light, chlorophyll, carbon dioxide, and water, a leaf manufactures simple carbohydrates like **glucose** (a sugar). This food-making process is called

photosynthesis. Carbon dioxide from the air enters the leaf through the stomata. Water from the soil enters the root through the root hairs and reaches the leaf through the xylem. Both carbon dioxide and water pass into the cells containing chlorophyll. There, with energy provided by light—especially sunlight (photosynthesis occurs in daytime)—the chlorophyll, carbon dioxide, and water react chemically. As a result of this reaction, glucose is formed and oxygen is given off as a by-product. The oxygen passes into the air spaces of the spongy region and out of the leaf through the stomata.

The overall, simplified reaction that takes place during photosynthesis is:

$$6\,CO_2 + 6\,H_2O \xrightarrow[\text{light (energy)}]{\text{chlorophyll (catalyst)}} C_6H_{12}O_6 + 6O_2$$

six molecules of carbon dioxide	six molecules of water	one molecule of simple sugar-glucose	six molecules of oxygen
(raw materials)		(product)	(byproduct)

Note that the chlorophyll acts as a *catalyst* in this reaction, enabling it to occur. The raw materials are carbon dioxide and water. The finished product is glucose and the by-product is oxygen. In the photosynthesis reaction, the light energy supplied by the sun is changed to chemical energy, which is then stored in the glucose.

Eventually, the glucose is oxidized within the cells of some organism that eats the plant. The process of oxidation releases the stored energy. This energy enables the organism to carry out its life activities. Thus, photosynthesis provides the primary source of food (glucose) to all organisms, and the main source of oxygen to the environment.

Formation of Starch. When conditions are favorable, a green plant makes more glucose than it needs to supply energy for its life activities. With the aid of numerous enzymes, the leaf can change the glucose to starch, which is then stored. **Starch** is a large molecule composed of a chain of many glucose sugar molecules. When energy is needed, enzymes (such as the diastase you used in the laboratory investigation in Chapter 2) change, or digest, the starch back to glucose (Fig. 13-6).

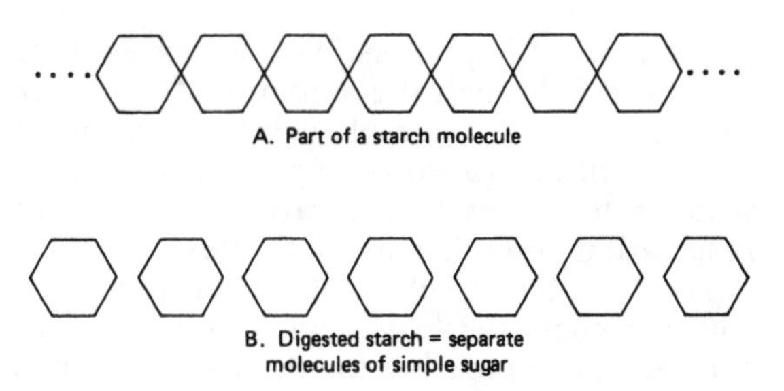

Fig. 13-6. A starch molecule and digested starch.

Formation of Other Nutrients. Besides making carbohydrates like starch, a green plant makes the following nutrients:

Proteins and Vitamins. The cells of the leaf, as well as the cells of other plant organs, remove nitrogen, sulfur, potassium, and phosphorus from the mineral salts that the roots absorb. These elements are then joined to the elements in glucose. The resulting substances are amino acids (which are then combined to form proteins) and vitamins. Sunlight and chlorophyll are not directly involved in making proteins and vitamins.

Fats and Oils. Various plant structures, including the leaf, change glucose into the fats and oils that the plant needs. Like glucose, the fats and oils are made up of carbon, oxygen, and hydrogen.

Respiration. During *respiration,* a plant takes in oxygen through the stomata, found mainly in its leaves. Food is oxidized and energy is released in the cells of the plant. At the same time, carbon dioxide and water vapor are formed as wastes. Both of these gases are released through the stomata.

Respiration and photosynthesis are opposite chemical processes. However, respiration goes on at all times; photosynthesis only occurs with sunlight.

Photosynthesis:

Energy (sunlight) + carbon dioxide + water → sugar + oxygen

Respiration:

Sugar + oxygen → carbon dioxide + water + energy (chemical)

Transpiration. The evaporation of water from the inside of a leaf is called **transpiration.** In your laboratory investigation at the beginning of this chapter, the water that appeared in the upper inverted cup came from inside the leaf by this process. The water escapes from a leaf through the stomata. Most water that is released from a plant by transpiration is water that was absorbed from the soil. Some water released by transpiration comes from respiration and other life activities.

Transpiration is of value to the plant because it cools the leaves and, together with osmosis, is responsible for the rise of liquids in the plant.

Value of Leaves. The existence of all land animals (including humans) depends on green leaves. These leaves, either directly or indirectly, supply all animals with food. In addition, leaves provide oxygen and remove excess carbon dioxide from the air.

The green leaves of only a few plants are used directly as food by people. Examples are spinach, lettuce, and cabbage. Some leaves are used by people for other purposes. For example, peppermint leaves are used for flavoring; digitalis leaves are used for making medicines.

INSECT-EATING PLANTS

In addition to carrying out photosynthesis, some plants also have modified leaves that trap and devour insects. These plants are said to be **insect-eaters.** Their special leaves secrete enzymes that digest the soft parts of the insects they trap. Examples of insect-eating plants are *Venus's flytrap* and the *pitcher plant* (Fig. 13-7).

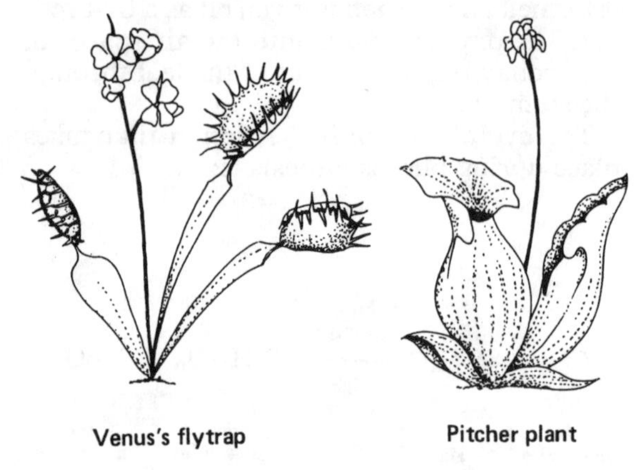

Fig. 13-7. Insect-eating plants.

Plant Diseases

It is clear that plants need food, water, light, and minerals to live. Plants also have to survive attack by various plant disease organisms. Just as animals are attacked by a variety of microscopic organisms, so are plants. Many types of fungi attack crops and trees. Examples are rusts and smuts, which can destroy grain and vegetable crops; and yeastlike fungi, which can cause Dutch elm disease, chestnut blight, and ergot disease of rye. Plants are also infected by various viruses, which can be specific to certain parts of a plant, like the leaves or stem only. The well-studied "tobacco mosaic" virus affects the leaves of the tobacco plant. Scientists are looking to nature for cures to some of these plant diseases. For example, a certain naturally occurring soil bacterium is being used to help fight a fungus that attacks the roots of wheat plants.

Biotechnology has been enlisted in the battle to develop better plants and plants more resistant to the environment. In plants, an early example of biotechnology has allowed scientists to develop tomatoes that are larger, sweeter, and do not rot as quickly. Genes that make agricultural plants resistant to disease can now be added to a crop's genes. Not only can these genes help the crops become more resistant to environmental factors, but the genes are capable of moving between plants. What if these genes moved to weeds; weeds that already cause problems for farmers by robbing the moisture from the soil and grow so large they shadow the sunlight from the crops? So, while on the surface genetically altered plants seem like a good idea, they could also prove harmful to our agriculture in the future.

CHAPTER REVIEW

Science Terms

The following list contains all of the boldfaced scientific terms found in this chapter and the page on which each appears.

chlorophyll (p. 154)
chloroplasts (p. 154)
glucose (p. 154)
guard cells (p. 154)
insect-eaters (p. 156)
leaf (p. 152)
phloem (p. 153)
photosynthesis (p. 155)
root (p. 152)
root hairs (p. 153)
starch (p. 155)
stem (p. 152)
stoma (p. 154)
transpiration (p. 155)
xylem (p. 153)

Matching Questions

On the blank line, write the letter of the item in column B which is most closely related to the item in column A.

Column A	Column B
____ 1. energy-releasing process in plants	*a.* white potato
____ 2. lets carbon dioxide into leaves	*b.* phloem
____ 3. a stem used as food	*c.* sunlight
____ 4. conducts liquids downward	*d.* respiration
____ 5. absorb soil minerals and water	*e.* photosynthesis
____ 6. a food-making process in plants	*f.* xylem
____ 7. conducts liquids upward	*g.* chloroplasts
____ 8. energy for photosynthesis	*h.* root hairs
____ 9. contain plant pigment	*i.* carrot
____ 10. an insect-eating plant	*j.* pitcher plant
	k. stoma

Multiple-Choice Questions

On the blank line, write the letter preceding the word or expression that best completes the sentence or answers the question.

1. Growth in the length of a root takes place in the region located at the
a. upper tip of the root *c.* center of the root
b. lower tip of the root *d.* surface of the soil 1 ____

2. In a beet plant, excess sugar is stored in the
a. stem *b.* leaf *c.* root *d.* petiole 2 ____

3. Air spaces in a leaf are found in the region called the
a. upper epidermis *b.* palisade layer *c.* spongy region *d.* vein 3 ____

4. Stems are important to plants because the stems help
a. the leaves get sunlight *c.* anchor the plant
b. the roots grow deeper *d.* absorb water 4 ____

5. The compound that the xylem in a corn root normally conducts to the upper part of the plant is
a. carbon dioxide *b.* protein *c.* fat *d.* water 5 ____

6. At night, the source of nourishment for a green plant is material located in the
a. xylem *b.* epidermis *c.* storage tissue *d.* palisade tissue 6 ____

7. The xylem of a stem is continuous with the
a. phloem of the root and the vein
b. phloem of the stem and the xylem of the leaf
c. xylem of the root and the xylem of the leaf
d. phloem of the leaf and the vein 7 ____

8. A plant gets the proteins, fats, and vitamins it needs by
a. absorbing them from the soil *c.* absorbing them from the air
b. making them itself *d.* absorbing them from sunlight 8 ____

9. The root of a maple tree
a. holds the tree in the soil *c.* takes in carbon dioxide for the leaves
b. takes in oxygen for the leaves *d.* manufactures food for the leaves 9 ____

10. Which plant stem is used as a food source?
a. sugar beet *b.* lettuce *c.* sweet potato *d.* sugar cane 10 ____

11. The part of the leaf in which guard cells are found is the
a. spongy layer *b.* vein *c.* palisade layer *d.* epidermis 11 ____

12. Substances manufactured in leaves reach the roots by way of the
a. phloem *b.* xylem *c.* epidermis *d.* root caps 12 ____

13. Most water absorption in a plant occurs through the
a. stomata *b.* leaf *c.* root hairs *d.* bark 13 ____

14. An example of a conducting tissue is the
a. xylem *b.* granule *c.* chloroplast *d.* stoma 14 ____

15. The gas released during photosynthesis is
a. carbon dioxide *b.* chlorophyll *c.* starch *d.* oxygen 15 ____

16. During photosynthesis, chlorophyll acts as
a. a raw material *b.* an energy source *c.* a by-product *d.* a catalyst 16 ____

17. The leaves of a plant give off excess water during the process of
a. respiration *b.* photosynthesis *c.* transpiration *d.* absorption 17 ____

18. A tree trunk grows in thickness as a result of cell divisions in the
a. storage cells *b.* growing layer *c.* xylem *d.* phloem 18 ____

19. A plant gets energy for its life activities as a result of the process of
a. respiration *b.* transpiration *c.* osmosis *d.* mitosis 19 ____

20. Transpiration aids the rise of water in a plant and also
a. cools the stem *c.* cools the leaves
b. warms the root *d.* warms the leaves 20 ____

21. From which of the following plants is the leaf used in the human diet?
a. bean *b.* cabbage *c.* pea *d.* carrot 21 ___

22. Organisms that cause plant diseases include the following *except*
a. rusts and smuts
b. yeastlike fungi
c. viruses
d. digitalis 22 ___

23. Four factors, or conditions, essential for the making of food in a leaf are
a. chlorophyll, carbon dioxide, water, light
b. chlorophyll, oxygen, water, darkness
c. carbon dioxide, oxygen, water, darkness
d. glucose, oxygen, carbon dioxide, light 23 ___

24. The pitcher plant is an example of a plant that eats
a. algae *b.* insects *c.* fungi *d.* mice 24 ___

Modified True-False Questions

In some of the following statements, the italicized term makes the statement incorrect. For each incorrect statement, write the term that must be substituted for the italicized term to make the statement correct. For each correct statement, write the word "true."

1. The chief organs of a rose plant are the root, the stem, and the *thorns*. 1 ___
2. Xylem tissue conducts fluids *upward* in a stem. 2 ___
3. Phloem tissue conducts fluids *upward* in a leaf. 3 ___
4. The leaf can store extra glucose as *protein*. 4 ___
5. Most of the chlorophyll in a leaf is located in the cells of the *palisade* region. 5 ___
6. A plant absorbs water through its *chloroplasts*. 6 ___
7. Plants take in carbon dioxide through the *stomata*. 7 ___
8. The energy that the leaf uses in making food comes from *chemicals*. 8 ___
9. As a result of respiration, a leaf gives off *carbon dioxide* and water. 9 ___
10. As a result of photosynthesis, a leaf gives off *carbon dioxide*. 10 ___

Testing Your Knowledge

1. State four uses of roots to plants. ___

2. State three uses of stems to plants. ______________________________

____

____

3. State five uses of leaves to plants. ______________________________

____

____

4. Twelve radish seeds were planted in a box of soil, moistened regularly, and kept near a window. In 7 days, young plants about an inch tall were visible.

a. The energy for the early growth of the seeds came from
(1) light (2) stored food in the seed (3) water (4) minerals ____

b. The radish plant became independent after the development of
(1) amino acids (2) chlorophyll (3) symbiotic algae (4) bark ____

c. Seven days after planting, three of the radish plants were dug up and finely chopped. They were then tested with Benedict's solution. The red color showed the presence of
(1) fats and oil (2) minerals (3) protein (4) simple sugars ____

5. It is often said that the sun is the source of all energy on earth, even the energy that is produced in your body. Explain this statement. ______________________________

____

____

____

6. A green leaf is very much like a factory. Show how a leaf is like a dress factory by completing the following table:

	Dress Factory	*Leaf*
Finished product	dresses	carbohydrates
Raw materials	cloth, thread	*a.* ______ *b.* ______
Materials delivered by	elevator	*c.* ______ *d.* ______ *e.* ______
Machinery	cutting machines and sewing machines	*f.* ______
Energy that runs machines	electricity and manpower	*g.* ______
By product	cloth scraps sold to paper manufacturer	*h.* ______ for respiration of plants and animals

7. Show that you understand the differences between respiration and photosynthesis by completing the following table:

	Photosynthesis	*Respiration*
When it occurs	only when light is present	*a.* ______________
Where it occurs	*b.* ______________	in all living cells
Substances used	*c.* ______________ *d.* ______________	oxygen, glucose
Substances formed	*e.* ______________ *f.* ______________	*g.* ______________ *h.* ______________
Energy changes	energy absorbed and then stored in carbohydrates	*i.* ______________ ______________ ______________

8. Why does chlorophyll make plants green? ______________________________

____

____

9. Explain why green plants produce carbon dioxide as well as oxygen. ______________

____

____

10. What is happening to a plant when leaves change their color in the autumn? ______________

____

14

Energy for Keeping Alive: Nutrition

LABORATORY INVESTIGATION

LOCATING STARCH IN FOOD

A. Your teacher will supply you with the following food items: a piece of bread; crackers; potato slices; sugar; grated cheese; baking power. Fill six test tubes about 1/3 full with each of these items. Add five drops of Lugol's (dilute iodine) solution to each tube. [Recall from Chapter 2 Laboratory Investigation that Lugol's solution tests for the presence of starch. A blue-black color shows that starch is present.] (See Fig. 14-1.)

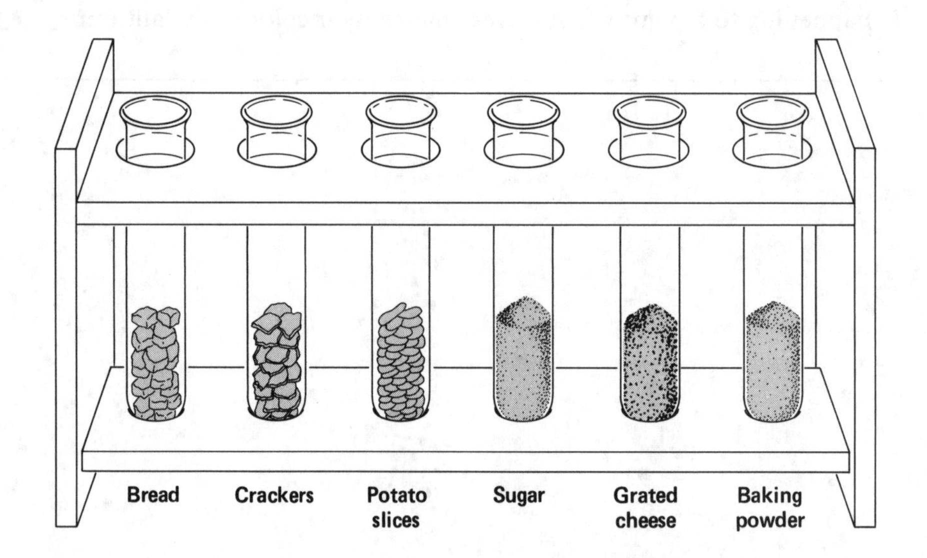

Fig. 14-1. Setup for locating starch in food.

1. What color is the Lugol's solution before it is placed in each of the test tubes? ________

2. What color is the Lugol's solution in the test tube containing
a. bread? ______ *b.* crackers? ______
c. potato? ______ *d.* sugar? ______
e. cheese? ______ *f.* baking powder? ______

3. Which food items have starch in them? ______

4. What kind of a nutrient is starch? ______

(**Caution:** Be extremely careful when working with iodine. Iodine is poisonous if swallowed; it can stain clothing if spilled on it.)

Food for Living Things

The energy that living things need is chemical energy. An organism needs a constant supply of this energy to carry out its life activities (see Ch. 2). Should an organism lack energy for a short time, it may weaken and die. The energy supply of animals and protozoans comes from the foods they eat. The energy supply of plants and algae comes from the foods they make with the aid of sunlight. In all organisms, energy is released when food—in a simple form that cells can use—is oxidized. In addition to serving as an energy source, food also provides the substances needed by all cells for growth and repair, and for maintaining health by regulating life processes.

NUTRIENTS

Foods are composed of one or more chemical compounds that cells use to carry out the life activities. These compounds, called **nutrients,** provide the cells with a source of energy and with the materials for their growth, maintenance, and repair. The nutrients are *carbohydrates*, *fats* and *oils*, *proteins*, *vitamins*, *water*, and *mineral salts*. Carbohydrates, fats and oils, proteins, and vitamins are organic compounds. Water and mineral salts are inorganic compounds.

Carbohydrates. The **carbohydrates** include such organic compounds as sugar, starch, and cellulose. *Glucose,* a type of sugar, is the major source of energy for cells. Every carbohydrate contains the elements carbon, hydrogen, and oxygen. The proportion of hydrogen atoms to oxygen atoms in carbohydrates is 2 to 1, the same as that in water (H_2O). For example, the chemical formula for table sugar is $C_{12}H_{22}O_{11}$. The proportion of hydrogen atoms to oxygen atoms in this compound is 22 to 11, which is 2 to 1.

Humans and many other animals get the carbohydrates they need from green plants and the food products made from them. Good sources of carbohydrates are fruits, vegetables, potatoes, and grains. About 2% of your body is made up of carbohydrates.

The process of digestion changes many complex carbohydrates to glucose. When the glucose is oxidized in the cells, energy is released. In the human body, excess glucose is carried by the blood to the liver, which changes the glucose to **glycogen** (also called "animal starch") and stores it. Under some conditions, carbohydrates may be changed into fat, which is also stored. When the body lacks glucose, the liver changes the glycogen back to glucose. When the supply of glycogen is exhausted, the body is able to change the stored fat into glucose. In these ways a steady supply of energy is made available to the body.

In green plants, excess glucose is often converted to starch. (That is why plants are a good source of this nutrient.) The potato plant, for example, stores starch in the form of granules in large storage-tissue cells. When the other cells of the plant need glucose, the starch granules are digested to glucose, which is then transported to the cells.

Test for Starch. *Lugol's (iodine) solution,* which is light brown in color, turns blue-black when it reacts with starch. To test a food for starch, mix the food with Lugol's solution. If starch is present, a blue-black color appears. If starch is absent, the solution's color remains brown.

Test for Simple Sugar. *Benedict's solution,* which is blue, turns an orange-to-red color when it reacts with a simple sugar. To test a food for a simple sugar, boil together equal quantities of the food and Benedict's solution. If a large amount of simple sugar is present, an orange-to-red color appears. If sugar is absent, the solution remains blue. A green or yellow color indicates the presence of only a small amount of sugar.

Fats and Oils. *Fats* and *oils* are of similar chemical composition, but **fats** are solid at room temperature, whereas **oils** are liquid. Examples of these organic compounds are lard (an animal fat) and olive oil. Most fats are composed of these small chemical units: 3 *fatty acids* and one *glycerol.* These units are composed of the same basic elements as carbohydrates—carbon, hydrogen, and oxygen—but not in the same proportions.

People obtain fats and oils by eating both plants and animals. Good sources of these nutrients are animal fats (including meats, milk products, and eggs), olive oil, corn, butter, margarine, and peanuts.

When equal quantities of fats and carbohydrates are oxidized, the energy released by fats is more than double that released by carbohydrates. However, the energy in a meal that is rich in carbohydrates is more rapidly available to the body than is the energy in a meal that is rich in fats. This is because it takes more time for fats to be digested to a form in which they can be oxidized than it does for carbohydrates.

About 20% of your body is made up of fats. In animals, fat that is not immediately needed by the body is deposited in the vacuoles of special fat cells. These cells are located under the skin and around such organs as the heart and intestine. In these areas, the fat serves as a protective cushion for organs, as insulation against the cold, and as a reserve supply of energy. Fats serve to dissolve certain vitamins that are not soluble in water. Fats are also used to build cell parts, like membranes. When too much fat is stored in the body, an individual's health may suffer. One health risk results when too much fat causes the body to produce high levels of cholesterol. Fatty deposits formed by the cholesterol have been linked to strokes and heart disease. In plants, excess fats or oils are deposited in the vacuoles of storage-tissue cells.

Test for Fats. When dipped in fat, a piece of heavy paper that light normally cannot pass through becomes *translucent,* letting some light pass through it. To test a food for fat, rub the food on heavy paper and carefully warm the paper over a source of heat, like a radiator. If fat is present, a long-lasting translucent spot will appear. If fat is not present, a temporary spot or no spot will appear.

Proteins. The **proteins** include such compounds as gelatin, keratin, and albumen. Proteins provide the body with the materials needed for growth and repair of tissues. Special proteins called *enzymes* control all chemical activities in the cells. When fats or carbohydrates are not available, the body uses proteins for energy.

All proteins are complex organic compounds made up of simple compounds called **amino acids.** The amino acids consist of carbon, hydrogen, oxygen, and nitrogen. Sulfur, phosphorus, iron, and other elements are also present in some amino acids. There are 20 different amino acids. Animals can make some of these amino acids in their bodies; the rest they have to get from their food. Some foods that are rich in amino acids (and, thus, in proteins) are eggs, fish, meat, cheese, nuts, and beans.

The process of digestion breaks down proteins to amino acids. In the process of assimilation, tissue cells rebuild the amino acids into the various proteins needed to make up the body's protoplasm. About 20% of your body is made up of thousands of different proteins.

Test for Proteins. When *biuret reagent,* which is a blue liquid, is mixed with a protein, the mixture turns violet or pink-violet. To test a food for protein, mix equal quantities of the food and the biuret reagent in a test tube. If protein is present, a violet or pink-violet color appears. If protein is absent, the solution remains blue.

Water. *Water* (H_2O), an inorganic compound, is composed of the elements hydrogen and oxygen. In the body, water is neither digested nor oxidized. Water is essential because it is the liquid in which all chemical reactions of the body take place. Water is also the substance that carries nutrients throughout the body, dissolves certain vitamins, carries away waste products, and helps regulate body temperature. A lack of water can be fatal. The bodies of many organisms contain as much as 80% water. The human body is over 50% water. An average person requires the equivalent of 8 glasses of water daily. Tap water is not the only source. Beverages, soups, fruits, and vegetables are also good sources of water.

Test for Water. Water, which is ordinarily a liquid, becomes a vapor, or gas, when heated (evaporation). Upon cooling, water vapor becomes a liquid (condensation). To test whether a substance contains water, place the substance in the bottom of a long test tube that is clean and dry. Heat the bottom of the tube. If water is present, condensed moisture collects on the upper, cooler part of the test tube. If water is absent, the upper part of the test tube remains dry.

Mineral Salts. The elements that make up **mineral salts** are needed by the body for building various tissues and for aiding the action of certain enzymes. Some of these elements are sodium, chlorine, calcium, phosphorus, potassium, magnesium, copper, iodine, sulfur, and iron. Sodium, potassium, and magnesium are necessary for muscles and nerves to work properly; chlorine, for the formation of hydrochloric acid in the stomach; calcium and phosphorus, for building strong bones and teeth; potassium, for aiding growth and maintaining a chemical balance; and iron, for the production of hemoglobin in red blood cells. People who have "iron-poor blood" develop a condition called *anemia*, in which the red blood cells cannot transport enough oxygen to all the body's cells. The mineral iodine is necessary for the proper functioning of the thyroid gland (which controls growth and chemical reactions); too little iodine results in the condition known as goiter. About 6% of the human body is composed of minerals. Foods that are rich in mineral salts include table salt (iodized), seafood, fruits, green vegetables, milk, and meats. (See Table 14-1.)

Test for Mineral Salts. When mineral salts are burned at the temperature of an ordinary

TABLE 14-1. MINERALS NEEDED BY THE BODY

Mineral	*Food Sources*	*Needed by Body For*
Calcium	Milk, dairy products, green vegetables	Building bones and teeth; regulating heart and nerve activity; blood clotting.
Copper	Liver, mushrooms, shellfish, green vegetables, cocoa, bran	Forming oxygen-carrying compounds in red blood cells.
Iodine	Seafood, iodized salt	Manufacture of thyroxin by the thyroid gland.
Iron	Liver, red meat, eggs, leafy vegetables, whole grains	Forming red blood cells; carrying of oxygen by red blood cells.
Magnesium	Vegetables	Muscle and nerve action.
Phosphorus	Milk, dairy products, meat, eggs, whole-grain cereals, vegetables	Building bones, teeth, and nerve tissue; energy production by cells.
Potassium	Vegetables, citrus fruits, bananas, apricots	Normal functioning of muscles and nerves; maintaining chemical balance.
Sodium	Table salt, vegetables	Normal functioning of nerves and muscles.

gas flame, they appear as gray ashes after they have cooled. To determine whether a food contains mineral salts, carefully burn the food in a gas flame. If minerals are present, a gray ash will remain. If minerals are absent, no ash remains.

Vitamins. Young laboratory animals that have been fed diets consisting of purified (overprocessed or overcooked) carbohydrates, fats, proteins, water, and mineral salts become ill and fail to grow properly. When such animals are fed more natural (less processed) foods, these nutrients can be properly utilized, and the animals remain healthy and grow normally. It has been discovered that the natural foods contain very tiny amounts of certain nutrients that are extremely important for maintaining health. When these substances are lacking in the diet, various diseases develop. These important nutrients are the **vitamins.**

All vitamins are organic compounds composed of various elements. Most vitamins that are taken into the body eventually become parts of different enzymes. If a particular vitamin is lacking in the diet, then a specific enzyme cannot be formed in the body. As a result, some chemical activity within the body stops, and signs of a deficiency disease appear. A **deficiency disease** is an illness that occurs because the body lacks a special substance. (Goiter and anemia are actually types of deficiency diseases, too.) In general, a state of poor health that is caused by a lack of adequate nutrients is called **malnutrition.**

Animals get most of their vitamins from the foods they eat—either directly from plants, or from other animals that eat plants. Some vitamins can also be made in the animal's body. Green plants manufacture all the vitamins they require.

The most important vitamins for human health are: (See Table 14-2.)

Vitamin A (Carotene). Foods that contain vitamin A include carrots, yellow vegetables, green leafy vegetables, liver, butter, whole milk, and eggs. Vitamin A is essential for good vision in dim light and for healthy skin. A deficiency of this vitamin can result in *night blindness.* Vitamin A is also thought to help prevent certain cancers.

Vitamin B_1 (Thiamin). Sources of this vitamin are meat, milk, whole grains, beans, soy, and legumes. A lack of vitamin B_1 leads to weakness, loss of appetite, and eventually to *beriberi,* a disease of the nerves that can lead to paralysis. B_1 is important for growth and for functioning of the heart, nerves, and muscles.

Vitamin B_2 or G (Riboflavin). Found in the same foods as vitamin B_1, in addition to eggs, vegetables, and yeast. Vitamin B_2 helps in cell respiration. When this vitamin is lacking, an inflammation (soreness and swelling) of the tongue and lips may develop. A person's vision and growth may also suffer.

Vitamin B_3 (Niacin). This vitamin is found in meats, peanut butter, vegetables, and whole grains. It is important for growth and metabolism (chemical reactions in cells). A lack of this vitamin can lead to skin irritations and digestive problems. Eventually, the nervous system can be damaged, leading to mental disorders. In its extreme form, this condition is called *pellagra.*

Vitamin B_{12} (Cobalamin). This vitamin is found in meat (liver), fish, and green vegetables. It is, along with the mineral iron and folic acid, essential in blood formation. A lack of vitamin B_{12} and folic acid leads to pernicious anemia and disorders of the nervous system.

B-Complex. Sometimes all the B vitamins are grouped together into one category. They are then referred to as the B-complex vitamins. *Folic acid* is a member of the B-complex group.

Vitamin C (Ascorbic Acid). Foods rich in vitamin C are citrus fruits, leafy green vegetables, and unpasteurized milk. This vitamin is rapidly destroyed by heat, oxygen, and baking soda. Vitamin C helps to keep the walls of the capillaries strong. A lack of vitamin C can result in a disease of the gums, called *scurvy.*

Vitamin D (Calciferol). Foods rich in vitamin D include milk, eggs, and fish. If directly exposed to sunlight (or an artificial source of ultraviolet light), the skin of many organisms is able to manufacture vitamin D. This vitamin, together with calcium and phosphorus, helps build bones and teeth. A lack of vitamin D can result in soft bones, a condition called *rickets.*

Vitamin E (Tocopherol). Foods that contain vitamin E are whole grains, milk, butter, leafy vegetables, wheat germ, and meat. The need for vitamin E by humans is not fully established. However, laboratory rats that lack this vitamin are unable to reproduce.

TABLE 14-2. VITAMINS NEEDED BY THE BODY

Vitamin	*Food Sources*	*Needed by Body For*
Vitamin A (carotene)	Milk; butter, cheese, egg yolk, fish-liver oils, liver, green and yellow vegetables	Normal growth and proper development of bones and teeth; formation of normal skin; normal vision, especially in dim light.
Vitamin B_1 (thiamin)	Meat, soybeans, milk, whole grain, eggs, nuts, green vegetables	Appetite, oxidation of foods; normal action of nervous system, heart, and muscles.
Vitamin B_2 (riboflavin)	Meat, fowl, soybeans, milk, nuts, green vegetables, eggs, yeast, whole grain	Healthy skin, especially around lips; functioning of the eyes; oxidation of foods.
Vitamin B_3 (niacin)	Meat, fowl, fish, liver, eggs, milk, potatoes, whole grain, tomatoes, leafy vegetables	Proper functioning of the stomach, intestines, and the nervous system; healthy skin.
Vitamin B_{12} (cobalamin)	Green vegetables, liver, kidney, fish	Normal blood formation; growth; normal action of nervous system.
Vitamin C (ascorbic acid)	Citrus fruits (oranges, grapefruits, lemons, limes), leafy green vegetables, tomatoes	Growth; strength of the blood vessels; development of teeth; health of gums, bones, and muscles.
Vitamin D (calciferol)	Fish-liver oils, milk, eggs, liver	Growth of strong teeth and bones.
Vitamin E (tocopherol)	Wheat germ, leafy green vegetables, meat, whole-grain cereals, milk, butter	Prevents damage to cell membranes and destruction of vitamin A by oxygen; normal reproduction.
Vitamin K	Leafy green vegetables, egg yolk, milk	Normal clotting of blood.

Vitamin K. Normally, vitamin K is made by bacteria found in the body's large intestine. Leafy green vegetables also contain vitamin K. This vitamin is essential for proper blood clotting. A deficiency of vitamin K can result in excessive bleeding from minor wounds.

Tests for Vitamins. There is a different test for each vitamin. For some vitamins, such as ascorbic acid, a chemical test is available. For others, such as vitamin D, the effect of the vitamin on the growth and health of laboratory mice has been used as a test.

Vitamin-Enriched Foods. During the processing of foods such as milk and flour, the vitamins that are normally present in them are usually destroyed. To make up for this loss, the vitamin content of such foods is artificially increased. A food to which vitamins have been added is called an **enriched food.** For example, flour is enriched by the addition of the B vitamins thiamin, riboflavin, and niacin. The vitamin D content of milk is usually increased by adding vitamin D to the processed milk.

Conserving Vitamins in Food. Many foods, particularly fruits and vegetables, lose some vitamins and minerals when cooked or prepared improperly. A few simple rules to help conserve these nutrients are:

1. Use only a small amount of water when cooking green vegetables, because some minerals, as well as vitamin C and the B vitamins, dissolve in water and become part of the juices that are usually discarded. Save these juices and add them to soups or gravies.
2. Since oxygen (from the air) and heat rapidly destroy vitamin C, cook foods containing this vitamin as quickly as possible in a covered pot. Refrigerate fruit juices in sealed containers.
3. Since baking soda chemically destroys vitamin C, do not use baking soda in cooking.
4. To prevent the oxidation and destruction of vitamins present in fresh vegetables, eat salads as soon as they are made.

CALORIES

Oxidization of nutrients in cells releases the energy required for the life processes. The total amount of energy that a food can release is found by completely oxidizing (burning) the food and measuring the amount of heat energy given off during the process. The unit used to measure heat energy is the *calorie.* A **calorie** is the amount of heat needed to raise the temperature of 1 gm of water 1° Celsius. The instrument in which this measurement is carried out is called a *calorimeter* (Fig. 14-2). The amount of energy that a food can provide is expressed in terms of **kilocalories,** or *Calories* (with a capital *C*). A Calorie is equal to 1,000 calories.

The amount of daily energy that an individual requires depends on such factors as age, sex, climate, metabolism, and physical activity. Younger, more active people require more Calories than older people do. Males usually need more Calories than females do. A person in a cold climate needs more energy (Calories) than a person in a warm climate who does the same amount of work. Table 14-3 below shows the approximate number of Calories required by people engaged in different activities.

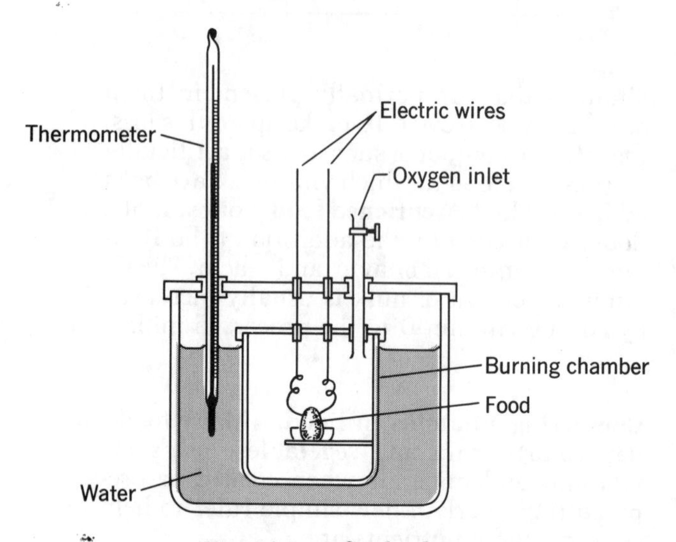

Fig. 14-2. A food calorimeter.

TABLE 14-3. CALORIE REQUIREMENTS OF PEOPLE.

Person/Activity	*Approximate Calorie Need per Day*
Athlete or Laborer	4000–4500
Man at desk job	2600–3000
Woman at desk job	2200–2500
Average teenage boy	2500–3600
Average teenage girl	2200–2600
Average adult—inactive	1900–2100
Average child to 10 years	1000–2200

GOOD NUTRITION

A **balanced diet** consists of a selection of foods that provides an adequate amount of energy and that includes a variety of nutrients. Such a diet provides all the substances needed for the work of our cells and for growth. As a guide for physical fitness, the United States Department of Agriculture recommends the daily selection of foods from the following four groups (Fig. 14-3):

1. **Meat and Beans Group.** Two or more servings of meat, poultry, fish, or eggs. Dried beans, peas, soy products, and nuts also may be used.
2. **Fruit and Vegetable Group.** Four or more servings of a citrus fruit or other fruit or vegetable containing vitamin C; a dark green or deep yellow vegetable for vitamin A; other vegetables and fruits, including potatoes.
3. **Milk and Dairy Group.** Children and teenagers should have 3 to 4 cups of milk; adults, about 2 cups. Ice cream, cheese, and butter also provide important nutrients from this food group.
4. **Bread and Cereal Group.** Four or more servings of whole grain foods, such as bread, cereal, and pasta, should be eaten.

a. Meat and beans group

b. Fruit and vegetable group

c. Milk and dairy group

d. Bread and cereal group

Fig. 14-3. The four food groups.

In addition, people require *fiber* in their diets. **Fiber** is the hard, cellulose part of food plants and is used by our bodies as **roughage,** the substance that helps our intestines eliminate waste products. Although not a nutrient, fiber helps to keep our bodies healthy.

CHAPTER REVIEW

Science Terms

The following list contains all of the boldfaced scientific terms found in this chapter and the page on which each appears.

amino acids (p. 164)
balanced diet (p. 168)
calorie (p. 168)
carbohydrates (p. 163)
deficiency disease (p. 166)
enriched food (p. 167)
fats (p. 164)
fiber (p. 169)
glycogen (p. 163)
kilocalories (p. 168)
malnutrition (p. 166)
mineral salts (p. 165)
nutrients (p. 163)
oils (p. 164)
proteins (p. 164)
roughage (p. 169)
vitamins (p. 166)

Matching Questions

On the blank line, write the letter of the item in column B which is most closely related to the item in column A.

Column A

____ 1. test for starch

____ 2. test for proteins

____ 3. test for mineral salts

____ 4. keratin and albumen

____ 5. lack of vitamin A

____ 6. glycerol and fatty acids

____ 7. lack of vitamin D

____ 8. stored glucose in the body

____ 9. lack of mineral iron

____ 10. vitamin K deficiency

Column B

a. make up fats
b. rickets
c. glycogen
d. Lugol's solution
e. excessive bleeding
f. biuret solution
g. anemia
h. heat in flame to ashes
i. proteins
j. scurvy
k. night blindness

Multiple-Choice Questions

On the blank line, write the letter preceding the word or expression that best completes the sentence or answers the question.

1. Which group contains only carbohydrates?
a. sugar, starch, glucose, glycogen
b. cellulose, starch, amino acid, fatty acid
c. sugar, starch, glycerol, glucose
d. glucose, fatty acid, glycogen, cellulose — 1 ____

2. A good source of iodine in the diet is
a. white cereal *b.* white flour *c.* fish *d.* vitamin D — 2 ____

3. Many green plants store excess glucose in the form of
a. glycogen *b.* fatty acids *c.* starch *d.* protein — 3 ____

4. Which nutrient is composed of the elements carbon, nitrogen, oxygen, and hydrogen?
a. fat *b.* sugar *c.* starch *d.* protein — 4 ____

5. Lemon juice turns orange when boiled with Benedict's solution. This shows that lemon contains
a. sugar *b.* citric acid *c.* protein *d.* starch — 5 ____

6. In the human body, fat serves as a reserve supply of
a. proteins *b.* energy *c.* starch *d.* vitamins — 6 ____

7. The nutrient a child needs for tissue building is
a. fats *b.* proteins *c.* carbohydrates *d.* vitamins — 7 ____

8. A rich source of vitamin D is
a. meat *b.* vegetables *c.* whole grains *d.* milk — 8 ____

9. Raw onion turns blue-black when soaked in a solution of iodine. This shows that the onion contains
a. glucose *b.* cane sugar *c.* protein *d.* starch — 9 ____

10. Soybeans are a rich source of
a. protein *b.* vitamin K *c.* vitamin B_{12} *d.* calcium — 10 ____

11. The nutrient we all need but that is never oxidized is
a. protein *b.* glucose *c.* water *d.* fat — 11 ____

12. Iron is used by the body in
a. forming hemoglobin
b. the growth of hard bones
c. making ptyalin
d. digesting amino acids — 12 ____

13. In all carbohydrates, the proportion of hydrogen atoms to oxygen atoms is
a. 1 to 1
b. 1 to 2
c. 2 to 1
d. 2 to 2 — 13 ____

14. Which vitamin affects the ability to see in dim light?
a. A *b.* B_{12} *c.* C *d.* D — 14 ____

15. The two vitamins and the mineral that are directly needed for the normal formation of blood are
a. vitamin B_{12}, vitamin B_1, and sodium
b. folic acid, vitamin D, and calcium
c. folic acid, vitamin B_{12}, and iron
d. vitamin A, vitamin D, and chlorine — 15 ____

16. Calorie requirements differ according to all of the following *except*
a. age *b.* intelligence *c.* climate *d.* activity — 16 ____

17. The compounds used in greatest amount to make protoplasm during the process of assimilation are
a. fatty acids *b.* amino acids *c.* sugars *d.* vitamins — 17 ____

18. An enriched food is one that has added
a. cream *b.* butter *c.* vitamins *d.* oil — 18 ____

19. The vitamin that is usually destroyed by contact with oxygen and heat is
a. B *b.* C *c.* D *d.* K — 19 ____

20. The vitamin that can prevent and cure scurvy is
a. A *b.* B *c.* C *d.* D — 20 ____

21. The best source of vitamins for teenagers is
a. a varied diet
b. vitamin pills
c. vitamin injections
d. ultraviolet light — 21 ____

22. In planning high-protein meals, beans can be substituted for
a. citrus fruit *b.* vegetables *c.* milk *d.* meat — 22 ____

23. Citrus fruits are important in nutrition because of their high content of vitamin
a. K *b.* E *c.* D *d.* C — 23 ____

24. The vitamin that helps build strong bones and teeth is
a. A *b.* B_2 *c.* C *d.* D — 24 ____

25. Whole grain foods supply large quantities of
a. water *b.* vitamin A *c.* B vitamins *d.* iodine — 25 ____

Modified True-False Questions

In some of the following statements, the italicized term makes the statement incorrect. For each incorrect statement, write the term that must be substituted for the italicized term to make the statement correct. For each correct statement, write the word "true."

1. The major food source of energy for plant and animal cells is *glycogen*. 1 ______

2. The useful chemical compounds found in foods are called *nutrients*. 2 ______

3. A state of ill health caused by a lack of nutrients is called *enriched*. 3 ______

4. The nutrient that can produce more energy than any other is *protein*. 4 ______

5. A translucent spot on a shopping bag can be a sign that a food rich in *fat* is present in the bag. 5 ______

6. Biuret testing solution is used to test for *sugar*. 6 ______

7. Most people need about 8 glasses of *milk* per day. 7 ______

8. The unit used to measure heat energy is the *enzyme*. 8 ______

9. A disease that can be prevented or cured by Vitamin B_1 is *beriberi*. 9 ______

10. *Pellagra* is the disease associated with a lack of vitamin B_3. 10 ______

Testing Your Knowledge

1. Explain the difference between the two terms in each of the following pairs:

a. food and nutrient ______

______

b. amino acid and fatty acid ______

______

c. protein and amino acid ______

______

d. glucose and glycerol ______

______

e. starch and glycogen ______

______

2. Give a scientific explanation for each of the following statements:

a. It may be necessary to include vitamin D in the diet during winter more than during the summer. ______

______

b. A bottle of orange juice should be kept tightly capped and refrigerated. ____________

____

c. Vitamin pills are unnecessary if food selection follows that suggested by the United States Department of Agriculture.

____

____

d. A person who exercises a lot probably has less glycogen and fat in his/her body than a person who eats a lot and exercises little. ____________

____

3. An unknown food is to be tested in the laboratory. When different testing agents are used, some color changes can be expected if certain nutrients are present in the food.

a. What color can we expect to result when Benedict's solution is used if the food contains
(1) simple sugar ____________
(2) starch ____________

b. What color can we expect when Lugol's iodine solution is used if the food contains
(1) simple sugar ____________
(2) starch ____________

c. What color can we expect when biuret solution is used if the food contains
(1) simple sugar ____________
(2) starch ____________
(3) protein ____________

4. When discussing food and its effect on living things, what do we mean by the "Basic Four?"

____

____

5. Why should you drink milk with every meal? ____________

____

____

6. Name the *Basic Four* and list three foods from each of the categories. ____________

____

____

Making Foods Better: Is It Really Worse for You?

People have been doing it for thousands of years. "It" is the treatment of foods so that they last longer, are in greater supply, better condition, and are more profitable to produce.

All of this can be achieved in a number of ways: by preserving fresh foods; by boosting supply through the use of pesticides on crops or special chemicals that increase growth or prevent disease in animals; or by adding flavors and colors that make foods more appealing.

If this were as far as it went, few people might argue with such efforts to provide an abundance of attractive food for consumers. However, critics of these processes say it goes much farther . . . right into your body! And that might *not* be so good for you.

Let's take a look at what you may be eating along with the food in a typical meal. If the food is processed, that is, treated in some way before it reaches the supermarket, it will have a label on it. If you read the label carefully, you'll notice two words that should set you thinking. The words are "preservatives" and "artificially."

Preservatives tend to make foods last longer by retarding the growth of organisms that spoil food, such as mold and bacteria. Unfortunately, some preservatives have been shown to be harmful to laboratory animals, like rats. Representatives of the food industry are quick to point out that huge amounts of a preservative are fed to experimental animals. No human being, they maintain, would ever eat that much of a preservative in proportion to his or her body size.

Other preservatives can cause allergic reactions in people. Some of these reactions can even be fatal. But preservatives reduce the amount of money a family must spend on food, because the food lasts longer and there is less waste.

How about the word "artificial?" What follows it on a label? Usually the name of a substance that enhances the flavor or color of a product. Yet some of these artificial ingredients also have been shown to harm animals. Not long ago, a food dye was found to cause cancer in laboratory animals. And some flavoring substances are known to trigger allergic reactions, ranging from hives to breathing difficulties.

And then there are the unprocessed foods like red meats, poultry, vegetables, and fruits. Don't they come to you straight from the farm? Perhaps, but that doesn't mean that they are pure foods.

Cattle are given certain hormones to increase their proportion of meat to fat. This seems like a good idea, given the dangers to our arteries from eating too much animal fat. But the hormones stay in the meat. So you may get a dose of hormones along with your hamburger. Are the hormones bad for you? Some people say no, especially in the small amounts you eat. Of course, other people disagree.

Fruits and vegetables are produced in greater quantity and free of rot or insects because crops are sprayed with chemicals, called pesticides, that kill insects and fungi. However, many of these chemicals are also poisonous to people. How poisonous? Not enough to do you harm, insist fruit and vegetable producers. Yet consumer advocates disagree and advise people to buy produce grown with little or no pesticides on them.

Can we do without preservatives? Artificial substances? Pesticides? Do we have to? What do you think?

1. *The term "artificially" on a food label usually refers to*

a. pesticides and flavors.
b. hormones and colors.
c. flavors and colors.
d. hormones and pesticides.

2. *Pesticides*

a. improve crop flavor.
b. improve fruit color.
c. kill insects and fungi.
d. are harmless to people.

3. *Preservatives*

a. slow down spoiling.
b. speed up spoiling.
c. increase meat content.
d. decrease allergies.

4. *Describe a positive and a negative effect of using pesticides.*

___

___

___

___

___

___

___

5. *A bacterium that can cause food poisoning is sometimes found in poultry. In order to protect consumers, farmers may give antibiotics to poultry to kill the bacterium. Do research on this topic and write a brief report concerning the pros and cons of this practice. Include your own opinions.*

15

Digestion and Absorption in Animals

LABORATORY INVESTIGATION

DIGESTION OF PROTEIN BY GASTRIC JUICE

A. Your teacher will present you and your lab partner with four 4-cm-long tubes filled with coagulated (hard-boiled) egg white, a nearly pure source of protein.

B. Take four glass bottles, number them, and fill them 3/4 full as follows: (See Fig. 15-1.)

Bottle 1: Artificial gastric juice (equal parts of 1% pepsin solution and 1% hydrochloric acid).
Bottle 2: 1% pepsin solution.
Bottle 3: 1% hydrochloric acid.
Bottle 4: Water.

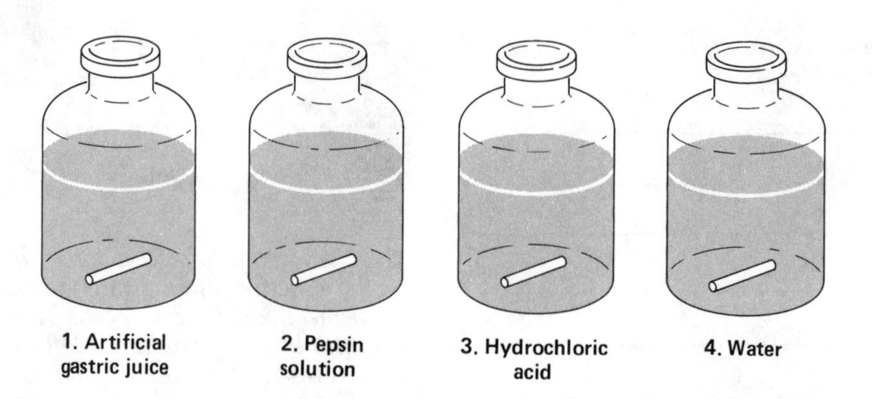

Fig. 15-1. Setup for gastric juice investigation.

C. Carefully place one tube of coagulated egg white into each of the bottles. Be sure that the liquid covers both open ends of the tubes. (**Caution:** Do not touch the HCl solutions with your bare hands.) Stopper the bottles and leave them in a warm place overnight.

D. Observe the tubes on the following day to detect any changes that may have occurred in the egg white.

1. *a.* Describe what changes, if any, you can see in the amount of egg white in the tube in

Bottle 1: ____________________

____

b. Bottle 2: ____________________

____

c. Bottle 3: ____________________

d. Bottle 4: ____________________

2. What has happened to some of the egg white? ____________________

____

3. What is the role of gastric juice in your stomach? ____________________

____

4. *a.* Can the enzyme pepsin alone digest protein? ____________________

b. As well as the gastric juice does? ____________________

5. *a.* Can the hydrochloric acid alone digest protein? ____________________

b. What does it need to help it work? ____________________

6. What is the role of water in this investigation? ____________________

____

Food-Getting and Digestion in Complex Animals

Animals seek, gather, and capture the foods that supply energy for their life activities. Animals usually have specialized organs, such as teeth, beaks, claws, or tentacles, that assist in food-getting and ingestion. Once an animal has gotten and eaten its food, special organs within the body digest the food.

The process of **digestion** prepares the food for use by all the body cells of the animal. In this process, food is changed from its natural complex form to simpler compounds that cells can use. The digestive process is much the same in all animals. Although the shape and size of the organs of digestion differ according to the diet and size of an animal, the digestive systems of humans and most other complex animals are similar. In this chapter, you will study the digestive system of humans.

The Human Digestive System

The digestive system, illustrated in Fig. 15-2, consists of two main divisions: the *alimentary canal* and the *digestive glands.* The **alimentary canal** is the tube through which all food passes. The **digestive glands** are the organs that secrete juices consisting of water, enzymes, and other compounds. Each juice plays a specific part in digesting, or breaking down, food. Tubes, or **ducts,** connect the digestive glands to the food tube.

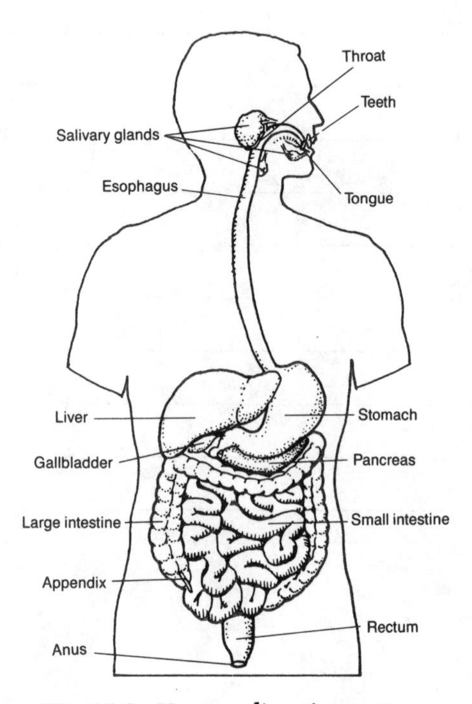

Fig. 15-2. Human digestive system.

PARTS OF THE DIGESTIVE SYSTEM

The food tube consists of the *mouth,* the *throat* (or *pharynx*), the *esophagus* (or *gullet*), the *stomach,* the *small intestine,* and the *large intestine* (or *colon*).

The digestive glands are the *salivary glands,* the *gastric glands,* the *pancreas,* the *liver,* and the *intestinal glands.*

There are two types of digestion: *mechanical* and *chemical.* **Mechanical** (or *physical*) **digestion** is the grinding and softening of the food which occurs in the mouth and stomach. **Chemical digestion** is the breakdown of food molecules by enzymes which occurs in the mouth, stomach, and small intestine.

Mouth. Food enters the food tube at the mouth, where the teeth and tongue, aided by the juice of the salivary glands, begin the digestive process.

Teeth. Human adults have 32 teeth. Different types of teeth carry out different activities (Fig. 15-3). The *incisor* and *canine* teeth are used in biting and tearing. The *premolar* and *molar* teeth are used in chewing. This activity grinds the food into small pieces, which are then ready for the action of the digestive juices.

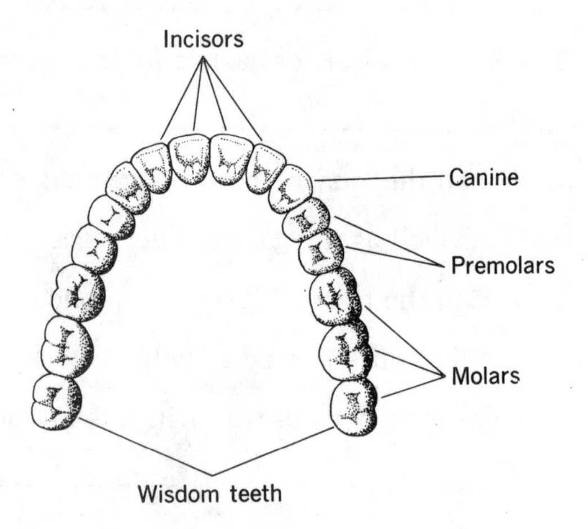

Fig. 15-3. Teeth of human adult.

The four types of teeth vary in shape but have similar parts. As shown in Fig. 15-4, a typical tooth is divided into three main regions: the crown, neck, and root. The *crown* is the part of the tooth visible above the gum line. The *neck,* the part under the margins of the gum, joins the crown with the root. The *root*

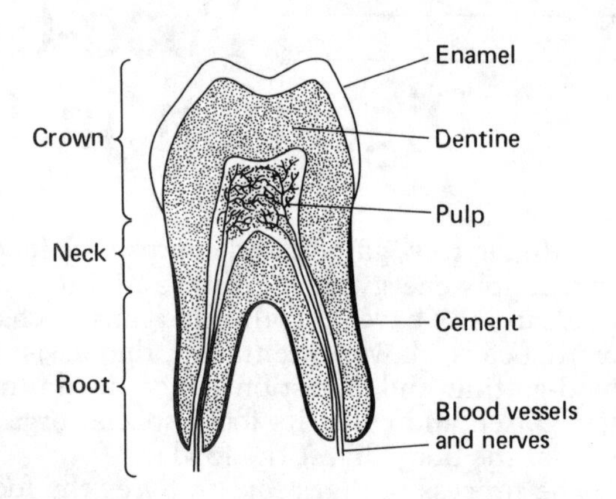

Fig. 15-4. Structure of a typical tooth.

is the part that holds the tooth securely in the bone of the jaw. The center of a tooth is the *pulp,* which consists mainly of blood vessels and nerves. *Dentine,* which is a hard bonelike substance, surrounds the soft pulp. Covering the dentine of the crown is the *enamel,* a layer of calcium phosphate that is even harder than bone, and which protects the layers below it. Covering the dentine of the root is *cement,* a substance that helps to hold the root in its socket within the jawbone.

Tongue. The surface of the tongue bears **taste buds,** which enable us to distinguish four taste sensations—sweet, sour, salty, and bitter. The inner region of this organ consists mainly of muscles. These muscles enable the tongue to move food around the mouth (to assist the teeth in chewing) and to push food down the throat (to assist in swallowing).

Salivary Glands. As food is chewed, **saliva,** the juice of the **salivary glands,** flows into the mouth cavity. *Salivary amylase,* a digestive enzyme, is present in saliva. This enzyme changes starch into *malt sugar* (or *maltose*), a simpler compound than starch. However, the body cells cannot use malt sugar, so it is changed to a more usable form later in the digestive process. The fluids in saliva moisten food, make it slippery, and enable it to be swallowed easily.

Throat (Pharynx). The **throat** (or **pharynx**) connects the mouth and esophagus. No new digestive action starts in the throat. The digestive process begun by saliva just proceeds further. Together with muscles near the base of the tongue, the muscular wall of the throat pushes food down into the esophagus in the process called *swallowing.*

Esophagus (Gullet). The **esophagus** is a tube that connects the throat and the stomach. The wall of the esophagus, or **gullet,** is composed of two sets of involuntary muscles. These muscles contract regularly, causing wavelike movements along the length of the tube. This motion, called **peristalsis,** forces food down the esophagus and into the stomach. Peristalsis also occurs throughout the rest of the food canal. *Mucous glands* that line the inner surface of the esophagus secrete **mucus,** a slimy substance that aids the passage of food. As in the throat, no additional digestion begins in the esophagus.

Stomach. The esophagus empties into the upper end of the baglike, J-shaped **stomach.** The opening in the lower end of the stomach, which empties into the small intestine, is controlled by a muscular valve. The wall of the stomach consists of three sets of muscles, which contract in three different directions. These muscles churn food, mix it with the stomach fluid, and, after a few hours, push the partially digested food into the small intestine.

Gastric Glands. The fluid in the stomach is called *gastric juice.* This juice is secreted by **gastric glands,** located in the inner lining of the stomach. Gastric juice consists of two enzymes, called *rennin* and *pepsin,* water, and *hydrochloric acid.*

Rennin curdles the protein in milk called *casein.* This action prepares the casein for digestion by pepsin. Rennin is present in the gastric juice of human babies and other young mammals, like calves, which live on a diet of milk.

Pepsin, like rennin, can curdle milk. Pepsin changes casein and other proteins into simpler proteinlike substances.

When food reaches the stomach, the hydrochloric acid softens fibers and kills many of the bacteria normally present in food. Both water and hydrochloric acid dissolve mineral salts. However, as you learned in your laboratory investigation, water and hydrochloric acid cannot change proteins to simpler substances. When water, hydrochloric acid, and pepsin are mixed with proteins, all three compounds together begin to digest the proteins.

Small Intestine. When food from the stomach enters the **small intestine,** the valve at the lower end of the stomach closes and prevents partially digested food from flowing back into the stomach. The coiled-up small intestine is about 7 meters in length and nearly 4 cm. in diameter. Like the wall of the esophagus, the wall of the small intestine is composed of two sets of involuntary muscles that push food along by peristalsis. The pancreas, liver, and intestinal glands secrete juices that continue and complete the digestion of all food. It takes food several hours to pass through the small intestine.

Pancreas. The **pancreas** produces *pancreatic juice.* By way of a duct, pancreatic juice reaches the small intestine. There the juice mixes with and acts on food. Pancreatic juice contains several enzymes, three of which are very important. One of these enzymes changes any starch that has not already been digested by saliva into maltose. Another enzyme

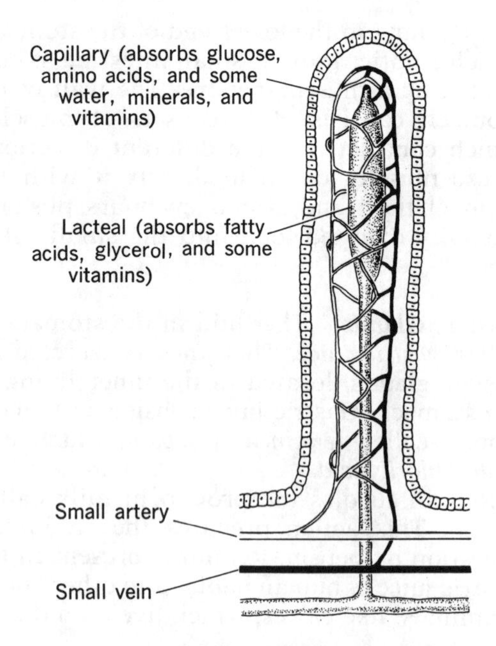

Fig. 15-5. A villus (much enlarged).

changes the proteinlike substances into amino acids and also acts on any proteins that have not yet been broken down by pepsin. The third enzyme breaks down fats into fatty acids and glycerol, thus completing the digestion of fat.

Liver. The **liver,** the body's largest gland, secretes the juice **bile** into the gall bladder. The **gall bladder** temporarily stores the bile. A bile duct from the gall bladder joins the pancreatic duct, so that one common duct carries bile and pancreatic juice together into the small intestine. Bile contains no enzymes. However, bile breaks down fat into very tiny particles. These tiny fat particles are then readily acted on by the pancreatic enzyme that digests fat.

Intestinal Glands. These glands are located in the inner lining of the small intestine. The juice secreted by the **intestinal glands** is called *intestinal juice.* There are four major enzymes in intestinal juice. One of these enzymes, like the pancreatic enzyme mentioned, changes all proteinlike material into amino acids, thereby completing the digestion of proteins. The other three enzymes act on various carbohydrates and change all of them into the simple sugar glucose, thus completing the digestion of carbohydrates.

End Products of Digestion. Before digestion takes place, the nutrients (carbohydrates, proteins, and fats) are composed of complex compounds that cannot enter cells. After they have been digested, the nutrients are called **end products.** These end products (glucose, amino acids, and fatty acids) are simpler compounds that are in solution. All of them are found in the lower portion of the small intestine. In solution, these compounds can pass through cell membranes. Once in the cells, the end products can be used by the cells to perform their life activities.

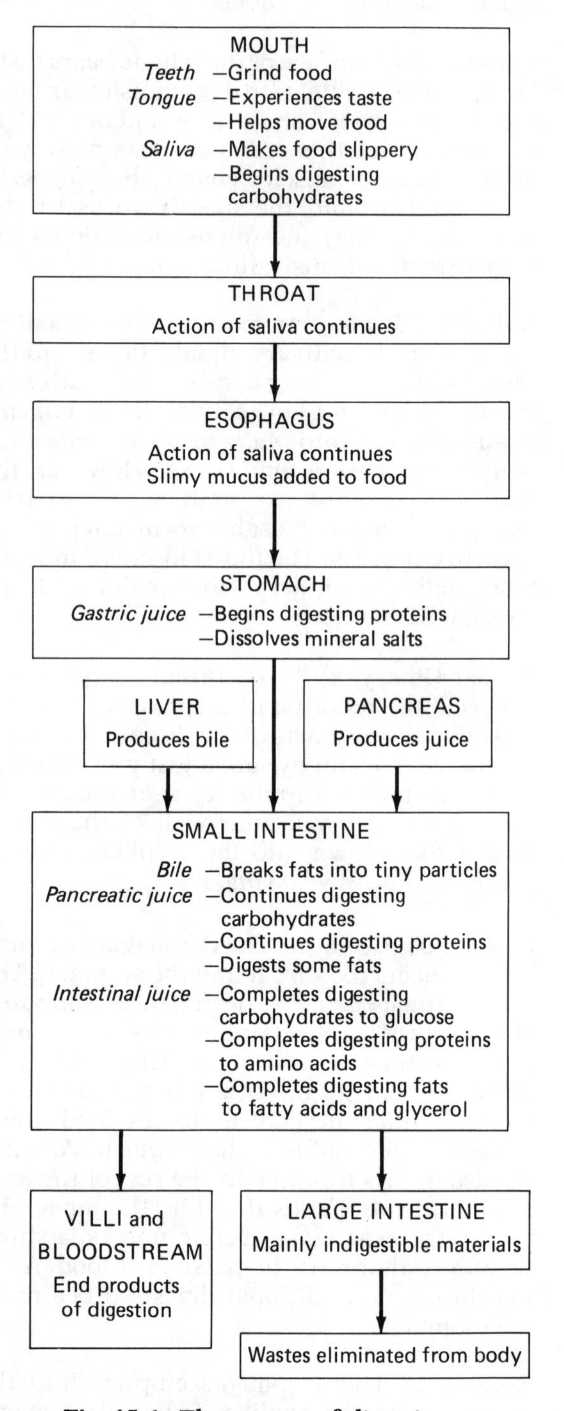

Fig. 15-6. The process of digestion.

Absorption in the Small Intestine. The process by which the end products of digestion pass through the membrane lining the small intestine and enter the bloodstream is called **absorption.** The bloodstream then distributes the nutrients to the body's cells.

The lining of the small intestine has a very large surface area in contact with the end products, so that it is particularly well-suited to absorb them. This large surface is the result of three major features: (1) the great length of the small intestine; (2) the folds and creases in its internal lining; and (3) the presence of millions of **villi,** each of which is a tiny fingerlike projection into the cavity of the small intestine.

A *villus* is microscopic in size. Located in the center of each villus (see Fig. 15-5) are *capillaries,* which are very tiny blood tubes, and a *lacteal,* which is a very tiny lymph tube. (You will learn about lymph in Ch. 16.)

The end products of digestion, which are present in the cavity of the small intestine, bathe the villi. By the process of diffusion, the capillaries absorb the glucose, amino acids, some water, dissolved minerals, and vitamins. The lacteals absorb the fatty acids, glycerol, and some other vitamins. All of the substances absorbed by the lacteals eventually reach the bloodstream.

Large Intestine (Colon). Undigested matter (such as fiber) and water pass from the small intestine into the **large intestine** (or **colon**). Although it is shorter than the small intestine, the large intestine is greater in width. The large intestine is about 1.5 meters in length and about 8 cm. in diameter. It resembles an upside-down letter "U."

As peristalsis moves the undigested fluid mass along this organ, the intestinal lining absorbs much of the water and some of the minerals from the mass. The absorbed substances enter the bloodstream. This process takes several hours to be completed. Peristalsis moves the remaining mass, the *solid wastes* (or *feces*), to the lower end of the large intestine. This portion of the large intestine is called the *rectum.* The solid wastes contain bacteria as well as undigested materials. The colon's naturally occurring bacteria help make vitamins B_{12} and K. Solid wastes are stored in the rectum until they are eliminated from the body through the *anus.*

Appendix. The *appendix* is a small tube, located in the lower right-hand side of the abdomen. The appendix is attached to the large intestine near the area where the large and small intestines join. The appendix takes no part in digestion in humans now. Sometimes it becomes infected and has to be surgically removed. Refer to Fig. 15-6 to review the process of digestion.

Care of the Digestive System

Do not stuff yourself at mealtimes. In addition to causing *obesity* (excessive fatness), overeating strains the digestive and circulatory systems.

Form the habit of regularly eliminating solid wastes. The excessive accumulation of solid wastes is harmful. *Roughage,* which is mainly indigestible vegetable fiber, stimulates peristalsis. By eating roughage and drinking plenty of water, you help your body eliminate solid wastes.

Clean your teeth and visit a dentist regularly. Proper chewing, which depends upon healthy teeth, prepares food for thorough enzyme action.

Eat when you are relaxed. A relaxed atmosphere at mealtime aids digestion because it permits the digestive glands to secrete juices. Tension and anxiety, on the other hand, interfere with normal secretion.

When eating causes discomfort regularly, consult a doctor. Although many digestive upsets may be minor in nature, it is best to have a physician investigate repeated discomfort.

CHAPTER REVIEW

Science Terms

The following list contains all of the boldfaced scientific terms found in this chapter and the page on which each appears.

absorption (p. 181)
alimentary canal (p. 178)
bile (p. 180)
chemical digestion (p. 178)
colon (p. 181)
digestion (p. 177)
digestive glands (p. 178)
ducts (p. 178)
end products (p. 180)
esophagus (p. 179)
gall bladder (p. 180)
gastric glands (p. 179)
gullet (p. 179)
intestinal glands (p. 180)
large intestine (p. 181)
liver (p. 180)
mechanical digestion (p. 178)
mucus (p. 179)
pancreas (p. 179)
peristalsis (p. 179)
pharynx (p. 179)
saliva (p. 179)
salivary glands (p. 179)
small intestine (p. 179)
stomach (p. 179)
taste buds (p. 179)
throat (p. 179)
villi (p. 181)

Matching Questions

On the blank line, write the letter of the item in column B which is most closely related to the item in column A.

Column A

____ 1. grinding and softening of food
____ 2. secrete enzymes in the mouth
____ 3. muscular movements along food tube
____ 4. tube through which all food passes
____ 5. undigested food passes through this
____ 6. breakdown of food by enzymes
____ 7. digested nutrients in small intestine
____ 8. tube that connects mouth and gullet
____ 9. secretes enzymes into small intestine
____ 10. secrete enzymes into the stomach

Column B

a. pancreas
b. chemical digestion
c. throat
d. mechanical digestion
e. gastric glands
f. salivary glands
g. appendix
h. alimentary canal
i. large intestine
j. peristalsis
k. end products

Multiple-Choice Questions

On the blank line, write the letter preceding the word or expression that best completes the sentence or answers the question.

1. The process of digestion in animals changes
a. simple compounds to complex compounds
b. complex compounds to simple compounds
c. mineral compounds to vitamin compounds
d. water to mineral compounds — 1 ____

2. Digestion in the mouth is begun by the
a. teeth, tongue, and saliva
b. lips, tongue, and teeth
c. cheeks, lips, and tongue
d. tonsils, lips, and salivary glands — 2 ____

3. The teeth used in biting into an apple are the
a. premolars *b.* molars *c.* incisors *d.* tricuspids — 3 ____

4. The passage of food through the gullet is aided by the slippery secretion called
a. bile oil *b.* ptyalin *c.* mucus *d.* gastric juice — 4 ____

5. Which part of a tooth is covered by enamel?
a. root *b.* crown *c.* neck *d.* pulp — 5 ____

6. The part of the food tube that connects the throat with the stomach is the
a. gullet *b.* pharynx *c.* tonsil *d.* intestine — 6 ____

7. Food is moved along the food tube by the process called
a. osmosis *b.* diffusion *c.* peristalsis *d.* absorption — 7 ____

8. In the mouth, a piece of food is crushed by the action of the
a. molar teeth and jaw muscles
b. lips and tongue muscles
c. tongue and throat muscles
d. pulp and enamel — 8 ____

9. As the muscles of the stomach wall move, food in the stomach is
a. mixed with gastric juice
b. usually made more solid
c. mixed with minerals
d. mixed with bile — 9 ____

10. Gastric glands are located in the
a. small intestine *b.* large intestine *c.* stomach *d.* esophagus — 10 ____

11. Which two enzymes are present in gastric juice?
a. pepsin and amylase
b. diastase and amylase
c. rennin and amylase
d. pepsin and rennin — 11 ____

12. Protein cannot be digested in the stomach in the absence of
a. amylase *b.* lactic acid *c.* casein *d.* hydrochloric acid — 12 ____

13. Mineral salts in food are dissolved in the stomach by
a. water *b.* citric acid *c.* Benedict's solution *d.* hydrochloric acid — 13 ____

14. The enzyme in gastric juice that curdles milk protein is
a. casein *b.* rennin *c.* saliva *d.* bile — 14 ____

15. The large intestine is about 1.5 meters long; how long is the small intestine?
a. 1 meter *b.* 7 meters *c.* 20 meters *d.* 3 meters — 15 ____

16. Three juices that help in digesting proteins are
a. saliva, gastric, and pancreatic
b. bile, saliva, and pancreatic
c. intestinal, bile, and saliva
d. gastric, intestinal, and pancreatic — 16 ____

17. Bile is secreted into the gall bladder by the
a. pancreas *b.* liver *c.* salivary glands *d.* intestinal glands 17 ___

18. Most chemical digestion in the food tube occurs in the
a. liver *b.* pancreas *c.* stomach *d.* small intestine 18 ___

19. The final usable form into which food is changed by the body is called
a. an end product *b.* a waste product *c.* an excretion *d.* a raw material 19 ___

20. Glucose that is present in the small intestine is
a. digested to sugar
b. changed to glycerol
c. absorbed into the bloodstream
d. excreted from the large intestine 20 ___

Modified True-False Questions

In some of the following statements, the italicized term makes the statement incorrect. For each incorrect statement, write the term that must be substituted for the italicized term to make the statement correct. For each correct statement, write the word "true."

1. The two main divisions of the human digestive system are the digestive glands and the *respiratory* canal. 1 ______

2. The root of a tooth is held in the jawbone by *enamel.* 2 ______

3. The enzyme in saliva that changes starch to a sugar is called salivary *maltose.* 3 ______

4. Normally, the food mass in the small intestine cannot flow back into the stomach because of a muscular *gland.* 4 ______

5. In addition to acting on proteins and fats, pancreatic juice acts on *starches.* 5 ______

6. *Bile* helps break large fat particles into tiny fat particles. 6 ______

7. As a result of the chemical changes in digestion, all proteins are changed to *fatty* acids. 7 ______

8. Glucose is the end product of the digestion of *carbohydrates.* 8 ______

9. A slimy secretion that aids movement of food in the gullet is *mucus.* 9 ______

10. Villi, which absorb end products, are located in the *large intestine.* 10 ______

Testing Your Knowledge

1. Describe the function that each of the following performs in the body:

a. bile ____

____

b. tooth enamel ____

____

c. roughage ____________________

____

d. peristalsis ____________________

____

e. villi ____________________

____

2. Fill in the missing parts of the table.

Glands	*Juices*	*Important Enzymes*
Salivary	Saliva	*a.* ________
Gastric	*b.* ________	*c.* ________ *d.* ________
e. ________	Pancreatic	Amylase, trypsin, lipase
f. ________	Bile	None
Intestinal	*g.* ________	Lactase, sucrase, maltase

3. The diagram below represents a villus. On the line before each description, write the number from the diagram that fits the description. [Numbers can be used more than once.]

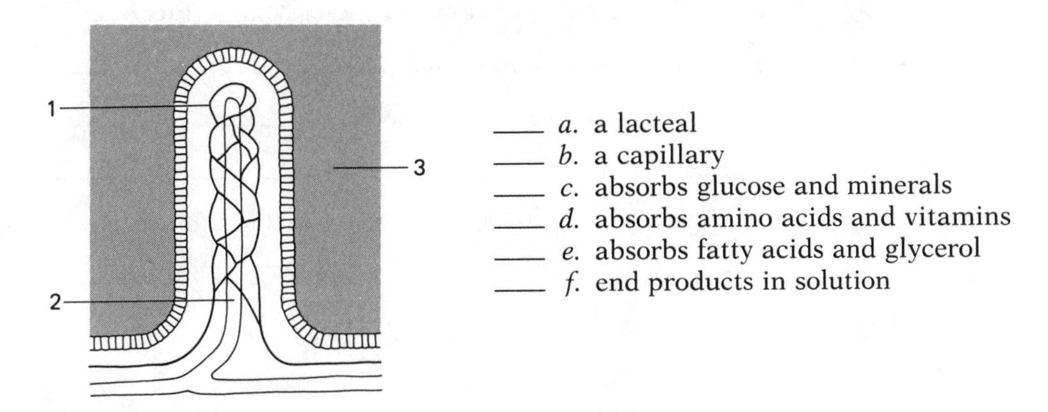

____ *a.* a lacteal
____ *b.* a capillary
____ *c.* absorbs glucose and minerals
____ *d.* absorbs amino acids and vitamins
____ *e.* absorbs fatty acids and glycerol
____ *f.* end products in solution

4. Pieces of hard-boiled egg white are placed in a test tube with some water. A membrane is securely fastened over the mouth of the tube. Then, the tube is inverted into a beaker of water. An equal quantity of hard-boiled egg white is placed in a second tube containing water, pepsin, and hydrochloric acid. A membrane is securely fastened over the tube and then this tube is inverted into a beaker of water. Both sets of apparatus are placed in an incubator at 98.6°F. The next day, the water in both beakers is tested for the presence of a simple protein.

a. What predictions can you make about the results for each beaker setup?

____

____

____

b. Explain your answer to *a.* ______________________________

____

____

____

c. What two body processes are illustrated in the second setup? ______________________________

____

d. Explain why you chose the two processes you did for *c.* ______________________________

____

____

____

5. A person ate a chicken sandwich that was made of two slices of wheat bread, lettuce, butter, and chicken.

a. Which of these foods is richest in fat? ______________________________

b. Which of these foods is richest in protein? ______________________________

c. Which of these foods is richest in carbohydrate? ______________________________

d. In which part of the body does the digestion of carbohydrate begin? ______________________________

e. In which organ does the first chemical change in the chicken occur? ______________________________

f. In which organ is the chicken fat changed to small droplets of fat? ______________________________

g. Which digestive juice will most likely act on the minerals in lettuce? ______________________________

h. Which food is most likely to provide roughage? ______________________________

i. Explain why roughage is not absorbed in the body. ______________________________

____

j. Name two parts of the food tube in which carbohydrate digestion occurs. ______________________________

____

16

Circulation and Blood

LABORATORY INVESTIGATION

TAKING YOUR PULSE

A. You will work with a classmate to time each other's *pulse rate* before and after mild exercise. (**Caution:** Students with health problems should be excused from this exercise.)

1. Using the index and middle fingers of your right hand, press gently to find your pulse on the inside of your left wrist (near the base of your thumb). Sit quietly and count the number of beats, while your partner times you for exactly one minute. Record the result in the first column of the chart below (see Table 16-1).

TABLE 16-1. TIMING YOUR PULSE RATE

Trials (Counting beats per minute)	*Partner 1*	*Partner 2*
After sitting quietly		
After running in place 2 minutes		
After resting for 2 minutes		
After resting for another 2 minutes		

2. Repeat this same procedure in reverse, that is, with your partner counting his/her pulse, while you time it for one minute. Record the result in the second column of the chart. Were the number of beats approximately the same for you and your partner? ____________

B. Run in place for two minutes. Sit down afterward and count the number of beats on your pulse again, while your partner times you for one minute. Record the result.

3. *a.* Was your pulse rate faster or slower this time? ____________

b. Hypothesize why the result may be different. ____________

C. Rest in your seat for two minutes, then take your pulse again, as your partner counts to one minute. Record your result.

4. *a.* Is this rate faster or slower than the pulse rate before exercise? ________________

b. Immediately after exercise? ________________

D. Rest in your seat for another two minutes, then take your pulse again, as your partner counts to one minute. Record your result.

5. *a.* Is this rate faster or slower than the previous pulse rate?

____

b. Explain why you think you got this result. ________________

____

E. Now repeat steps B through D, with your partner running in place for two minutes, then resting in the seat afterward, and then taking his/her pulse each time while you time it for one minute. Record and compare the results.

6. Are there any differences between your pulse rates and those of your partner? Explain.

____

7. What is the overall effect of exercise on pulse rate? ________________

____

8. What is the effect of rest on pulse rate? ________________

9. What do you think causes the actual "beat" of the pulse? ________________

____

Transport: The Process of Circulation

Circulation is the process by which oxygen, digested nutrients, and waste materials are transported around the body of an organism. Almost every part of a simple aquatic organism's body is near its environment—the water. Such organisms take in the materials they need directly from their environment and, usually, do not have special circulatory structures. For example, in the paramecium, which is unicellular, the entire cytoplasm continuously moves around the cell and distributes the substances that the organism has taken in. (This is called cyclosis.)

However, larger complex organisms have circulatory systems composed of a network of tubes, a pump, and a fluid that moves. By means of this system, nutrients, oxygen, and other materials are delivered to the cells, and wastes are removed from them. The circulatory systems of most complex animals are similar, although they differ in detail. You will study the human circulatory system.

The Human Circulatory System

THE CIRCULATORY SYSTEM

The circulatory system is composed of the *heart* (pump), *blood vessels* and *lymph vessels* (tubes), and *blood* and *lymph* (moving fluids).

THE HEART

As shown in Fig. 16-1 below, the **heart** is a saclike, muscular organ that is divided into four hollow chambers, or sections. The two upper chambers are called **atria** (or *atriums*), also known as *auricles*. Atria receive blood from veins and discharge it into the two lower chambers, called **ventricles.** Ventricles pump blood out of the heart into arteries. **Valves,** which are flaps of tissue that prevent the backward flow of blood, are located between each atrium and ventricle and at the base of the arteries that are attached to the heart. The walls of the heart are composed mainly of involuntary muscle tissue.

The right side and left side of the heart are separated by a wall, called a **septum.** Although the atria contract at the same time, and the ventricles pump at the same time, oxygen-poor blood on the right side does not mix with oxygen-rich blood on the left side of the heart.

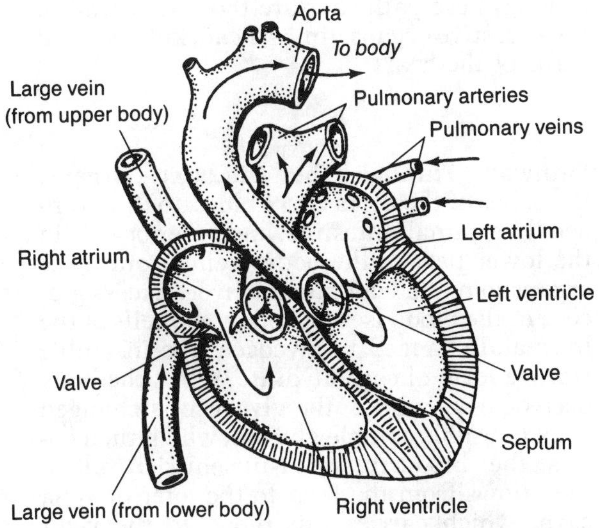

Fig. 16-1. The heart (cutaway view).

Walls of the Heart. The thickness of the muscle wall of each chamber varies according to the work that the muscle wall does.

The atria have thinner walls than the ventricles. Atria receive blood, contract, and push blood into the ventricles.

The right ventricle has thicker walls than the atria. It has to pump blood out to the lungs.

The left ventricle has the thickest walls of all the chambers of the heart. The left ventricle is three times as muscular as the right ventricle. It pumps blood the longest distance—throughout the rest of the body.

Heartbeat. The heart contracts and relaxes regularly about 70 times a minute in an average adult at rest. This rate, called the **pulse rate,** varies according to the age, sex, state of health, and activity of an individual. At each heartbeat, the two atria contract and force blood into the ventricles. Both ventricles then contract and push blood into the two large arteries attached to them. At each relaxation of the heart, blood from three large veins enters the atria. With the next contraction, these events are repeated. Blood is transported to and from the heart and throughout the body by three types of tubes known as **blood vessels.**

BLOOD VESSELS

Arteries. The vessels (tubes) in which blood flows away from the heart are called **arteries.** Every minute, the heart pumps about 5 liters of blood through the body's arteries. Blood is pumped from the left ventricle into the **aorta,** which is the only artery that leads out of the left ventricle. It is the largest artery in the body. The aorta carries oxygen-rich blood to branches of arteries, which distribute blood to all parts of the body except the lungs. Blood from the right ventricle is pumped into the *pulmonary arteries,* which lead to the lungs. Arteries branch repeatedly, becoming smaller and smaller until they are microscopic in size. These small vessels end in still smaller vessels called *capillaries.*

The walls of an artery consist mainly of thick layers of smooth muscle and connective

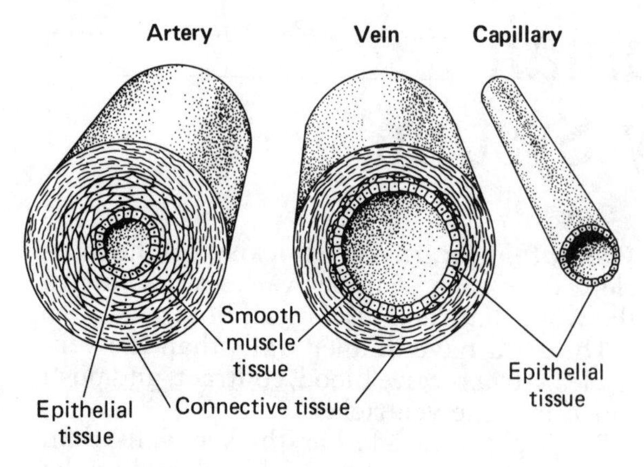

Fig. 16-2. The three kinds of blood vessels (cross-sections).

tissue (Fig. 16-2). Arterial muscle helps pump the blood. Arteries carry blood that is rich in oxygen and nutrients. The only exceptions are the large pulmonary arteries, which carry blood poor in oxygen and rich in carbon dioxide to the lungs.

Veins. The vessels in which blood flows toward the heart are called **veins.** Blood from several capillaries flows into a very small branch of a vein. This branch unites with other small branches, which join with other veins, becoming larger and larger. Two of the largest veins in the body are the *caval veins:* the superior (upper) vena cava and the inferior (lower) vena cava. The *superior vena cava* carries blood from the head and arms to the right atrium. The *inferior vena cava* carries blood from the lower part of the body to the right atrium.

The walls of a vein consist of thin layers of smooth muscle and connective tissue. Valves within the veins keep blood from flowing backward. Veins carry blood that is rich in carbon dioxide and other cell wastes. The only exceptions are the other very large veins, the *pulmonary veins,* which carry blood rich in oxygen from the lungs to the left atrium of the heart.

Capillaries. The abundant microscopic tubes that carry blood through all tissues and organs, connecting the small arteries and small veins, are called **capillaries.** Made of epithelial tissue, the walls of a capillary are only one cell thick. Exchange of materials (nutrients and wastes) between the bloodstream and body cells takes place by diffusion through the capillary walls.

BLOOD PATHWAYS

Although other people had been aware that blood moves in the body, it was the English scientist *William Harvey* (1578–1657) who first proved experimentally that blood flows through the body repeatedly in closed pathways. The major branch pathways through which the blood flows are shown in Fig. 16-3. The pathway of blood between the heart and the lungs is called the *pulmonary circulation.* The pathway of blood between the heart and the rest of the body is called the *systemic circulation.*

Pathway of Blood Through the Lungs. The circuit from the heart to the lungs and back to the heart again is separate from all other blood pathways. Blood containing nutrients that have been absorbed from the intestinal villi passes into the right atrium of the heart by way of a vena cava. The blood that enters the right atrium also contains carbon dioxide and other dissolved wastes, which the bloodstream has carried away from the body cells. From the right atrium, blood passes to the right ventricle, is pumped into the pulmonary arteries, and then flows to the capillaries of the air sacs in the lungs. Here the blood gives off carbon dioxide and some water and takes in oxygen. Now, loaded with oxygen, the blood flows to the pulmonary veins, into the left atrium, and then to the left ventricle. When the left ventricle contracts, the blood is pumped into the aorta.

The numerous arteries that branch off the aorta carry blood to various organs and systems. Then veins return the blood to the heart. Among these pathways are those that lead to the digestive organs, limbs, head, kidneys, and walls of the heart.

Pathway Through the Digestive Organs. Branches of the aorta carry blood to the stomach, liver, small intestine, and other organs in the lower part of the body. Veins from these organs join and lead into the liver. Excess glucose in the blood is removed by the cells of the liver and changed to glycogen, the insoluble storage form of carbohydrate. When the body needs carbohydrate, the glycogen is changed back to usable soluble glucose, which then enters the bloodstream. Nutrient-rich blood then flows from the liver to the inferior vena cava, which carries this blood to the right atrium of the heart.

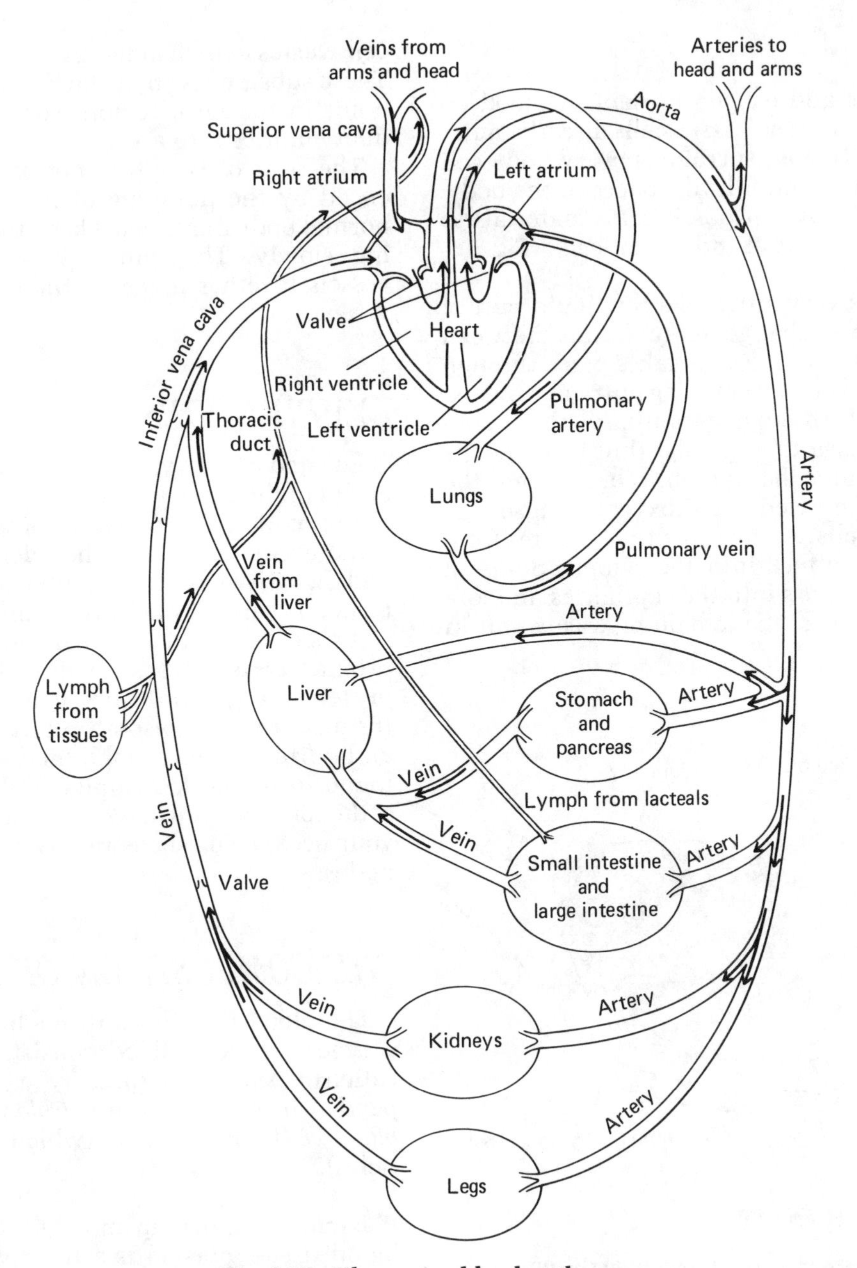

Fig. 16-3. The major blood pathways.

Pathway Through the Limbs and Head. Separate arteries branch from the aorta and carry blood rich in oxygen and nutrients to the cells of each limb and the head. In each of these organs, capillaries and small veins carry waste-laden blood away from the cells. This blood is then transported to the right atrium of the heart via the superior vena cava.

Pathway Through the Kidneys. Branches from the aorta deliver blood to the capillaries in each kidney. Dissolved wastes in this blood, such as urea and excess water, pass out of the capillaries and into certain cells of the kidney, which excrete the wastes. The waste-free blood in the capillaries then enters the vein leading out of the kidney. The veins from each kidney join the inferior vena cava, which empties into the right atrium.

Pathway Through the Walls of the Heart. A branch from the aorta penetrates the heart walls, bringing nutrients and oxygen to the heart muscle cells. Capillaries and small veins among these cells gather the waste-laden blood and pass it to a large vein that empties the blood into the right atrium.

LYMPH

Nutrients and oxygen present in blood do not reach individual tissue cells directly. A liquid, called **lymph,** surrounds every body cell and capillary, and fills all spaces in the body. This liquid makes it possible for materials to pass into and out of individual cells (Fig. 16-4).

Lymph is very much like the liquid part of blood. The exchange of materials between blood and the tissue cells takes place through the lymph. As blood carrying nutrients and oxygen passes through a capillary, these substances diffuse through the thin walls of the capillary into the lymph. Then, from the lymph, the nutrients and oxygen diffuse into the tissue cells. At the same time, wastes from these cells diffuse into the lymph. Some of these wastes pass into the capillaries and join the bloodstream; the remaining wastes stay in the lymph.

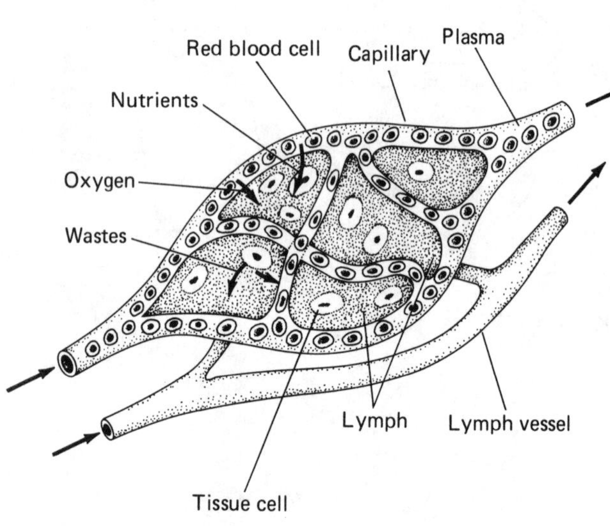

Fig. 16-4. Passage of materials to and from tissue cells.

LYMPH VESSELS

Lymph, containing wastes, enters **lymph vessels.** The structure of these vessels is similar to that of blood capillaries and veins. Small lymph vessels join and form larger ones. The largest of these is the *thoracic duct,* which lies in the left-hand side of the body. This duct joins a vein near the base of the neck. **Lacteals,** the tiny lymph vessels of the intestinal villi, join together, forming larger lymph vessels. Eventually these vessels also lead into the thoracic duct. Thus, absorbed fats as well as some cell wastes can be found in the lymph stream. These substances enter the bloodstream at the point in the neck region where the thoracic duct empties into a vein.

The flow of lymph is not especially influenced by the pumping of the heart. Rather, normal body movements keep the lymph flowing slowly. The lymph vessels and blood vessels together make up the body's *vascular system.*

LYMPH NODES

Located at intervals along the lymph vessels are glandlike masses of spongy tissue called **lymph nodes.** The tissue of these nodes is the same as that found in the adenoids, tonsils, and digestive system. Lymph nodes filter bacteria and viruses out of the lymph before it is returned to the bloodstream. And the nodes produce certain types of white blood cells and proteins, called *antibodies,* which help fight off the diseases and poisons produced by bacteria and viruses. The lymph nodes are mostly found in the neck, armpits, and groin. When a doctor says you have "swollen glands" in your neck, he or she is referring to the lymph nodes.

THE COMPOSITION OF BLOOD

The **blood** that flows through your body is a type of tissue. Blood consists of a liquid, called *plasma,* and three types of cells suspended in it. These are *red blood cells, white blood cells,* and *platelets* (which are cell particles).

Plasma. Making up more than half of the blood tissue, **plasma** is a transparent, straw-colored liquid. It is the fluid that transports the blood cells, platelets, and numerous dissolved substances. Among the major substances in the plasma are:

1. *Water*—makes up over 90% of the plasma. Numerous substances are dissolved in the water.
2. *Dissolved nutrients*—glucose, amino acids, fatty acids, mineral salts, and vitamins—are absorbed from the digestive system and storage centers in the body. All of these nutrients are transported to the tissue cells.
3. *Dissolved wastes*—carbon dioxide, urea, and other cell wastes—are absorbed from tissue cells. These wastes are carried to organs that eliminate them from the body.

4. *Antibodies*—protein substances made by certain white blood cells—destroy bacteria, viruses, and their poisons.
5. *Hormones*—compounds secreted by ductless glands—regulate many body processes.
6. *Fibrinogen*—a protein made in the liver—is essential in blood clotting.

Red Blood Cells. The **red blood cells** are disc-shaped and thicker around the edge than in the center (Fig. 4-6*a*, page 40). Nearly 45% of the blood is made up of red blood cells. A small drop of blood may contain as many as five million red blood cells. Mature red blood cells lack nuclei. Instead, the protein molecule **hemoglobin** fills the thin center of the red blood cell.

Red blood cells carry oxygen to the body's tissues by means of this hemoglobin molecule. A lack of either hemoglobin or red blood cells results in a condition called *anemia*. The iron present in hemoglobin gives blood its red color when it combines with oxygen.

Red blood cells wear out in 90 to 120 days and are then broken down in the liver and the *spleen* (a blood-storage organ attached to the circulatory system). The hemoglobin is reused in new red blood cells. Other products of the old red blood cells become part of the bile, which is important in the digestion of fat. New red blood cells are constantly produced in the *marrow* of the body's long bones.

White Blood Cells. The **white blood cells** vary in shape (Fig. 4-6*b*, page 40), are larger than red blood cells, and have large nuclei. Many of them move about and change shape like an ameba. White blood cells can pass through small blood vessel walls, moving between the bloodstream and tissues. Normally, there are nearly one thousand red blood cells for each white blood cell. White blood cells live only one to three days.

One type of white blood cell, called the *phagocyte,* can engulf and destroy bacteria that invade the tissues. Another type of white blood cell, called the *lymphocyte,* is very important in fighting bacteria and viruses. When germs infect the body, the number of white blood cells usually increases.

Most types of white blood cells are produced and stored in the bone marrow. Other types are produced in the lymph nodes.

Platelets. Colorless cell pieces, **platelets** are smaller than either red or white blood cells (Fig. 4-6*c*, page 40), and they lack nuclei. Platelets are produced from cells in the bone marrow, and last less than 10 days. Platelets act only in *clotting,* the process that stops the flow of blood from a cut. When a blood vessel is injured, the platelets stick to the broken vessel and release an enzyme. This starts a reaction that results in the formation of a clot. During this reaction, *fibrinogen,* a liquid protein normally present in plasma, is changed to *fibrin.* This substance consists of threads that become tangled and form a spongy mesh. By holding back the blood cells, the mesh forms a plug, called a *blood clot.*

The blood of some people clots very slowly, which can cause excessive bleeding. This condition is called *hemophilia.* It is a hereditary condition in which the platelets do not act normally. Hemophilia occurs more frequently in males than in females.

Slow blood clotting that is not hereditary may afflict people who lack either vitamin K or the mineral calcium. This condition can be remedied by improving the diet or by using medications containing vitamin K and calcium. Both of these substances are necessary in the chemical reactions that lead to clotting.

CHANGES IN THE COMPOSITION OF CIRCULATING BLOOD

As blood passes through the organs of the body, the cells of each organ exchange certain materials with the blood. As a result, the composition of the blood changes according to the special activity of the organ.

Table 16-2 shows the chief changes in the composition of blood in a few selected organs.

MODERN USES OF BLOOD

Control of Bleeding in Hemophilia. People who suffer from hemophilia are in danger of bleeding to death even from minor injuries. Such people can be helped either by giving them blood transfusions from healthy (nonhemophiliac) people or by the use of fibrin sponges. These sponges, which are made from natural fibrinogen, can be used as artificial clots to plug wounds.

Control of Certain Infectious Diseases. The antibodies in blood are found in the part of blood plasma called *gamma globulin* (or GG). Individuals whose bodies cannot make gamma globulin are unable to fight off some

TABLE 16-2. CHANGES IN THE COMPOSITION OF THE BLOOD IN VARIOUS PARTS OF THE BODY

Organ	*Organ Takes From Blood*	*Blood Takes From Organ*
Small intestine	Materials for secretion of juices by the intestinal glands, hormones, oxygen	Food, cell wastes including carbon dioxide and excess water
Lungs	Carbon dioxide, water, food	Oxygen, cell wastes
Muscles	Food, oxygen, hormones	Cell wastes
Kidneys	Cell wastes, food, oxygen	Carbon dioxide
Ductless glands	Raw materials for secretion of hormones, oxygen, food	Hormones, cell wastes
Liver	Glucose (for storage), amino acids, worn-out red blood cells, oxygen	Glucose (when needed), urea, and other cell wastes
Marrow of long bones	Iron for making hemoglobin, food, oxygen	Platelets, new red blood cells, some types of white blood cells, cell wastes

infectious diseases. Such people are given an injection of gamma globulin either before or shortly after they have been exposed to these diseases. (See Chapter 23 for more information on how the body fights disease.)

REPLACEMENT OF LOST BLOOD

Transfusions. Excessive loss of blood (hemorrhage) can be fatal to a person unless he or she receives a **blood transfusion.** In a transfusion, the patient receives a supply of blood either directly from a healthy person or, more likely, from a hospital's blood bank. But a transfusion is successful only if the person who donates the blood (the *donor*) has the same blood type as the patient who receives it.

Blood Types. One's **blood type** is inherited. There are four major blood types, called A, B, AB, and O. These blood types are determined by the different special proteins that are present on the red blood cells. In addition, a person with one of these four types may have either Rh positive or Rh negative blood.

Before a transfusion is given, the patient's blood type is determined. Then the patient is given blood of the same type. If a different type of blood is transfused, harmful reactions can occur—the two different types of red blood cells can clump together, which prevents proper circulation and may cause death. An example of an improper transfusion would be giving type A blood to a patient having type B blood. Another unsafe transfusion would be giving Rh positive blood to a patient having Rh negative blood. Table 16-3 shows the transfusions of different blood types that may be made safely.

Blood Banks. Institutions that collect and store blood are called *blood banks.* Blood can be stored as whole blood or as blood plasma. Before whole blood is stored, its type is determined. The blood is kept from clotting by means of the addition of chemicals. Whole blood can be stored under refrigeration for a few weeks.

Blood plasma, the liquid that remains after the blood cells have been removed from blood, can also be stored. By removing the water from such plasma, the plasma is changed to a powder. Powdered plasma can be kept for long periods in sterilized containers without refrigeration. When needed for a transfusion, the powder is mixed with sterilized distilled water. Although it is better to use whole blood for transfusions, plasma is used when whole blood of the proper type is either too inconvenient to use or is not available. Most transfusions, however, include the blood cells, not just the plasma.

TABLE 16-3. SAFE BLOOD TRANSFUSIONS

Blood Type	*May Give Blood to Type*	*May Receive Blood From Type*
A	A or AB	A or O
B	B or AB	B or O
AB	AB	A or B or AB or O
O	A or B or AB or O	O
Rh positive	Rh positive	Rh positive or Rh negative
Rh negative	Rh positive or Rh negative	Rh negative

There are also other factors that have to be considered when choosing blood for transfusions. The blood has to come from a healthy donor who is free of infectious diseases such as hepatitis B and AIDS, which can be transmitted through blood transfusions. Blood banks now screen their donors more carefully to prevent this from occurring.

Care of the Circulatory System

Medical reports show that diseases of the circulatory system are the leading causes of death among older people today. With proper care of the circulatory system, heart attacks, *atherosclerosis* (clogged arteries), high blood pressure, and strokes may be prevented, and life can be extended.

Get adequate rest. Periods of rest and relaxation relieve strain on the organs of circulation, especially the heart.

Exercise moderately, but regularly. Moderate, regular exercise is beneficial. It strengthens the heart and assists the flow of blood in veins and the flow of lymph in lymph vessels.

Eat moderate amounts of a balanced, varied diet. Excessive eating can lead to serious obesity, a condition that places a strain on the heart. Diets containing proteins, vitamins, and iron provide the raw materials for the manufacture of hemoglobin and red blood cells; diets containing calcium and vitamin K provide the materials for normal blood clotting; and diets low in saturated fats help prevent clogged arteries.

Avoid using alcohol and tobacco. Excessive use of these substances is often associated with such conditions as the bursting of blood vessels, blood pressure disturbances, and some types of heart disease.

Consult a physician regularly. Experience shows that early medical treatment is successful in curbing many diseases of the circulatory system.

CHAPTER REVIEW

Science Terms

The following list contains all of the boldfaced scientific terms found in this chapter and the page on which each appears.

aorta (p. 189)
arteries (p. 189)
atria (p. 189)
blood (p. 192)
blood transfusion (p. 194)
blood type (p. 194)
blood vessels (p. 189)
capillaries (p. 190)
circulation (p. 188)
heart (p. 189)
hemoglobin (p. 193)
lacteals (p. 192)
lymph (p. 192)
lymph nodes (p. 192)
lymph vessels (p. 192)
plasma (p. 192)
platelets (p. 193)
pulse rate (p. 189)
red blood cells (p. 193)
septum (p. 189)
valves (p. 189)
veins (p. 190)
ventricles (p. 189)
white blood cells (p. 193)

Matching Questions

On the blank line, write the letter of the item in column B which is most closely related to the item in column A.

Column A

___ 1. can donate to all blood types
___ 2. can receive from all blood types
___ 3. produced in the bone marrow
___ 4. tiniest blood vessels
___ 5. where lymph meets bloodstream
___ 6. excessive bleeding condition
___ 7. carry oxygen-rich blood
___ 8. upper chambers of the heart
___ 9. lower chambers of the heart
___ 10. central wall of the heart

Column B

a. hemophilia
b. ventricles
c. thoracic duct
d. atria
e. Rh positive
f. type AB
g. type O
h. septum
i. pulmonary veins
j. capillaries
k. red blood cells

Multiple-Choice Questions

On the blank line, write the letter preceding the word or expression that best completes the sentence or answers the question.

1. What prevents the backward flow of blood from the ventricles to the auricles?
a. muscles *b.* lymph nodes *c.* veins *d.* valves — 1 ____

2. Which is the largest artery in the body?
a. pulmonary artery *c.* kidney artery
b. aorta *d.* leg artery — 2 ____

3. Which are the largest veins in the body?
a. caval veins *b.* kidney veins *c.* aortas *d.* brain veins — 3 ____

4. Most arteries contain blood rich in
a. carbon dioxide *b.* oxygen *b.* wastes *d.* lymph — 4 ____

5. Capillaries carry blood from small
a. veins to arteries *c.* arteries to veins
b. lymph vessels to veins *d.* auricles to ventricles — 5 ____

6. The exchange of nutrients and wastes between tissue cells and the blood takes place by diffusion through the walls of
a. veins *b.* arteries *c.* valves *d.* capillaries — 6 ____

7. The general pathway of blood flow is
a. heart, capillaries, veins, arteries, heart
b. veins, heart, capillaries, arteries, veins
c. veins, heart, arteries, capillaries, veins
d. capillaries, arteries, veins, heart, capillaries — 7 ____

8. Blood rich in oxygen leaves the heart from the
a. left ventricle *c.* pulmonary artery
b. right ventricle *d.* pulmonary vein — 8 ____

9. As it passes through the liver, blood may gain
a. glucose *b.* red blood cells *c.* white blood cells *d.* hormones — 9 ____

10. Blood leaving the brain contains large quantities of
a. oxygen *b.* nutrients *c.* cell wastes *d.* glucose — 10 ____

11. Urea is removed from the blood in the
a. large intestine *b.* kidneys *c.* lungs *d.* bladder — 11 ____

12. If the flow of lymph were blocked, the most likely result would be an inability
a. to digest fat *c.* of cells to receive glucose
b. of cells to receive fats *d.* of the blood to clot — 12 ____

13. Most of the blood plasma is composed of
a. growth nutrients *b.* water *c.* vitamins *d.* fibrinogen — 13 ____

14. All the end products of digestion can be found in greatest quantity in the small intestine and in the
a. red blood cells *b.* white blood cells *c.* blood plasma *d.* platelets — 14 ____

15. Which blood cells are normally most numerous?
a. red blood cells *b.* white blood cells *c.* phagocytes *d.* platelets — 15 ____

16. Worn-out red blood cells are removed from the bloodstream by the
a. kidneys *b.* large intestine *c.* liver *d.* lymph glands — 16 ____

17. New red blood cells are constantly added to the bloodstream by the
a. liver *b.* villi *c.* lungs *d.* bone marrow 17 ____

18. The type of blood cell that aids the body in fighting infections is the
a. blood platelet *b.* fibrinogen *c.* red blood cell *d.* white blood cell 18 ____

19. An inherited disease of the blood is
a. hemophilia *b.* deficiency anemia *c.* leukemia *d.* pellagra 19 ____

20. The transfer of blood from one person into the bloodstream of another person is called
a. hemorrhage *b.* blood banking *c.* transfusion *d.* blood typing 20 ____

21. A person who can donate blood to any other person but can receive blood only of his or her own type has blood type
a. A *b.* B *c.* AB *d.* O 21 ____

22. The health of the circulatory system can be maintained by
a. good diet, rest, exercise *c.* rest and vitamin pills
b. 3 hours of daily weight lifting *d.* avoiding physicians 22 ____

23. Bleeding from small cuts soon stops because of
a. clotting by platelets *c.* hemophilia
b. the drying effect of air *d.* the white blood cells 23 ____

24. In passing through the air sacs of lungs, blood from the pulmonary artery gains
a. carbon dioxide *b.* nitrogen *c.* water *d.* oxygen 24 ____

Modified True-False Questions

In some of the following statements, the italicized term makes the statement incorrect. For each incorrect statement, write the term that must be substituted for the italicized term to make the statement correct. For each correct statement, write the word "true."

1. The four heart chambers are two ventricles and two *aortas.* 1 ________

2. The heart chambers that have the thickest walls are the *ventricles.* 2 ________

3. *Valves* are the smallest blood vessels in the body. 3 ________

4. The walls of the heart receive nutrients and oxygen from blood vessels that branch off the *aorta.* 4 ________

5. Materials can diffuse between capillaries and tissue cells by way of the *villus,* which is a liquid. 5 ________

6. The liquid part of the blood is called *lymph.* 6 ________

7. Blood platelets are important in blood *transfusions.* 7 ________

8. The protein in blood plasma that aids clotting is *fibrinogen.* 8 ________

9. Phagocytes and lymphocytes are types of *white* blood cells. 9 ________

10. The pulmonary arteries carry oxygen-poor blood to the *heart.* 10 ________

Testing Your Knowledge

1. In column *A* below, list five types of compounds (besides water) that are normally present in blood plasma. In column *B*, tell how or where these compounds enter the blood.

	Column A	*Column B*
a.		
b.		
c.		
d.		
e.		

2. Name three types of cells normally present in blood and state the role of each type of cell.

3. Fill in the missing parts of the table.

	Arteries	*Veins*	*Capillaries*
Relative thickness of walls	*a.*	*b.*	*c.*
Transport blood into	*d.*	*e.*	*f.*

4. When a person cuts an artery or a vein in an arm, pressure may be applied somewhere on the arm in order to prevent too great a loss of blood. If, for example, a wrist *artery* is cut, the pressure may be applied between the cut and the shoulder. If a *vein* in the upper arm is cut, the pressure may be applied between the cut and the elbow. In each case, explain why the pressure places mentioned should be used.

5. The diagram represents a heart and the blood vessels leading into and out of it. In the proper blank, name each structure shown in the diagram.

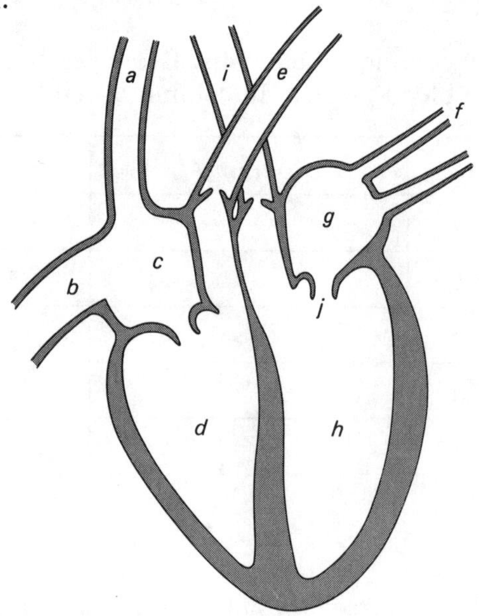

a. ______________________

b. ______________________

c. ______________________

d. ______________________

e. ______________________

f. ______________________

g. ______________________

h. ______________________

i. ______________________

j. ______________________

6. Name the four chambers of the heart and their functions.

a. ______________________

b. ______________________

c. ______________________

d. ______________________

7. Describe the path of the blood flow through the human heart.

____

____

____

8. What forces the blood through the arteries? ______________________

____

____

Respiration

LABORATORY INVESTIGATION

ANALYZING YOUR EXHALED AIR

The presence of water in the air can be indicated by condensing water from the vapor (gas) to the liquid state.

A. Exhale against a cool shiny surface, such as a mirror or glass plate. (Be sure the surface is clean and dry.)

1. Describe the appearance of the plate before and after breathing against it. ____________

____

B. Pump air from an empty atomizer against the cool shiny surface.

2. Describe the appearance of the plate before and after pumping air against it. ____________

____

3. *a.* Which contains more water vapor, the air you inhale or the air you exhale? ________

b. Explain your answer on the basis of your results. ____________________

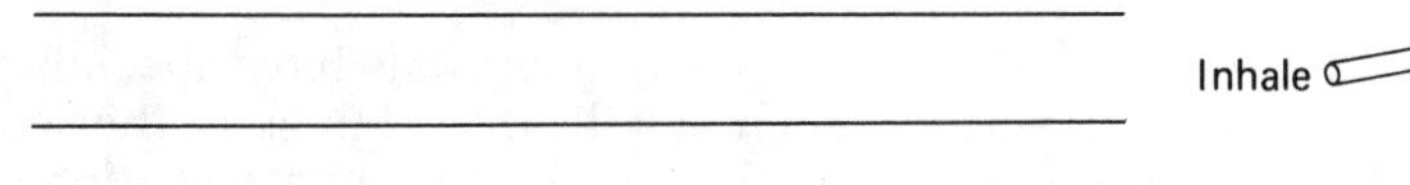

C. Your teacher will supply you with stoppers containing long and short L-shaped glass tubes. Half fill a bottle with limewater and close the bottle with the rubber stopper bearing the two tubes. Inhale (by mouth) through the shorter tube for 1 minute, to draw outside air into the limewater. (See Fig. 17-1.)

Fig. 17-1. Setup for analyzing exhaled air.

4. Describe what happens to the appearance of the limewater.

____

Exhale into the longer tube for 1 minute, to expel the air from your lungs into the limewater.

5. Describe what happens to the appearance of the limewater.

____

(*Note:* The change in appearance of fresh limewater, from being clear and colorless to having a milky color, is used as a test for the presence of carbon dioxide.)

6. *a.* Which contains more carbon dioxide, the air you inhale or the air you exhale? ______

b. Explain your answer on the basis of your results. ______________________

____

D. Compare the differences you have found in the composition of inhaled and exhaled air.

7. Hypothesize why the air you exhale is different from the air you inhale. ______________

The Process of Respiration

Respiration is the process by which living things release the stored energy present in nutrients (specifically glucose). Respiration in people and in most land animals take place in three stages—*external:* breathing and exchange of gases between the lungs and bloodstream; *internal:* exchange of gases between the blood and cells; and *cellular:* the reactions inside the cells.

The automatic process of **breathing** involves two steps: inhalation and exhalation. *Inhalation* is the taking in of air rich in oxygen. *Exhalation* is the giving off of air rich in carbon dioxide and water vapor. Your rate of breathing is controlled by the amount of carbon dioxide in your blood. Breathing also serves to regulate body temperature and body water content.

Oxygen that has been inhaled is distributed to all the cells of the body. The oxygen reacts with nutrients inside the cells in the process of *oxidation.* During oxidation, energy is released and carbon dioxide and water are formed as wastes. After they are carried away from the cells through the bloodstream, the wastes are exhaled.

Water-dwelling animals have special structures, such as gills or a thin skin, that enable these organisms to carry out respiration while underwater. These aquatic organisms absorb the oxygen that is dissolved in water and then give off carbon dioxide, which also dissolves in the water.

Land animals have either special air tubes or lungs, which enable these organisms to get their oxygen directly from the air. The respiratory system of insects includes air tubes. Respiratory systems of most other land animals include lungs, and are generally alike. You will learn about the respiratory system of humans in this chapter.

The Human Respiratory System

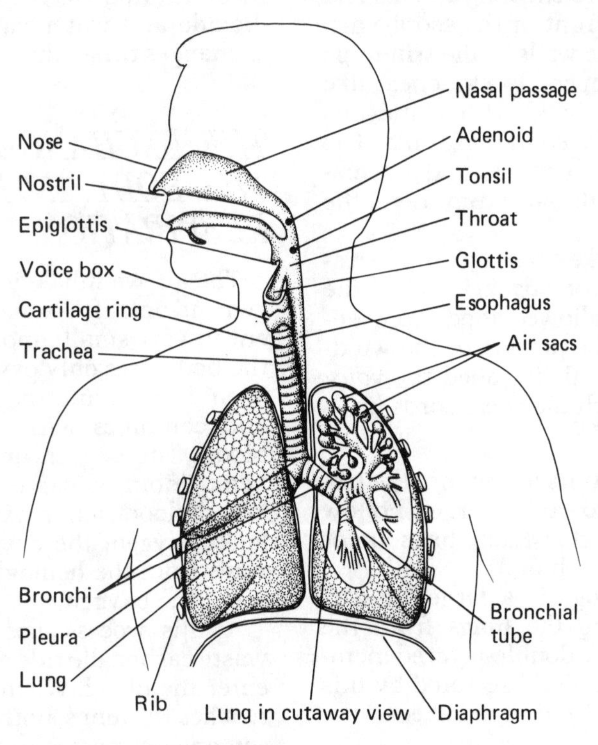

Fig. 17-2. Human respiratory system.

The respiratory system (Fig. 17-2) consists of two main divisions: an air pathway and a pressure-changing device that encloses most of this pathway. The air pathway consists of the *nose, throat, trachea* (*windpipe*), *bronchi,* and *lungs.*

PATHWAY OF AIR IN THE RESPIRATORY SYSTEM: EXTERNAL RESPIRATION

Nose. Two openings in the nose, the *nostrils* (or *nares*), admit air into the body. The nostrils open into the *nasal passages,* which are hollow cavities that extend from the nostrils to the throat. In the nose, inhaled air is cleaned, moistened, and warmed as follows:

1. *Coarse hairs* projecting from the inner wall of the nostrils strain some dust particles out of the air that enters the body.
2. The *mucous membrane,* a layer of specialized cells, lines the walls of the nasal passages. These cells secrete *mucus,* which is a moist, thick, gummy substance. The water in it moistens the incoming air. Very tiny particles such as bacteria and dust stick to the mucus. The base of the mucous membrane contains a network of numerous blood capillaries. The heat of the blood in these capillaries quickly warms the air to the temperature of the body.
3. **Cilia,** microscopic hairs on the surface cells of the mucous membrane, move back and forth at all times. By this means, the cilia push mucus, and any material in it, toward the outside of the nose.

Throat. It is better to inhale air through the nose than through the mouth because of the cleaning and warming that takes place in the nasal passages. Air from the nasal passages then enters the throat and passes over the *adenoids* and *tonsils.* The adenoids are embedded in the wall between the back of the nasal passages and the throat. The tonsils lie in the side walls near the junction of the mouth and throat. Both the adenoids and the tonsils are made up of a special tissue that normally helps protect the body against infection.

Trachea (or Windpipe). Air passes through the throat and then enters the **trachea,** or *windpipe,* a tube about 10 cm. long and 2.5 cm. in diameter. It lies in front of the esophagus. Rings of cartilage in the walls of the windpipe keep the air passage open. The trachea, like the nasal passages, is lined with a mucous membrane that has cells containing cilia. The cilia of the windpipe push mucus, and any material in it, toward the throat. From there, the mucus is coughed up.

The *glottis* is the slitlike opening into the trachea. A flap composed of cartilage, called the *epiglottis,* prevents swallowed food from entering the windpipe. The portion of the windpipe just below the glottis is called the *voice box* (or *larynx*), in which the vocal cords lie.

Bronchi and Lungs. At its lower end, the trachea is divided into two large hollow branches, called **bronchi.** Each branch, or *bronchus,* extends into a lung.

Each of the two **lungs** is a large, elastic, spongy sac. Separating the lungs from the chest wall is a protective double-layered membrane called the *pleura.* Fluid secreted by this membrane prevents friction between the lungs and the chest wall.

Inside each lung, the bronchus branches into smaller **bronchial tubes.** These bronchial tubes branch repeatedly into even smaller microscopic tubes, called **bronchioles.**

Each tiny bronchiole opens into thin-walled bulbs called **air sacs** (or *alveoli*). As shown in Fig. 17-3, each air sac is surrounded by a network of blood capillaries. As it passes through these capillaries, the blood gives up carbon dioxide and water vapor to the air sac, and exchanges these wastes for oxygen from the air sac.

THE EXCHANGE OF GASES IN THE BODY: INTERNAL RESPIRATION

The air we inhale is about four-fifths nitrogen and one-fifth oxygen. Other gases are present in very small amounts. Of all these gases, the body uses only oxygen. At the air sacs, oxygen diffuses into the blood of the capillaries and combines with the hemoglobin of the blood. The oxygen, in combination with the hemoglobin, is carried to all the body cells. When blood rich in oxygen reaches a cell that lacks oxygen, the oxygen in the blood separates from the hemoglobin. The cell then absorbs the oxygen.

As this process goes on, the cell gives off its waste carbon dioxide and water, both of which enter the blood. When this waste-laden blood reaches the lungs, both the carbon dioxide and some water pass into the air spaces of the air sacs. (The wastes are exhaled from the lungs.) Then the hemoglobin of the blood absorbs more oxygen from the air in the air sacs, and the cycle is repeated. The exchange of gases in the lungs, blood, and cells goes on continuously, thereby supplying all parts of the body

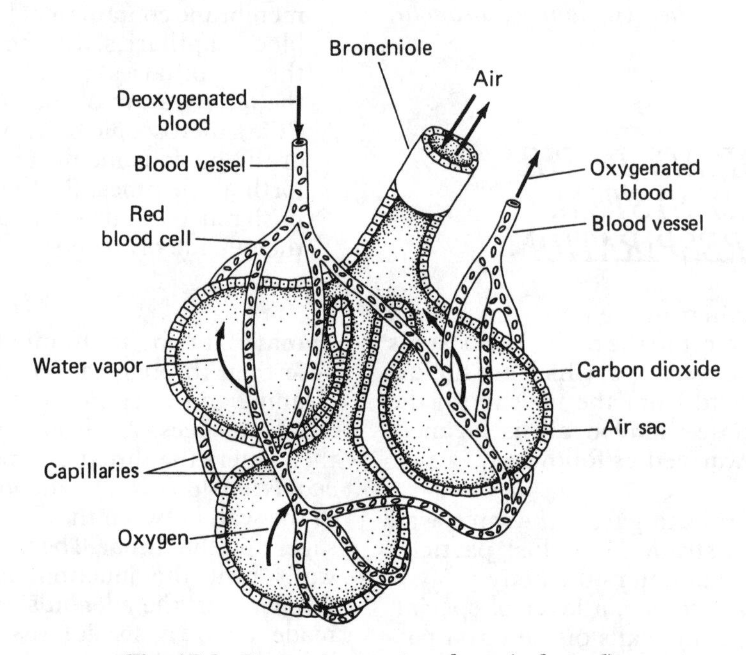

Fig. 17-3. Some air sacs in a lung (enlarged).

with fresh oxygen and removing the wastes of oxidation.

OXIDATION IN CELLS: CELLULAR RESPIRATION

Energy for carrying out the life activities is produced within the cells of an organism. This energy is the chemical energy present in the molecules of the nutrient glucose. Within a cell, the glucose acts as a fuel and undergoes oxidation—that is, it reacts with oxygen and releases its chemical energy.

The energy in glucose is not released all at once during oxidation. Rather, it is released gradually, a small quantity of energy at a time. This energy enters and becomes part of a molecule called **ATP.** The ATP stores the energy until it is needed for the life activities of a cell. For example, the energy of ATP is used in the contraction of muscle and in the movement of cilia. Actually, glucose is the indirect source of energy and ATP is the direct, immediate source of energy for every cell.

Numerous enzymes, called *respiratory enzymes,* help transfer the chemical energy from glucose to ATP. Since most of the enzymes that aid in releasing this energy are located inside the mitochondria of a cell (see Chapter 3), the reason for calling mitochondria the "powerhouses of the cell" becomes clear.

The final steps of oxidation (and respiration) in cells produce carbon dioxide and water, both of which are released as wastes from the cells. These wastes are carried away from the cells to the lungs by the blood. The lungs expel the carbon dioxide and some water vapor, as you found out in your laboratory investigation about exhaled air.

PRESSURE CHANGES IN THE RESPIRATORY SYSTEM

Air enters and leaves the body because of changes in pressure upon the lungs. These changes are the results of two breathing movements—*inhalation* and *exhalation* (see Fig. 17-4).

Inhalation. The act of inhaling air is called **inhalation,** or *inspiration.* When you inhale, one set of *rib muscles* and the **diaphragm,** which is a sheet of muscle below the lungs, become active. The rib muscles involved contract and cause the ribs to move upward and

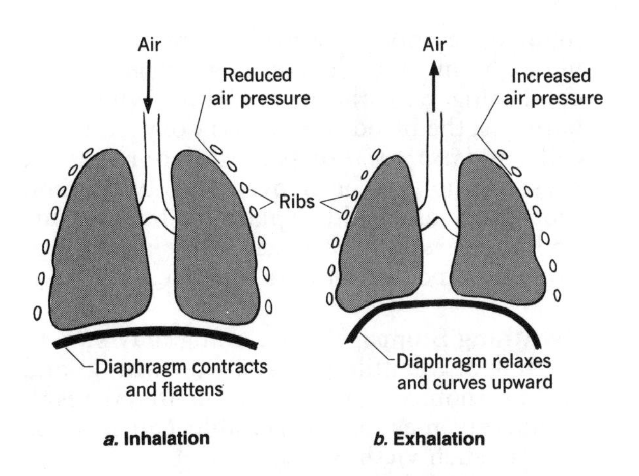

Fig. 17-4. Air going into and out of the lungs.

outward. At the same time, the diaphragm—which is in an arched position when relaxed—contracts, moves downward, and flattens. These two events, occurring at the same time, enlarge the chest cavity and lower the air pressure within the lungs. Then, the greater outside air pressure causes air to rush into the lungs and fill them.

Exhalation. The act of exhaling air is called **exhalation,** or *expiration.* When you exhale, the set of rib muscles that was contracted during inhalation relaxes. At the same time, a second set of rib muscles contracts. This set was relaxed during inhalation. These changes in the muscles cause the ribs to move downward and toward the center of the body. Also at the same time, the diaphragm, which contracted during inhalation, relaxes and moves upward to its former arched position. These two events cause pressure to be exerted on the lungs and cause the chest cavity to become smaller. The increased pressure squeezes the used air out of the lungs.

SERIOUS DISTURBANCES OF RESPIRATION

Certain chemicals, diseases, and accidents can disturb respiration so much that life is endangered. Some of the most serious of these disturbances are:

Carbon Monoxide Poisoning. Carbon monoxide is formed by the incomplete burning of the carbon in fuels such as gasoline, fuel oil, and coal. Carbon monoxide is extremely poisonous, even in small quantities. When inhaled, carbon monoxide combines with hemoglobin, thus preventing oxygen from combining with the hemoglobin. When this happens, the blood cannot carry oxygen to the cells, and *asphyxiation* (suffocation) results. A person who has been overcome by carbon monoxide gas should be placed in the open air and, if necessary, be given artificial respiration until medical attention arrives.

Breathing Stoppage. Breathing may stop in victims of infantile paralysis, drowning, and carbon monoxide poisoning. Artificial respiration, given as soon as possible, can save the lives of such victims.

In **artificial respiration,** pressure on the lungs is rhythmically increased and decreased by an outside force. Some methods of artificial respiration are the "iron lung," the pulmotor, mouth-to-mouth breathing, and the back pressure-arm lift.

Irritation of Respiratory Organs. Air that contains harmful substances is called *polluted air.* Such air irritates the lining of the respiratory system, affecting cilia action and the production of mucus. Repeated irritation may cause allergic attacks and numerous other respiratory diseases, such as lung cancer, emphysema, and bronchitis. Among the major causes of air pollution are smoke from open fires, exhaust fumes of heating systems and gasoline engines, wastes from factory smokestacks, and ash from the burning of fuels.

Many of the respiratory diseases caused by polluted air can be checked or prevented by reducing air pollution. (You will read more about air pollution in Ch. 25.)

Care of the Respiratory System

Get into the open air. Spend 2 hours or more each day outdoors. When indoors, be sure that the room is well-ventilated. In a well-ventilated room, fresh air comes in, circulates for a while, passes out of the room, and is then replaced by more fresh air. Most rooms can be properly ventilated by opening windows from the top and bottom.

Practice breathing deeply. Deep breathing helps get fresh air into the lower sections of the lungs. Brisk walking and other forms of exercise stimulate deep breathing.

Breathe with the mouth closed. The nose and nasal passages have structures that normally warm, clean, and moisten the air before it reaches the lungs, whereas the mouth has none of these structures.

Don't smoke. There is strong evidence from medical studies that cigarette smoking is a major cause of lung cancer and other respiratory illnesses. The numerous harmful substances in cigarette tobacco and smoke (especially when combined with air pollution) irritate the respiratory system and damage lung tissue.

Prevent the accumulation of carbon monoxide. Furnaces, stoves, and engines should be checked regularly to make sure that enough air is present for them to burn their fuels completely. Be sure that all such devices are used in well-ventilated rooms. Never run an automobile engine in a closed garage.

Prevent air pollution. Cooperate with the clean-air program in your community. Refrain from making open, smoky fires. Be sure that your furnace or incinerator is properly adjusted for complete combustion. Factory and automobile exhausts should be equipped with devices that prevent harmful fumes from entering the atmosphere.

Consult a physician regularly. Regular physical examinations that include X rays of the chest help detect respiratory diseases at an early stage when they are most curable.

CHAPTER REVIEW

Science Terms

The following list contains all of the boldfaced scientific terms found in this chapter and the page on which each appears.

air sacs (p. 204)
artificial respiration (p. 206)
ATP (p. 205)
breathing (p. 202)
bronchi (p. 204)
bronchial tubes (p. 204)
bronchioles (p. 204)
cilia (p. 203)
diaphragm (p. 205)
exhalation (p. 205)
inhalation (p. 205)
lungs (p. 204)
respiration (p. 202)
trachea (p. 204)

Matching Questions

On the blank line, write the letter of the item in column B which is most closely related to the item in column A.

Column A

____ 1. given off during exhalation
____ 2. release of stored energy
____ 3. can cause asphyxiation
____ 4. taking in of oxygen-rich air
____ 5. contain respiratory enzymes
____ 6. sheet of contracting muscle
____ 7. molecule that stores energy
____ 8. microscopic hairs in nose and trachea
____ 9. small branching tubes in lungs
____ 10. protective membrane around lungs

Column B

a. ATP
b. inhalation
c. bronchioles
d. air sacs
e. pleura
f. carbon monoxide
g. respiration
h. carbon dioxide
i. mitochondria
j. cilia
k. diaphragm

Multiple-Choice Questions

On the blank line, write the letter preceding the word or expression that best completes the sentence or answers the question

1. Which gas is present in greater quantity in exhaled air than in inhaled air?
a. carbon dioxide *b.* oxygen *c.* nitrogen *d.* carbon 1 ____

2. The chemical process that produces energy inside a muscle cell is
a. digestion *b.* oxidation *c.* diffusion *d.* secretion 2 ____

3. In animals that carry out respiration through the skin, the skin must always be
a. thick and dry *b.* thin and dry *c.* thin and moist *d.* thick and moist 3 ____

4. Many large dust particles entering the nostrils are stopped by the
a. blood *b.* tonsils *c.* hairs *d.* adenoids 4 ____

5. Besides trapping bacteria and small dust particles, mucus in the nose
a. dries the air *b.* warms the air *c.* cools the air *d.* moistens the air 5 ____

6. Although we may inhale air having a temperature of 32°F, the temperature of air in the lungs is close to
a. 35°F *b.* 50°F *c.* 70°F *d.* 98°F 6 ____

7. Microscopic hairs that clean air before it reaches the lungs are called
a. mucus *b.* adenoids *c.* cilia *d.* mucous membranes 7 ____

8. In which region are the tonsils located?
a. bronchi *b.* throat *c.* adenoids *d.* windpipe 8 ____

9. The windpipe, or trachea, is located
a. behind the esophagus *c.* to the right of the esophagus
b. to the left of the esophagus *d.* in front of the esophagus 9 ____

10. The windpipe does not collapse because its wall contains
a. bone *b.* mucus *c.* blood vessels *d.* cartilage 10 ____

11. The epiglottis prevents
a. food from entering the windpipe *c.* food from entering the gullet
b. air from entering the gullet *d.* air from entering the windpipe 11 ____

12. The exchange of carbon dioxide in the blood for oxygen in the air takes place in
a. each body cell *c.* the pleura
b. each air sac *d.* the bronchial tubes 12 ____

13. The main branches of the windpipe are the
a. arteries *b.* bronchi *c.* cilia *d.* vocal cords 13 ____

14. Air moving into the nostrils passes through several channels and eventually stops at the walls of the
a. bronchi *b.* windpipe *c.* air sacs *d.* voice box 14 ____

15. The vocal cords are located in the
a. larynx *b.* pharynx *c.* epiglottis *d.* bronchus 15 ____

16. A cell of the stomach that requires oxygen gets it from the nearest
a. red blood cells *b.* lung *c.* air sacs *d.* nostril 16 ____

17. Blood loses oxygen as it passes through the
a. nasal passages *b.* cavity of the windpipe *c.* air sacs *d.* tissues 17 ____

18. Blood loses carbon dioxide as it passes through the
a. nasal passages *b.* cavity of the windpipe *c.* air sacs *d.* tissues 18 ____

19. In organisms, the energy in glucose is released
a. by rapid oxidation *c.* a little at a time
b. by photosynthesis *d.* by digestion 19 ____

20. When the diaphragm moves downward and the ribs upward, the chest cavity
a. becomes smaller *c.* remains the same
b. becomes larger *d.* disappears 20 ____

21. When a muscle contracts, ATP is used up. This is an indication that
a. the muscle cells use energy
b. water stays out of the muscle
c. osmosis occurs
d. ATP leaves the environment
21 ____

22. Enzymes that aid in oxidation are found in the
a. mitochondria *b.* chromosomes *c.* vacuoles *d.* ribosomes
22 ____

23. When the diaphragm moves upward and the ribs downward, the chest cavity
a. becomes smaller
b. becomes larger
c. remains the same
d. disappears
23 ____

24. Which compound in cells makes energy available for their immediate use?
a. carbon dioxide *b.* hemoglobin *c.* DNA *d.* ATP
24 ____

25. Air moves into and out of the lungs because of differences in
a. blood pressure
b. blood circulation
c. quantities of oxygen and carbon dioxide
d. air pressure
25 ____

Modified True-False Questions

In some of the following statements, the italicized term makes the statement incorrect. For each incorrect statement, write the term that must be substituted for the italicized term to make the statement correct. For each correct statement, write the word "true."

1. Carbon dioxide in an air sac must pass through a(an) *bronchial tube* on its way out of a lung. 1 ________

2. The *pleural* membrane lines the nasal passages. 2 ________

3. The oxygen carried by red blood cells is temporarily combined with *carbon dioxide*. 3 ________

4. The energy that is released from glucose becomes part of a(an) *DNA* molecule. 4 ________

5. The energy needed for the life activity of digestion is provided by the life activity of *reproduction*. 5 ________

6. Both the *diaphragm* and rib muscles aid breathing movements. 6 ________

7. The poisonous gas that combines with hemoglobin and causes asphyxiation is carbon *dioxide*. 7 ________

8. An unconscious person who is not breathing should be given *artificial* respiration. 8 ________

9. Automobile exhaust causes *well-ventilated* air. 9 ________

10. Inhaled air passes from the throat to the *nasal passages*. 10 ________

Testing Your Knowledge

1. In correct order, name the parts through which air passes on its way to the air sacs. ______

____

2. Without looking at the list you have just made, in correct order name the parts through which air passes from the air sacs until it is outside the body. ______________

____

3. Both carbon monoxide and carbon dioxide are compounds of carbon and oxygen. A person who is exposed to air containing 0.3 percent of carbon monoxide for half an hour is in danger of dying. However, a person exposed to air containing 2.0 percent of carbon dioxide for an indefinite time may not feel any ill effects. Why is carbon monoxide so much more harmful than carbon dioxide? ______________________________

____

____

____

____

4. This diagram is of a model that shows how air gets in and out of the lungs. Study it and answer the questions below.

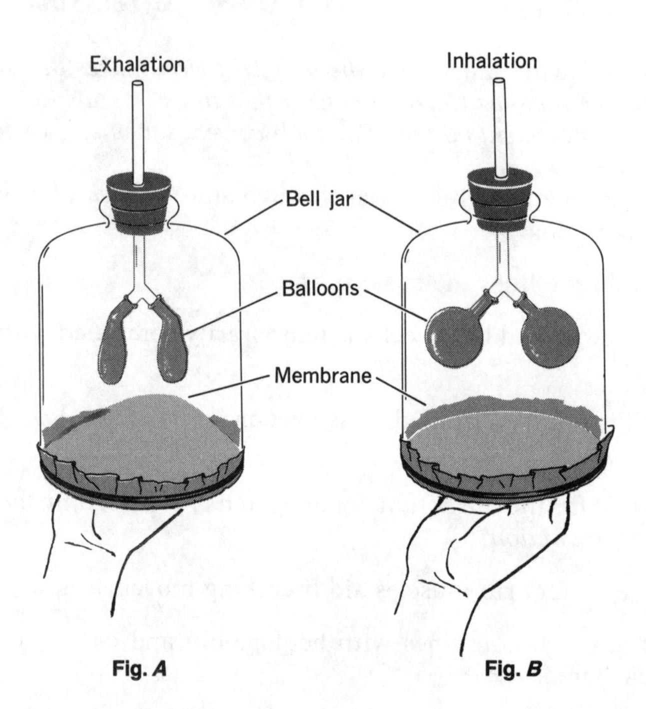

Fig. *A* Fig. *B*

a. What body organs do the balloons represent? ______________________

b. What body structure is represented by the combination of the membrane and fist? ____

____

c. What does the bell jar represent? ______________________________

d. In *A*, how does the air pressure inside the bell jar compare with the air pressure inside the balloons? ______

______

e. In *B*, how does the air pressure inside the balloons compare with the air pressure outside the bell jar? ______

______

f. In a short paragraph, explain what makes air go in and out of the balloons. ______

______

______

______

______

______

5. Name three sources of air pollution in your neighborhood and tell how each source can be either reduced or eliminated. ______

______

______

6. Describe the four phases of respiration:

a. breathing ______

______

b. external respiration ______

______

c. circulation ______

______

d. internal respiration ______

______

7. Distinguish between breathing and respiration.

______

______

8. Why is it impossible for a person to die by holding his or her breath? ______

______

______

Asbestos: How Dangerous Is It?

Asbestos is the name of a group of naturally occurring minerals (silicate) that can be separated into fibers. The fibers are resistant to fire and heat, are strong and durable. They are also thin, long, and flexible and can be woven into cloth. Because of these properties, asbestos has been used in thousands of products. Asbestos can be found in homes, schools, shipyards, insulation materials, shingles, siding, roof tiles, soundproofing materials, fireproof gloves, automobile brake pads, and commercial buildings.

The most dangerous of the asbestos fibers are the ones that are too small to see. They become airborne when material containing them is disturbed. When they are inhaled the fibers accumulate in the lungs. This increases the risk of lung cancer, mesothelioma (a cancer of the chest and abdominal linings), and asbestosis (irreversible lung scarring that can be fatal). Many of the symptoms of the inhalation of asbestos fibers do not appear until many years after exposure. Most people experiencing asbestos-related diseases were exposed to elevated concentrations on the job.

But what can we do about all those millions of tons of asbestos already installed in our homes and other buildings? Oddly, some scientists say "do nothing" while others say "get rid of the stuff." The "do nothing" supporters point to the fact that there is little evidence to link low-level exposure to asbestos with serious illnesses. They further argue that most of the asbestos in U.S. homes and buildings is of a "curly" type called chrysolite, which research shows is relatively harmless in low doses. The other asbestos type, called amphibole, may be more dangerous. It is needle-shaped, and thus more likely to penetrate and damage delicate tissues such as those found in the lungs.

Other scientists, however, have determined that chrysolite is more poisonous to cells than is amphibole. When asked to comment, an official of the EPA said in 1990 that the EPA had "yet to reach a definitive decision about one type of asbestos being less harmful" than the other.

Everyone agrees that asbestos of any kind in sufficiently high doses can kill. Ten thousand U.S. citizens die each year from asbestos-related diseases. Most of these people—such as asbestos miners, workers who manufacture or install asbestos products, maintenance workers like painters, carpenters, electricians, and building custodians—are exposed almost daily to the mineral. According to one expert, the rest of us are exposed to "no more than 1%" of the amount of asbestos considered safe for asbestos workers.

But 1% is still too much, the critics argue. They feel that even a small amount of such a dangerous substance floating around in the air is too much of a risk.

So what should be done about the problem? Uphold the ban and do not make any more asbestos products, the EPA would say. Continue to make the very useful products, say others, but educate people about the possible dangers.

And what can we do about the asbestos already out there? Some people think it's safer leaving it alone than trying to expose and remove it. Still others say it should be coated with materials that will keep the asbestos from getting into the air. What do you think?

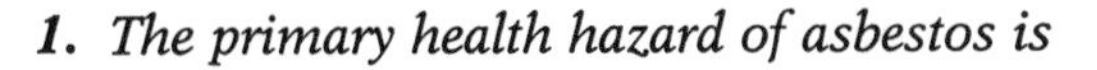

1. *The primary health hazard of asbestos is*

   a. heart disease.
   b. skin disease.
   c. respiratory disease.
   d. nervous disease.

2. *In 1990, the EPA said that chrysolite was*

   a. more harmful than amphibole.
   b. less harmful than amphibole.
   c. about as harmful as amphibole.
   d. might or might not be as harmful as amphibole.

3. *Why should the average person worry or not worry about asbestos?*

   ____
   ____
   ____
   ____
   ____
   ____

4. *On a separate sheet of paper, write one or two sentences of the beneficial uses of asbestos. Then write one or two sentences suggesting how other materials or methods might provide the same benefits with none of the drawbacks. Finally, write a concluding sentence or two describing the economic impacts of your suggestions. Consider loss or gain of jobs, costs of replacing asbestos, costs to consumers, etc.*

18

Excretion and Elimination

LABORATORY INVESTIGATION

STUDYING A SECTION OF KIDNEY TISSUE

A. Under the low power of a microscope, examine a stained slide of a thin slice of kidney tissue. Move the slide around and explore it. Note that the cells are arranged in a variety of patterns, most of which enclose a clear space.

1. Describe two of the main patterns you see. ______________________________

____

____

B. Find a portion of the slide which shows somewhat circular masses that are more darkly stained than the surrounding area. Each mass contains a mesh, or knot, of capillaries. Center one of these masses in the field of the microscope and then focus on it with high power. These capillaries are the main pathways through which wastes pass out of the bloodstream and into the kidney, as you will learn about in this chapter.

Compare your specimen with Fig. 18-1 on page 216 and see how many of the labeled structures you can identify in your specimen.

2. List the identifiable structures that you have located in your slide. ______________

____

The Processes of Excretion and Elimination

Excretion is the process by which dissolved wastes resulting from the activities of cells are removed from the body of an organism. These wastes come from the oxidation of nutrients, the breakdown of excess amino acids, and other life activities. If these wastes accumulate in the cell, they can poison the cell.

Simple organisms, such as the paramecium, excrete most of their dissolved wastes directly through their cell membrane by means of diffusion. Since they live in water, these organisms take in more water than their bodies need. A special structure of the paramecium, the *contractile vacuole,* continuously pumps out the excess water from the organism's body.

In complex organisms, such as humans, the circulating blood removes dissolved wastes from each cell. The blood carries the wastes to special organs that take wastes out of the bloodstream. These organs then discharge the wastes from the body. Among such organs are the kidneys and the skin.

Elimination is the process by which solid wastes are passed out of the body of an organism. These wastes, which are mainly the remains of food materials that could not be digested, never entered the protoplasm of any cell. In a paramecium, solid wastes remaining in a food vacuole are eliminated from a specific area, called the *anal spot,* on the surface of the organism's one-celled body. In complex animals, undigested materials pass into the large intestine and are discharged from there.

Excretion and Elimination in Humans

ORGANS OF EXCRETION

The parts of the body that aid in excretion are the *lungs, liver, urinary system,* and *skin.*

Lungs. As you learned in the previous chapter, when digested nutrients are oxidized in tissue cells, energy is released and carbon dioxide and water are formed as wastes. These wastes diffuse from the cells to the lymph and blood. Eventually, the wastes are carried by the blood to the air sacs of the lungs. During exhalation, carbon dioxide gas and some water vapor are excreted from the lungs.

Liver. Amino acids, the end product of the digestion of protein, are ordinarily used by the cells of the body for building and repairing tissue. However, excess amino acids that are not used by the cells produce a poisonous compound called *ammonia.* This nitrogen-rich product is not stored in the body, but is broken down. As blood carrying ammonia passes through the gland known as the **liver,** the cells of the liver change the ammonia into the less toxic compound *urea.* The urea dissolves in the blood plasma and is then carried to the kidneys. The kidneys excrete urea in the urine. Mineral salts that are produced as wastes in protein digestion are also excreted in urine.

Certain cells of the liver remove worn-out red blood cells from the blood. These blood cells are broken down and become part of the bile that the liver secretes. As you have learned, bile aids digestion by breaking large particles of fat into small droplets. Accordingly, since bile has a useful job in the body, it is often considered a *secretion* rather than a waste, or excretion. Eventually, used bile flows into the large intestine, from which it is eliminated, along with undigested matter.

Urinary System. Most of the excretion of wastes takes place through the **urinary system,** which consists of the *kidneys, ureters, urinary bladder,* and *urethra.* This system also maintains the correct water balance in the body.

Kidneys. There are two **kidneys,** each of which is between 10 and 15 cm. long and shaped like a bean. They are located a little above the waist, one on either side of the backbone. When cut open, a kidney is seen to be divided in two main regions—an outer, fairly solid region, and an inner, hollow region.

If it is examined under a microscope, as you did in your laboratory investigation, tissue from the outer region of a kidney is seen to consist of numerous long, twisted tubules (tiny tubes) and tiny capsules (pockets). Each capsule contains a set of twisted capillaries that resembles a knot. The capillary knot is surrounded by a fluid-filled cavity. (See Fig. 18-1.) Fig. 18-2 illustrates a **kidney unit,** or **nephron:** a capsule, its contents, and the tubule attached to it.

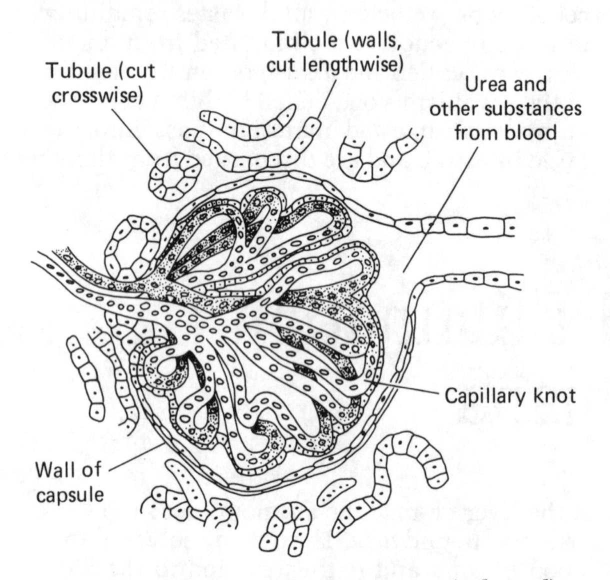

Fig. 18-1. Section of kidney tissue (enlarged).

There are about a million of these units in each kidney. The special job of the kidney units is removing dissolved wastes from the blood.

When blood from an artery passes through the capillary knot inside a capsule, a solution containing numerous substances filters out of the blood into the cavity of the capsule. Among these substances are water, urea, mineral salts, and glucose. As this fluid passes through the long, twisted tubule, the glucose and most of the water are reabsorbed back into the bloodstream by the capillaries around the tubule. These capillaries, containing waste-free blood, lead to the vein that leaves the kidney. The fluid remaining in the tubule is now called **urine.** The main substances in urine are excess water, urea, and excess salts. Urine from several tiny tubules flows into a larger collecting tubule. The collecting tubules join and empty into the hollow region of the kidney (Fig. 18-3).

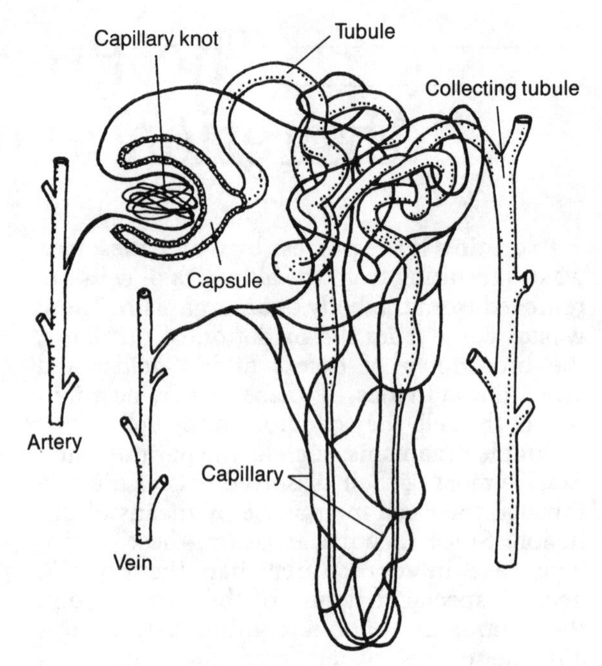

Fig. 18-2. A kidney unit, or nephron.

Ureters. Like the mouth of a funnel, the hollow region of each kidney leads to a narrower region and ends in a tube. This tube is called a **ureter.** Urine from the hollow region of each kidney passes into each ureter. Both ureters lead to the **urinary bladder.**

Urinary Bladder. This saclike organ temporarily stores the urine. At the base of the urinary bladder is a single opening that leads to a tube called the **urethra.**

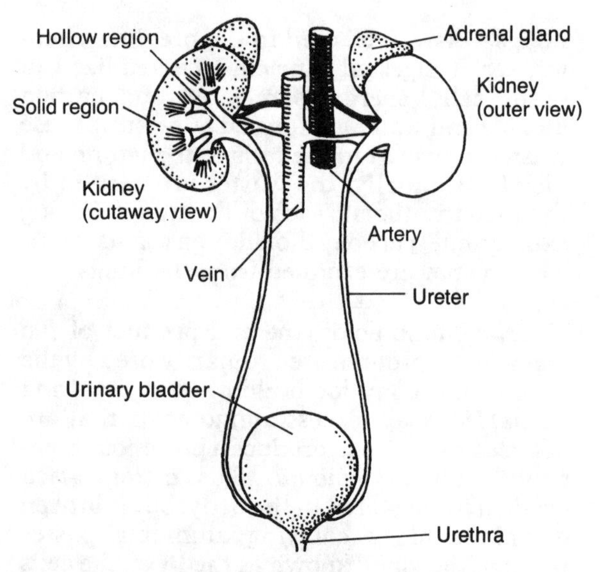

Fig. 18-3. Human urinary system.

Urethra. At fairly regular time intervals, when the urinary bladder is full, urine passes from the bladder into the urethra and then out of the body.

Skin. One of the most important functions of the **skin** is excretion. Other functions include sensation, cooling the body, and protecting the body from infection. The skin is actually the largest organ of the body. As shown in Fig. 18-4, there are two major layers in the skin. The outer layer is called the **epidermis;** the inner, the **dermis.**

Epidermis. Although it is thin, the epidermis (which has a top layer of dead cells) protects the dermis layer under it. Scattered throughout the epidermis are many microscopic openings called **pores.** Each pore leads down into a tubule in the dermis; each tubule ends in a coiled mass called a **sweat gland.**

Dermis. The dermis, which is thicker than the epidermis, contains these major structures:

1. *Capillaries* branch throughout the dermis and also surround each sweat gland.
2. *Nerve endings* are sensitive to changes that take place around the body, such as changes in temperature and pressure.
3. *Hair follicles* are pockets that contain growing hairs.
4. *Fat deposits* serve as a cushion against blows, as an energy reservoir, and as insulation against cold.
5. *Oil glands* secrete oil which keeps the skin and hair soft, and lubricates the skin.
6. *Sweat glands* excrete wastes in the form of perspiration, and thus help cool the body.

Excretion by the Skin. Perspiration (or **sweat**) is the liquid waste that is excreted by the skin. Perspiration is a solution of about 98% water and small quantities of urea, salt, and other substances. Sweat glands remove these wastes from the skin capillaries. The wastes pass into the sweat gland channel and reach the skin surface through the pore at the end of each channel.

The process of perspiration goes on almost continuously—more on the palms of the hands and in the armpits than elsewhere. Conditions such as heat, excitement, or vigorous physical activity increase the circulation of blood. The sweat glands then become more active, producing sweat at a faster rate than they ordinarily would.

Because body heat is used up when the water of perspiration evaporates, the surface of the skin is cooled. This cooling is essential in maintaining the normal temperature of the body. As the water of perspiration evaporates, urea and salts are left behind on the surface of the skin. These substances must be removed from the skin by bathing, so that the pores do not get clogged up and infected.

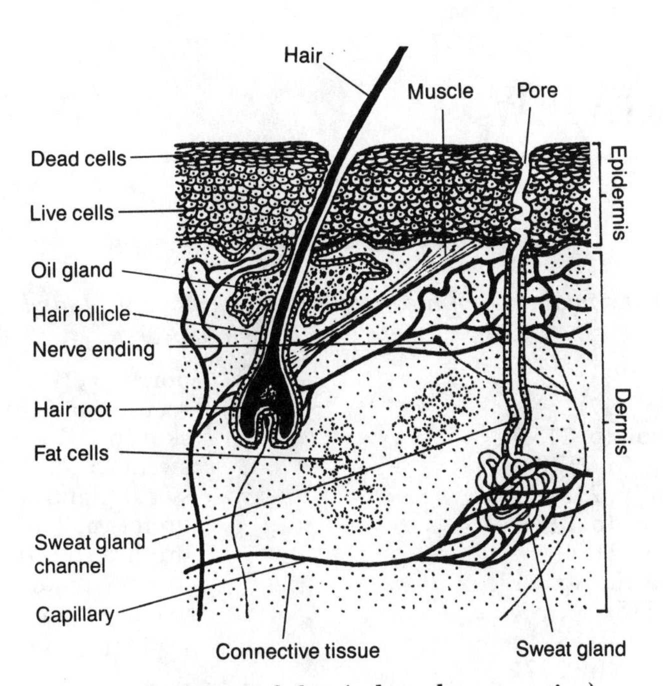

Fig. 18-4. Portion of skin (enlarged cross section).

ORGAN OF ELIMINATION—THE LARGE INTESTINE

Some materials in the food you eat cannot be digested. Among these are cartilage and vegetable fibers (roughage). Undigestable plant fiber is also known as *cellulose.* Such materials pass through the food tube and into the **large intestine** without being changed chemically. These undigested materials, in the form of solid wastes, are temporarily stored in the **rectum.** At intervals, these solid wastes are pushed out of the body through the opening of the rectum, called the **anus.** The large intestine also reabsorbs some water and aids in the removal of excess mineral salts.

Care of the Organs of Excretion and Elimination

Drink the equivalent of 8 glasses (2 liters) of water daily. Water is essential for digestion, circulation, and all cell processes. It is also needed to rid the body of wastes. In removing these wastes, the kidneys, skin, and large intestine also take about 2 liters of water per day out of the body. This quantity must be replaced by water in our foods and beverages.

Eliminate solid wastes regularly. Regular bowel movements rid the body of some wastes. Two natural aids for this process are eating foods containing roughage and drinking plenty of water. Roughage helps to stimulate the process of peristalsis in the food tube.

Bathe daily. The accumulation on the skin of dirt, dried perspiration, and oil provides a good environment for bacteria. These organisms produce bad odors and can cause skin infections. The regular use of soap and water can prevent such problems.

Wear appropriate clothing. Clothing made of loosely woven, porous fabrics allows perspiration to evaporate freely. Such clothing should be worn in warm weather. Clothing made of heavier, more closely woven fabrics prevents the loss of heat from the body. Such clothing should be worn in cold weather.

CHAPTER REVIEW

Science Terms

The following list contains all of the boldfaced scientific terms found in this chapter and the page on which each appears.

anus (p. 218)
dermis (p. 217)
elimination (p. 215)
epidermis (p. 217)
excretion (p. 215)
kidney unit (p. 216)
kidneys (p. 216)
large intestine (p. 218)
liver (p. 215)
nephron (p. 216)
perspiration (p. 217)
pores (p. 217)
rectum (p. 218)
skin (p. 217)
sweat (p. 217)
sweat glands (p. 217)
ureter (p. 216)
urethra (p. 216)
urinary bladder (p. 216)
urinary system (p. 215)
urine (p. 216)

Matching Questions

On the blank line, write the letter of the item in column B which is most closely related to the item in column A.

	Column A	Column B
___	1. secretion of bile	*a.* dermis
___	2. ridding of solid wastes	*b.* ureter
___	3. capsule, capillaries, and tubule	*c.* ammonia
___	4. ridding of dissolved wastes	*d.* epidermis
___	5. outer layer of the skin	*e.* pores
___	6. poisonous nitrogen-rich compound	*f.* sweat glands
___	7. inner layer of the skin	*g.* kidney unit
___	8. tube between kidney and urinary bladder	*h.* excretion
___	9. tiny openings in the skin	*i.* liver
___	10. excrete wastes from dermis	*j.* urethra
		k. elimination

Multiple-Choice Questions

On the blank line, write the letter preceding the word or expression that best completes the sentence or answers the question.

1. The most important wastes that must be removed from animal cells include
a. oxygen, water, amino acids *c.* urea, oxygen, hormones
b. carbon dioxide, water, ammonia *d.* hormones, enzymes, water 1 ___

2. The organ of higher animals that carries out a task similar to that of the stomata of leaves is the
a. stomach *b.* heart *c.* large intestine *d.* skin 2 ___

3. Solid waste materials that cannot be digested by the digestive system are discharged from the
a. large intestine *b.* kidneys *c.* contractile vacuole *d.* lungs 3 ___

4. Which two glands help in excretion?
a. gastric glands and liver *c.* heart and liver
b. tear glands and gastric glands *d.* liver and sweat glands 4 ___

5. Cell wastes such as carbon dioxide and water are excreted from the
a. lymph *b.* lungs *c.* liver *d.* veins 5 ___

6. Worn-out red blood cells are broken down in the organ called the
a. kidney *b.* large intestine *c.* liver *d.* pineal gland 6 ___

7. The kidneys are located
a. in front of the heart *c.* at waist level, in front of the stomach
b. above the lungs *d.* at waist level, in back of the stomach 7 ___

8. Dissolved wastes are removed from the blood as the blood passes through the
a. kidney units *b.* ureter *c.* urinary bladder *d.* urethra — 8 ____

9. Two important nutrients that are excreted from the body when they are present in excess are water and
a. plasma *b.* urea *c.* fat *d.* salts — 9 ____

10. The ducts that connect the kidneys with the urinary bladder are the
a. urethras *b.* ureters *c.* urinary tubules *d.* renal veins — 10 ____

11. The part of food that cannot be digested consists mainly of cartilage and
a. muscle fibers *b.* glucose *c.* cellulose *d.* protein fibers — 11 ____

12. After a mile run, an athlete may weigh several pounds less than before as a result of
a. water lost from tissues
b. excretion of carbon dioxide
c. a lack of oxygen
d. exhaustion — 12 ____

13. Two values of perspiration are
a. regulation of body temperature and loss of weight
b. opening of the pores and secretion
c. excretion of wastes and regulation of body temperature
d. elimination and sensitivity — 13 ____

14. Which tube leads urine out of the body?
a. ureter *b.* bladder *c.* kidney unit *d.* urethra — 14 ____

15. The rectum temporarily stores
a. solid wastes *b.* liquid wastes *c.* excess glucose *d.* urea — 15 ____

Modified True-False Questions

In some of the following statements, the italicized term makes the statement incorrect. For each incorrect statement, write the term that must be substituted for the italicized term to make the statement correct. For each correct statement, write the word "true."

1. Wastes are discharged from an organism in the process of *secretion*. — 1 ________
2. Urea is formed in the *liver* and then excreted by the kidneys. — 2 ________
3. The open end of a sweat gland is a(an) *capillary*. — 3 ________
4. The epidermis of the skin is *thicker* than the dermis. — 4 ________
5. The kidney unit is also known as a(an) *ureter*. — 5 ________
6. Excess *salt* is one type of waste excreted by the skin. — 6 ________
7. Undigested food fibers pass from the *small intestine* to the rectum. — 7 ________
8. The body normally requires about 8 *quarts* of water per day. — 8 ________
9. Liquid waste that is excreted by the skin is called *respiration*. — 9 ________
10. Food *roughage* aids elimination of solid wastes. — 10 ________

Testing Your Knowledge

1. Explain the difference between the two terms in each of the following pairs:

a. excretion and secretion ______________________________
____
____

b. excretion and elimination ______________________________
____
____

c. epidermis and dermis ______________________________
____
____

d. sweat glands and oil glands ______________________________
____
____

e. ureter and urethra ______________________________
____
____

2. Fill in the blanks in the following table:

Waste	*Where the Waste Enters the Blood*	*Organ Through Which the Waste Leaves the Body*
Carbon dioxide	*a.* ______________	*b.* ______________
Urea	*c.* ______________ *d.* ______________	*e.* ______________ *f.* ______________
Water	*g.* ______________ *h.* ______________	*i.* ______________ *j.* ______________ *k.* ______________

3. Give a scientific explanation for each of the following true statements:

a. Perspiration is beneficial. ______________________________
____
____

b. A person needs to drink more water during the summer than during the winter. ______
____
____

c. A person needs to eat more sodium chloride during the summer than during the winter.

___

___

___

d. Roughage is an aid to elimination. ___

___

___

e. A daily bath promotes a clear, healthy complexion. ___

___

___

___

4. Study the table and then answer the questions below it.

	Normal Blood Plasma	*Normal Urine*
Water	About 90%	95%
Proteins	About 8%	None
Glucose	0.1%	None
Urea	0.03%	2%

a. How many times greater is the percentage of urea in urine than in blood plasma?
(1) 6.6 (2) 66 (3) 15 (4) 1.5 ___

b. The organ in which urea is transferred from the blood into urine is the
(1) skin (2) urethra (3) bladder (4) kidney ___

c. The percent of glucose in normal urine is
(1) 0.1% (2) 2% (3) 95% (4) 0% ___

d. As blood passes through the kidney, glucose and water are removed from the blood. By the time urine is formed, the glucose concentration of urine is that shown in the table. A process that can account for this difference in glucose concentration is
(1) excretion of glucose (3) secretion of glucose
(2) body need (4) reabsorption of glucose ___

5. A person who completely covers his or her skin with body paint is likely to become very ill. This can happen because the body may become overheated, even at normal room temperature. How can this situation be explained?

___

___

___

___

___

19

The Skeleton and the Muscles

LABORATORY INVESTIGATION

MOVING A CHICKEN'S TOES

A. Obtain a fresh chicken leg and examine the end that has been cut. With forceps, touch the different tissues that lie under the skin. The hard material in the center is bone. The oval, tubelike structure that lies to one side of the bone is a *sheath* (or covering) that encloses several tendons. (See Fig. 19-1.)

B. Using dissecting scissors, slit the sheath lengthwise for a distance of about 2-1/2 cm. Look for a group of shiny white bands of tissue. With forceps, pick up these bands and fold them toward you. Each band is a *tendon.*

C. On the other side of the leg, opposite the tendons you have just exposed, slit the skin lengthwise for a distance of about 5 cm. Find at least one single tendon inside the slit skin and, with forceps, bring its end outside the slit.

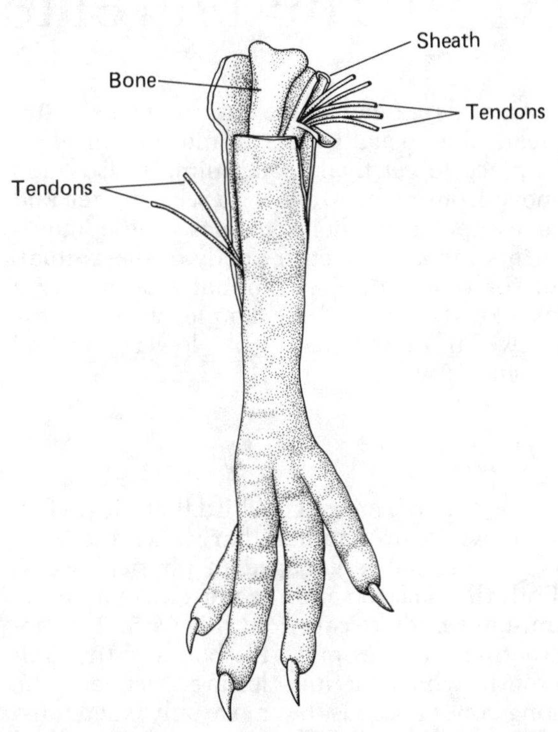

Fig. 19-1. Examining the tendons in a chicken leg.

D. Firmly grasp the single tendon with your forceps and pull the tendon upward.

1. What happens to the chicken's toes? ______________________________

____

E. Release the tendon.

2. What happens to the chicken's toes? ______________________________

____

F. Firmly grasp the group of tendons on the opposite side of the leg and pull all of them upward at one time.

3. Describe what happens. ______________________________

____

G. Pull each tendon of the group separately.

4. Describe what happens. ______________________________

____

5. In a live chicken, what body structures do you think would pull on these tendons? ______

H. Using your forceps, try to tear one of the tendons.

6. Describe one outstanding property of a tendon that you now observe. ______________

____

The Movement of Animals

Animals cannot make their own food. Instead, they usually have to move from place to place to get their food. Animals also often move from place to place to seek shelter and to escape from their enemies. These movements involve the entire body of the animal. In some activities, an animal moves only a part of its body—for example, when a cow chews, it moves mainly its lower jaw and tongue.

LEVERS

Whether the animal is a bird that flies, a fish that swims, or a human that walks, the body of the animal is engaged in physical work. Both the **skeleton** and the **muscles** enable the animal's body to carry out this work. By contracting, muscles move the bones of the skeleton to which the muscles are attached. The long bones are rods that move only when muscles pull them. A stiff rod that can be moved about some support or stationary point is called a **lever.** The point about which a lever moves is called a **fulcrum.** In the body of an animal, many different joints acts as fulcrums for different bony levers. The muscles attached to a bony lever provide the force that makes it move.

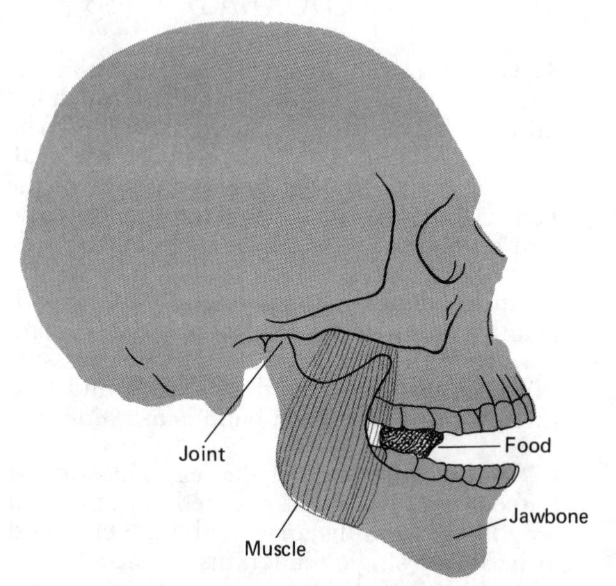

Fig. 19-2. How a lever (jawbone) and fulcrum (joint) work in chewing.

There are several types of levers in a body. One type is seen in the process of chewing. In this activity, the bone of the lower jaw is the lever (Fig. 19-2). The joints of the jaw are fulcrums for this curved lever. When muscles attached to the lower jaw and to the skull contract, they pull the lower jaw upward. This action crushes food between the teeth.

The Human Skeleton

IMPORTANCE OF THE SKELETON

The skeleton is important to the body in several major ways:

1. *For Motion.* The skeleton provides a surface for muscle attachment. Together with the muscles, the skeleton enables the body to carry out its movements.
2. *For Support.* Because the material that makes up the bones is stiff and strong, the skeleton acts as a framework that supports the rest of the body.
3. *For Storage.* The skeleton stores calcium and phosphorus, to be used later by cells in their chemical activities.
4. *For Protection.* Since the skeleton is hard and tough, it protects delicate organs inside the body from being injured by blows. The skull bones protect the brain; the rib bones protect the heart and lungs.
5. *For the Formation of Blood Cells.* The only place in the body where red blood cells are made is in the **marrow** of certain bones. Marrow is the soft tissue found in the center of the long bones. As red blood cells wear out, the marrow forms new ones. Certain white blood cells are also formed in the bones.

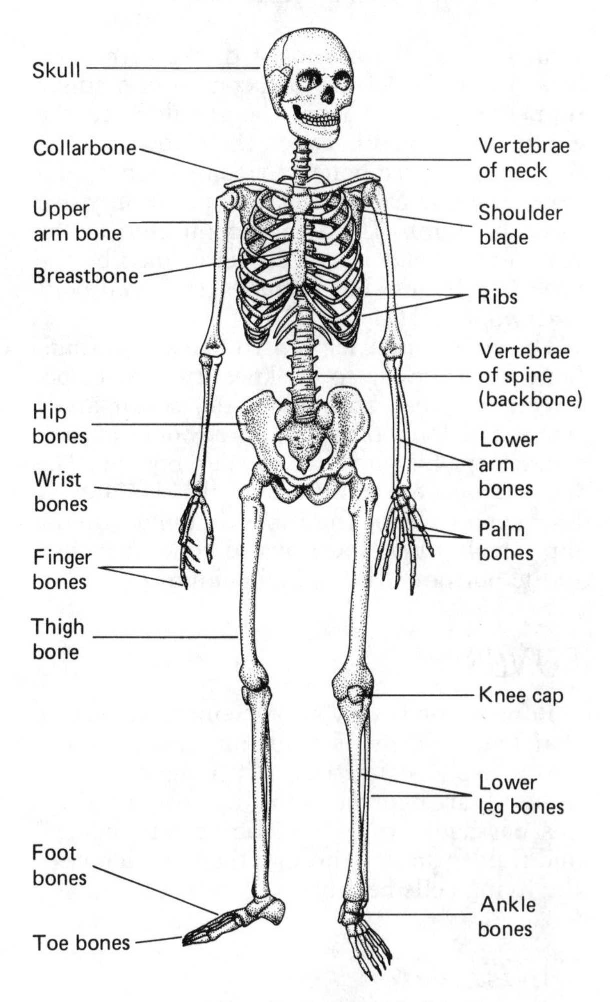

Fig. 19-3. The human skeleton.

MAIN DIVISIONS OF THE SKELETON

The human skeleton (Fig. 19-3) consists of 206 separate bones, in three chief divisions: the *skull,* the *spine,* and the *limbs.* The size and shape of the bones depend on their particular functions.

Skull. The **skull** consists of several curved flat bones that are interlocked at their edges. These lightweight bones form a hollow case that encloses the brain. Indentations in the front part of the skull form the cavities in which the eyes lie. Chambers in each side of the skull house the inner parts of the ears. A framework of bone and cartilage (a tissue that is softer than bone), and a bony passageway, form the nose. The lower jawbone is the one part of the skull that is movable. The skull rests on the upper end of the spine.

Spine. The **spine** (or *backbone*) consists of thirty-three bones, called **vertebrae.** These bones are separated from one another by disk-shaped cushiony pads of cartilage. This arrangement of bones and disks gives the spine some flexibility. The vertebrae and pads form a hollow tube that encases and protects the delicate nerve tissue of the spinal cord. The spine and the muscles attached to it keep the trunk and head erect.

Beginning at shoulder level, there are twelve pairs of **ribs** attached to the spine. These bones form the *rib cage.* Most of the ribs meet in front of the body at the **breastbone.**

The ribs and breastbone encase the chest cavity, in which the heart and the lungs lie. Two sets of muscles attached to the ribs raise and lower the ribs during breathing.

Limbs. The **limbs** consist of the arms and legs. The bones of the arm consist of a single (upper arm) bone from the shoulder to the elbow; two bones between the elbow and the wrist; eight wrist bones; five palm bones; and three bones in each finger, except for only two in each thumb. The arm is connected to the rest of the skeleton by a socket formed by the group of bones that includes the *collarbone* and *shoulder blade.*

The bones of the leg consist of a single thigh bone from the hip to the knee; two bones between the knee and the ankle; seven ankle bones; five foot bones; and three bones in each toe, except for only two in each big toe. The leg is connected to the rest of the skeleton by a socket formed by the bones that make up the hip. The **hip bones** are curved bones attached to the bottom of the spinal column.

BONES

Bone tissue is very strong connective tissue that has deposits of hard minerals (calcium and phosphorus) between its living cells. The minerals are bound together by proteins. Bone has passageways with blood vessels, nerves, and fluids running through them, which keep the living cells healthy.

CARTILAGE

Cartilage tissue is strong connective tissue, but it is softer than bone, slippery, and somewhat elastic. Cartilage supports the outer ear and the tip of the nose. This tissue also covers the ends of bones at movable joints, where its slipperiness aids movement. Cartilage serves as a cushion between the vertebrae of the backbone. Bands of cartilage connect the ribs to the breastbone.

JOINTS

A **joint** is the place where two bones meet (Fig. 19-4). There are two main types of joints—*movable* and *immovable.* The bones of the skull are immovable, except the jawbone. Movable joints act as fulcrums and enable bony levers to move around them. A *ball-and-socket joint,* such as the shoulder and hip joints, allows the limb to move in all directions. A *hinge joint,* such as the knee and elbow joints, allows the limb to move forward and backward. A *pivot joint* allows movement between the skull and the first vertebra.

LIGAMENTS AND TENDONS

The connective tissues that bind the individual bones of the skeleton together and that bind muscles to bones are composed of very dense, tough fibers. **Ligaments** are bands of tissue that bind one bone to another bone. **Tendons** are bands of tissue that bind a muscle to a bone. Both types of tissue aid in movement.

In your laboratory investigation, you discovered how tough the tendons are and how a force that pulls a tendon moves the bone to which it is attached.

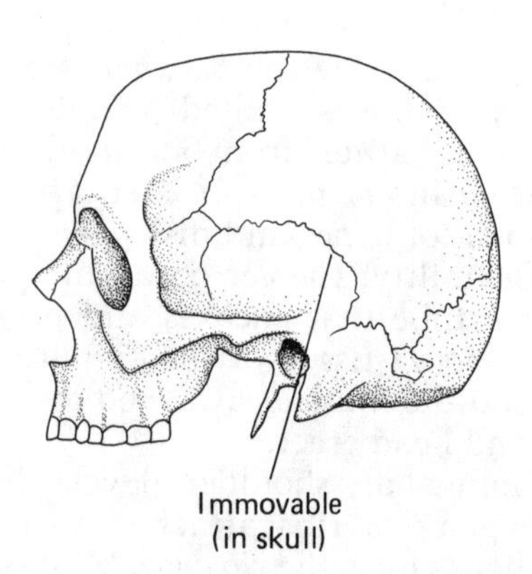

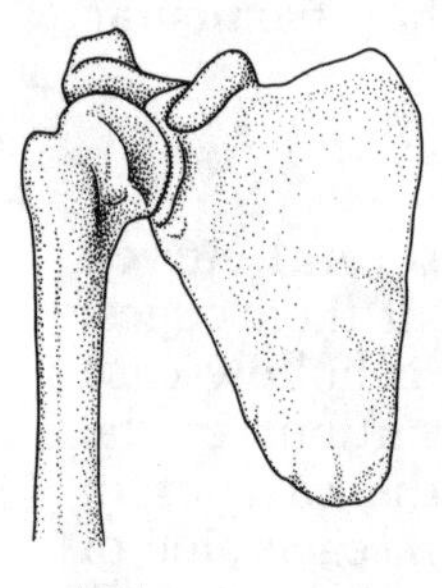

Hinge (at knee)

Fig. 19-4. Kinds of joints in the human skeleton.

The Muscles

IMPORTANCE OF THE MUSCLES

The human body has over 700 muscles in it. There are three types of muscle cells: *striped* (or *striated*), *smooth*, and *cardiac*. These different muscle cell types enable the body to carry out two types of movement: *involuntary movement* and *voluntary movement*.

Involuntary Movement. Movement that we cannot control is called **involuntary movement.** Usually we are not aware that such movements are taking place. Involuntary movements go on inside the body at all times—even during sleep. Examples of involuntary muscle movement include the pumping action of the heart and the pushing of food along the food canal. The muscles that make up the walls of the heart contain cardiac muscle cells, which look similar to striped cells, but function automatically. The muscles of the food tube organs are composed of smooth muscle cells. These cells work under involuntary control.

Voluntary Movement. Movement that we can control is called **voluntary movement.** Such movement does not occur unless you want it to occur. You can control this type of movement and are usually aware of it. Examples of voluntary movement include opening a book and doing a dance step. Voluntary movements are carried on by striped muscle cells. The muscles that are attached to the skeleton are all composed of this type of muscle tissue.

HOW MUSCLES MOVE BONES

A bone can move from a given position only when the muscle attached to the bone contracts. Muscle tissue contracts when proteins in it receive stimulation from the nervous system. When the muscle contracts, it shortens and pulls on the bone. The bone then moves, with its joint serving as its fulcrum. When the muscle relaxes, it lengthens and releases its pull on the bone. At the same time that this muscle relaxes, another muscle on the opposite side of the bone contracts. These actions then restore the bone to its original position.

As shown in Fig. 19-5, when a person bends an outstretched arm upward, the **biceps** muscle, located on the upper side of the arm between the elbow and the shoulder, contracts. At the same moment, the **triceps** muscle, located on the opposite side of the arm, relaxes. When a person straightens a bent arm, the triceps muscle contracts as the biceps muscle relaxes. Many muscles, such as these, work in pairs.

The action of muscles on bones was demonstrated in your laboratory investigation. As you pulled on different tendons, the toes of the chicken curled or straightened. Pulling on these tendons produced the same toe movements that the muscles of a live chicken would ordinarily produce.

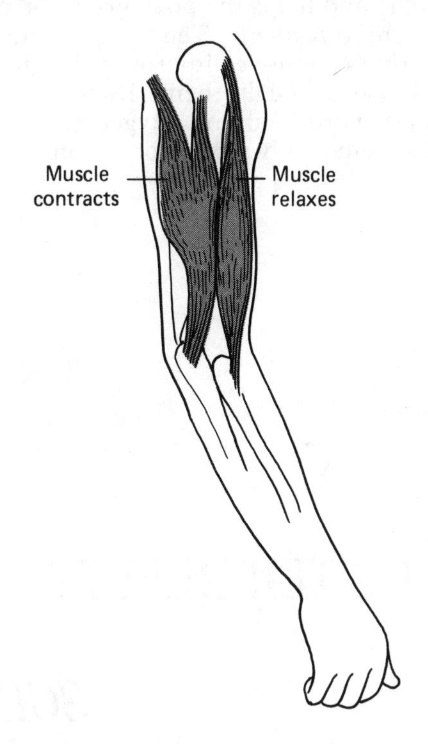

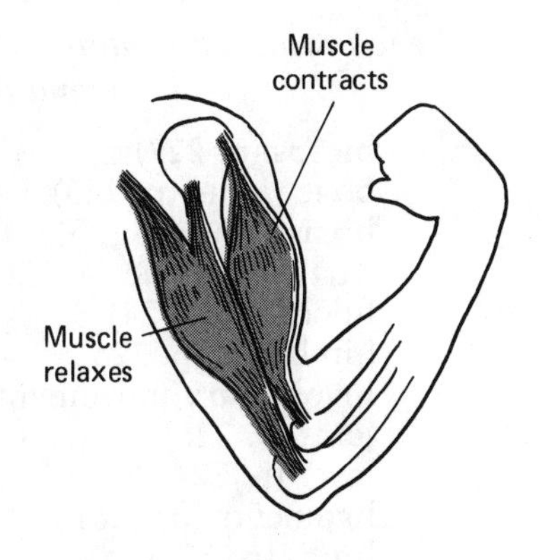

Fig. 19-5. Muscle action in the upper arm.

Care of the Skeleton and the Muscles

Eat properly. A well-balanced diet, which includes all the nutrients, helps normal growth and keeps the skeleton and muscles in good condition. Proteins, calcium and phosphorus, and vitamin D are required for the normal growth of bone. Calcium and protein are required for muscles to contract normally. Proteins also supply the building material for muscle growth.

Practice good posture. Good posture (holding the head and trunk erectly at all times) lessens fatigue and helps the body grow correctly.

Exercise regularly. The contraction and relaxation of muscles improves the circulation of blood through them. Better circulation brings more food and oxygen to the muscle cells, removes wastes from them, and thus aids their growth. Regular exercise also keeps the muscles and joints in good condition.

Relax regularly. All muscles must have periods of rest. During such periods, accumulated wastes are removed from muscle cells, and energy-producing nutrients are restored to them.

Get adequate sleep. The body is most relaxed during sleep. Most people require about 8 hours of sleep each night. The bed should be comfortable and firm enough to keep the muscles and skeleton from being cramped.

Wear shoes that fit. Excessive standing and walking in shoes that do not fit well may throw the skeleton out of balance. If this happens, poor posture, foot discomfort, pain, and excessive fatigue can result.

CHAPTER REVIEW

SCIENCE TERMS

The following list contains all of the boldfaced scientific terms found in this chapter and the page on which each appears.

Matching Questions

On the blank line, write the letter of the item in column B which is most closely related to the item in column A.

Column A	Column B
____ 1. A lever moves around this point	*a.* cartilage tissue
____ 2. where red blood cells are made	*b.* tendon
____ 3. bony case that encloses the brain	*c.* ligament
____ 4. where most pairs of ribs join	*d.* bone tissue
____ 5. the place where two bones meet	*e.* hipbones
____ 6. elastic connective tissue	*f.* fulcrum
____ 7. mineral-rich connective tissue	*g.* vertebrae
____ 8. tissue that connects bone to bone	*h.* skull
____ 9. tissue that connects muscle to bone	*i.* joint
____ 10. bones that make up the spine	*j.* breastbone
	k. marrow

Multiple-Choice Questions

On the blank line, write the letter preceding the word or expression that best completes the sentence or answers the question.

1. Which two body systems, acting together, enable an animal to move?
a. muscular and digestive systems
b. digestive and skeletal systems
c. skeletal and respiratory systems
d. skeletal and muscular systems
1 ____

2. The lungs are protected by the part of the skeleton called the
a. vertebrae *b.* spine *c.* rib cage *d.* shoulder blade
2 ____

3. Blows to the head usually do not harm the eyes because they are protected by the
a. upper eyelids *b.* lower eyelids *c.* bones of the skull *d.* eyebrows
3 ____

4. The force that enables body levers to move is supplied by the
a. brain *b.* kidneys *c.* heart *d.* muscles
4 ____

5. Long bones manufacture the cells of the tissue called
a. epithelial *b.* heart muscle *c.* blood *d.* smooth muscle
5 ____

6. The bones of the skeleton can be grouped into three main divisions, which are
a. spine, skull, hips
b. skull, spine, limbs
c. limbs, pelvis, skull
d. legs, collarbone, shoulder blade
6 ____

7. Bones in the body move only when the
a. bones attached to ligaments contract
b. muscles attached to bones contract
c. nerves attached to muscles contract
d. bones attached to muscles contract
7 ____

8. The nerve tissue of the spinal cord is covered by a
a. solid bony tube
b. bony tube composed of separate vertebrae and disks of cartilage
c. bony tube composed of only separate vertebrae
d. disks of cartilage only 8 ____

9. In the human, the number of bones found between the right elbow and the right wrist is
a. 2 *b.* 5 *c.* 8 *d.* 13 9 ____

10. Involuntary muscle is located in the
a. heart and food tube
b. arm, leg, and ear
c. heart, knee, and toe
d. finger, tongue, and lip 10 ____

11. Several immovable joints are present in the
a. thumb *b.* wrist *c.* shoulder *d.* skull 11 ____

12. The joint at the elbow is of the type called
a. hinge *b.* ball-and-socket *c.* lever *d.* immovable 12 ____

13. The beating of the heart during sleep is an example of
a. voluntary movement
b. peristalsis
c. triceps action
d. involuntary movement 13 ____

14. The number of bones in the left leg between the hip joint and the knee joint is
a. 1 *b.* 2 *c.* 5 *d.* 7 14 ____

15. Which type of muscle is directly used in throwing a ball?
a. involuntary *b.* voluntary *c.* heart *d.* automatic 15 ____

Modified True-False Questions

In some of the following statements, the italicized term makes the statement incorrect. For each incorrect statement, write the term that must be substituted for the italicized term to make the statement correct. For each correct statement, write the word "true."

1. A joint between two movable bones often acts as a *fulcrum*. 1 ________

2. The framework of the human body consists mainly of *cartilage*. 2 ________

3. Marrow is located in the center of many *teeth*. 3 ________

4. The space enclosed by the breastbone and the ribs is the *chest* cavity. 4 ________

5. The outer ear is supported by *tiny ear bones*. 5 ________

6. The bones of the hand are held together by *tendons*. 6 ________

7. For the skeleton to stay healthy, two mineral elements needed are phosphorus and *copper*. 7 ________

8. *Vitamin* B_{12} is needed for the building of strong bones. 8 ________

9. *Tendons* bind the muscles of the leg to its bony levers. 9 ________

10. Regular exercise and relaxation improve circulation and help remove *nutrients* from muscle cells. 10 ________

Testing Your Knowledge

1. Explain the difference between the two terms in each of the following pairs:

a. bone and cartilage ______________________________

____

____

b. hinge joint and ball-and-socket joint ______________________________

____

____

c. tendon and ligament ______________________________

____

____

d. lever and fulcrum ______________________________

____

____

e. biceps muscle and triceps muscle ______________________________

____

f. voluntary movement and involuntary movement ______________________________

____

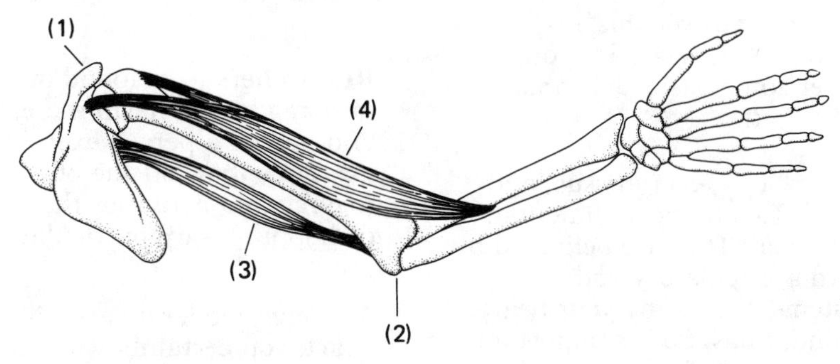

2. This is a diagram of an arm and the shoulder.

a. What type of joint is present at (1)? ______________________________

b. What type of joint is present at (2)? ______________________________

c. When muscle (3) contracts, what happens to muscle (4)? ______________________________

d. When muscle (3) contracts, what happens to the arm? ______________________________

e. When muscle (4) contracts, what happens to muscle (3)? ______________________________

f. When muscle (4) contracts, what happens to the arm? ______________________________

Exercise: Can You Do Too Much of It?

Want to live longer? Lose weight? Be in a better mood? Have stronger bones? Eat better? Reduce your chances of getting cancer, colds, or a heart attack?

Then walk briskly 30–45 minutes a day. Or maybe jog for one hour every day. Or do aerobic exercises 30 minutes, once a day. Or should it be for ten minutes, three times a day? Or three times a week? How about dancing away the night? Or dance for just a few minutes before school, after school, and between TV shows in the evening?

And does it matter if you are male or female? Older or younger? In shape or out of shape? With a heart condition or without one? Overweight or underweight? There are many answers to these questions. And some of them trigger heated arguments among physical fitness experts, scientists, and doctors. So what's a person to do?

Before attempting to answer this all-important question as it relates to you, you should remember some basic facts about the human body.

Every cell in your body, every tissue, and every organ needs two things to function: nutrients and oxygen. They are delivered by the actions of your circulatory and respiratory systems, that is, by your heart, blood vessels, and lungs. So it's important to keep these systems in tip-top shape.

One way to do this is to exercise. That's because exercise tends to strengthen muscles like the heart. Exercise also makes organs function more efficiently, that is, they do more work for you with less effort.

You might easily conclude that the more you exercise, the more efficient and strong your organs will become. Not so, say some researchers. There's a point, insist these scientists, beyond which exercise does little to improve your health and may actually harm it. What exactly is that point?

The answer actually varies from person to person and from one environmental condition to another. But there may be certain points beyond which no one should go. You should not perform exercises that increase the risk of injury to your body. That makes sense. But what kinds of exercises can do this? How and to whom?

One target for injury is your *musculoskeletal system*—your muscles, bones, and joints. Anyone who is active in sports has at one time or another sprained a joint, injured a muscle (gotten a "Charley horse"), or just become sore.

Researchers have found out that the chances of doing bodily harm increase the longer and harder a person exercises. What's more, exercises that jolt the bones and joints—like jogging and activities that involve jumping or hopping—are particularly damaging.

Although moderate exercise is good for your heart, you certainly wouldn't want to overwork it, especially if you had any kind of prior heart problem. You also would not want to work out for a long time during a hot and humid day. On such days, your body loses a lot of water through perspiration. But on humid days the perspiration doesn't evaporate readily, which would normally cool your body. So you can overheat and pass out from heat exhaustion. That's why marathon runners are not happy when races come on hot, humid days.

So how much and what kind of exercise is good for *you*? What do you think?

1. *Organs that function more efficiently*

a. require less effort to do the same amount of work.
b. require more effort to do the same amount of work.
c. require no effort to do work.
d. require work to have energy.

2. *The musculoskeletal system includes the*

a. heart, lungs, and blood vessels.
b. heart, muscles, and bones.
c. muscles, bones, and joints.
d. heart, joints, and bones.

3. *Heat exhaustion is more likely to occur on days that are:*

a. hot and dry.
b. hot and humid.
c. cool and dry.
d. cool and humid.

4. *Explain your answer to Question 3.*

____

____

____

____

____

____

5. *A number of questions were posed in the first three paragraphs of this feature. Choose one from each paragraph, do library research on them, and discuss your findings in a brief written report.*

The Nervous System and Behavior

LABORATORY INVESTIGATION

STUDYING REACTIONS OF THE EYE

A. Work with a partner to observe and record the reactions of eyes to light. Look into one of your partner's eyes. Note the size of the pupil. Shine the beam of a small flashlight into the eye as you continue to observe the size of the pupil.

1. Describe what happens to your partner's pupil when the light is shining on it. ________

____

2. What happens to the pupil when you switch off the light?

____

B. Repeat steps 1 and 2, with your partner shining the light into one of your eyes.

3. Record your results. ________________________________

____

C. Repeat each of the preceding steps, but now each of you should try to keep your pupil from changing size.

4. Describe your results. ________________________________

____

5. What conclusions can you draw about the behavior of your pupil? ________________

____

D. Close your eyes for 1 minute, holding a hand over them. Have your partner observe any changes in pupil size when you open your eyes. Repeat this procedure, with your partner closing and covering his or her eyes for 1 minute. Observe your partner's eyes.

6. Record your results. ______________________________

E. Try to explain the changes you have observed.

7. *a.* Do pupils get larger or smaller in more light? ______________________________

b. in less light? ______________________________

c. Explain why you think this is so. ______________________________

Stimuli and Responses

Sensitivity is the ability of a living thing to detect changes that take place both outside and inside of its body. These changes are called **stimuli** and the reactions to them are called **responses.**

An organism is affected by many stimuli. Some stimuli can harm the organism; others can be either harmless or beneficial to it. Stimuli that start outside the body are called *external stimuli.* Among important external stimuli are light, heat, water, and the presence of other organisms. Stimuli that start inside the body are called *internal stimuli.* Among important internal stimuli are the presence of food in the stomach, the accumulation of carbon dioxide in the blood, and the secretions of various glands.

An organism can respond to changes that may be harmful only if it has the means to detect and react to such stimuli. Simple organisms, such as the paramecium, can detect and swim away from harmful substances like strong acids. Although the paramecium has no nervous system, it has the ability to respond in a manner that aids its survival.

In more complex organisms, such as the human, there are two special systems that enable the body to adjust to changes, and to regulate its life activities. These are the *nervous system* and the *endocrine system,* which is a group of ductless glands. The **nervous system** is sensitive to both external and internal stimuli and controls the responses of the body to both types of stimuli. For example, the stimulus of pain—from a burn or a stomachache—leads a person to react in ways that protect the body against further injury. The endocrine system responds only to internal stimuli. These stimuli come from the nervous system and from changes in the quantities of certain chemicals within the body. The endocrine system will be studied separately in Chapter 22. In this chapter, you will study the human nervous system and behavior.

The Human Nervous System

The major structures that make up the nervous system are the *nerves,* the *brain,* and the *spinal cord.* All of these structures are composed mainly of nerve tissue.

NERVE CELLS

The basic unit of nerve tissue is the **nerve cell** (or *neuron*). The protoplasm of this cell is more sensitive than that of any other type of cell. Some nerve cells are specialized to detect stimuli; others send along **impulses,** or messages, from one part of the body to another. The system of nerve cells is so complex that, ultimately, every part of the body is connected to every other part.

Parts of a Nerve Cell. As shown in Fig. 20-1, a nerve cell has the following specialized parts:

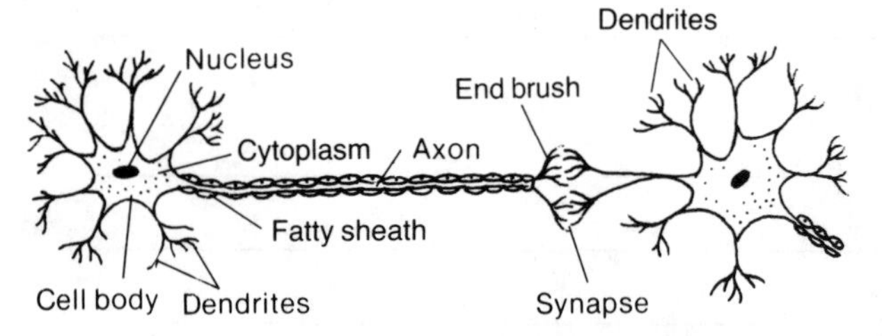

Fig. 20-1. Parts of a nerve cell.

1. The **cell body** (or *cyton*) is the control center of the cell, and includes the nucleus of the cell and the cytoplasm that immediately surrounds it.
2. The **dendrites** are the numerous short, branching fibers that project from the cell body. Dendrites receive stimuli and transmit messages to the cell body.
3. The **axon** is the single, long fiber projecting from the cell body. The axon usually extends in a direction opposite that of the dendrites, and sends messages to other nerve cells.
4. The **end brush** consists of numerous tiny branches found at the end of the axon.
5. The **fatty sheath** is a layer of fatlike material around the axon. The fatty material is enclosed by a delicate membrane. Some axons lack the fatty sheath but have the delicate membrane around them.

The dendrites of a nerve cell pick up impulses and pass them to the cell body which, in turn, leads the impulse to the axon. The axon then leads the impulse away from the cyton to the end brush. The end brush passes the impulse to the dendrites of another nerve cell or to a muscle or a gland.

Types of Nerve Cells

1. *Receptor nerve cells,* which are present in the **receptors,** or *sense organs,* receive stimuli from the environment. When a change in the environment occurs, impulses begin in the receptor nerve cells.
2. *Sensory nerve cells* pick up impulses from receptor nerve cells in the sense organs and pass the impulses to the brain or spinal cord. In that area, the sensory nerve cells make contact with many other nerve cells.
3. *Motor nerve cells* carry impulses to muscles or glands. These organs then respond, that is, the muscles move or the glands secrete. Since their activities bring about an effect, muscles and glands are called **effectors.**
4. *Associative nerve cells* in the brain and spinal cord connect sensory nerve cells with motor nerve cells. (See Fig. 20-2.)

Synapse. In the region where the end brush of one nerve cell lies very close to the dendrites of another nerve cell, there is a tiny gap called a **synapse.**

Nerve Impulse. The impulse that passes along a nerve cell is an electrochemical signal. *Electrochemical* means the signal is sent by electrical impulses and by chemicals. The signal moves from the dendrites to the cell body, and then to the axon and end brush. When the impulse reaches the end brush of an axon, the nerve cell releases special chemicals. These

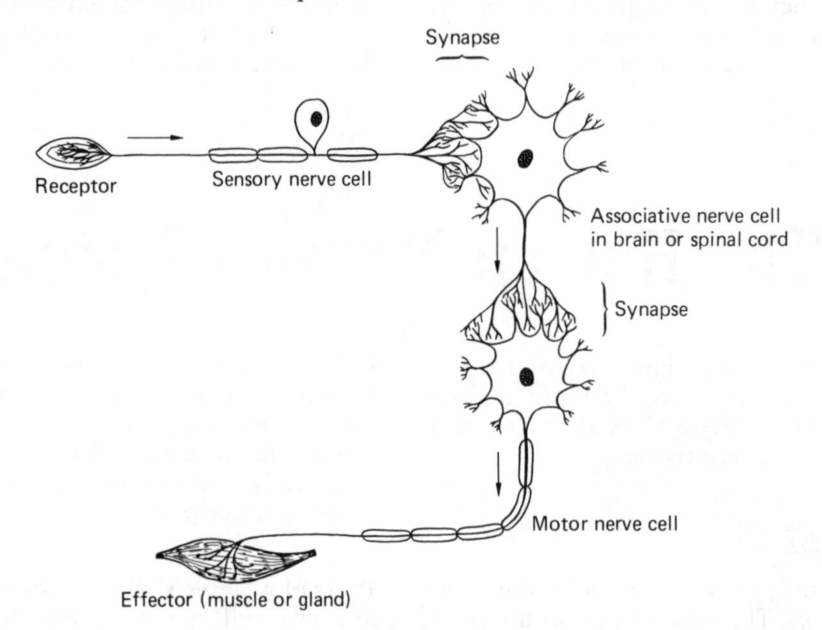

Fig. 20-2. How associative nerve cells connect sensory and motor nerve cells.

chemicals start the impulse up again in the dendrites of the nerve cell across the synapse gap.

Nerves and Ganglions. A **nerve** consists of a bundle of axons of several nerve cells that lie side by side. The nerve cells are arranged such that messages travel in only one direction along the nerve. A covering of connective tissue encloses the bundle. The cell bodies of associative nerves and motor nerves lie within the brain or spinal cord. The cell bodies of sensory nerves are located in clusters along the nerves, called **ganglions** (Fig. 20-3). Ganglions are located in the trunk and neck on either side of the spinal cord, and in the head, below the brain. All impulses that are transmitted to the brain and spinal cord are sent through the sensory nerve cells in the ganglions.

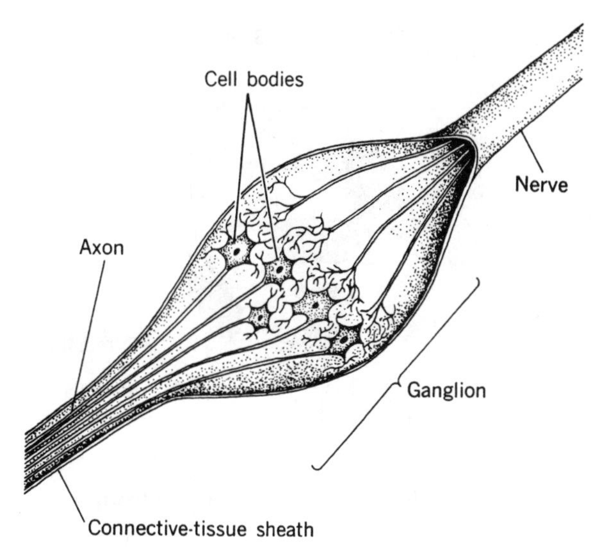

Fig. 20-3. A nerve and a ganglion.

The Central Nervous System

The **central nervous system** is composed of the *brain* and the *spinal cord.* Impulses are received by and sent out from the central nervous system to all parts of the body. The billions of nerve cells in the brain and spinal cord are organized into numerous interconnected *nerve centers,* or networks of nerve cells.

THE BRAIN

Three membranes, called **meninges,** tissue fluid, and the bones of the skull protect the large mass of delicate nerve cells that make up the **brain.** (Inflammation of the meninges of the brain or spinal cord is called *meningitis,* a serious disease.) There are three main sections to the brain: the *cerebrum, cerebellum,* and *medulla oblongata* (Fig. 20-4).

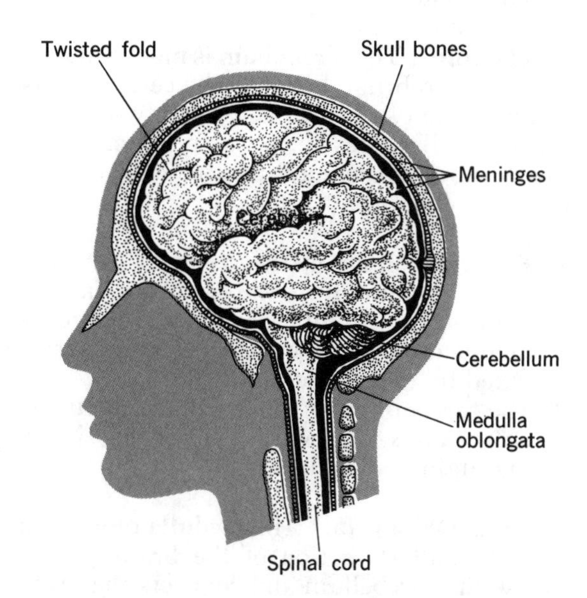

Fig. 20-4. The brain and its protective parts.

Cerebrum. The **cerebrum** is the largest section of the brain. It consists of two halves, called *cerebral hemispheres,* the bottoms of which are joined by millions of nerve fibers. There are two layers in the cerebrum. The outer layer consists mostly of cell bodies and synapses. This layer is often referred to as "*gray matter,*" because of the grayish color of the cell bodies. The outer layer of the cerebrum has many twisted folds, which increase its surface area. The inner layer of the cerebrum, consisting mainly of axons, is referred to as "*white matter,*" because of the whitish color of the fatty sheath around the axons. The cerebrum has three major activities:

1. The cerebrum interprets the impulses, or sensations, received from the sense organs. Nerves from the sense organs reach particular portions of the cerebrum, where the impulses register as sound, light, taste, or other sensations (Fig. 20-5).

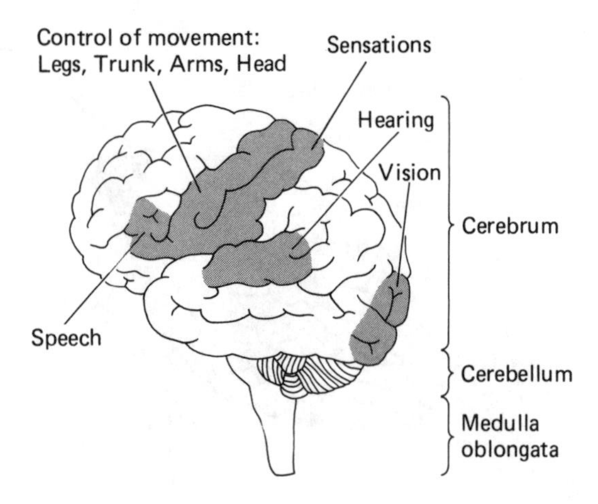

Fig. 20-5. Activity areas of the cerebrum.

2. The cerebrum is the center of consciousness and thinking. Such mental activities as reasoning, deciding, remembering, speaking, and learning are associated with special areas of the cerebrum.
3. The cerebrum controls all voluntary and learned acts.

Cerebellum. The **cerebellum** is much smaller than the cerebrum, but, like the cerebrum, is composed of two hemispheres and two layers. The cerebellum lies below the rear portion of the cerebrum. There are two major activities of the cerebellum:

1. The cerebellum is the pathway through which impulses from the cerebrum pass on their way to voluntary muscles. The movements of voluntary muscles are coordinated by the cerebellum.
2. The cerebellum, together with the *semi-circular canals* of the ear (see page 252), helps maintain body balance.

Medulla Oblongata. The **medulla oblongata** is the smallest section of the brain. It lies below the cerebellum and connects the brain with the spinal cord. The medulla has two major activities:

1. The medulla oblongata controls many automatic processes that are necessary to maintain life. It contains nerve centers that regulate such involuntary activities as breathing, the heartbeat, digestion, and glandular secretions.
2. The medulla oblongata regulates such actions as the contraction of the pupil and the secretion of saliva.

The medulla oblongata is the coordinating center of the autonomic nervous system. The *autonomic nervous system* regulates many involuntary essential activities.

THE SPINAL CORD

The **spinal cord** is the tubelike mass of nerve tissue that extends downward from the medulla. The bones of the spine (the vertebrae), tissue fluid, and three membranes (the meninges), which are continuations of the membranes around the brain, protect the spinal cord. Like the brain, the spinal cord has two layers. But, unlike the brain, the white matter is the outer layer and the gray matter, the inner. There are two major activities of the spinal cord:

1. The spinal cord connects the brain with nerves located in all parts of the body below the head.
2. The spinal cord controls many reflex acts. **Reflexes** are quick, automatic body movements, made in response to stimuli.

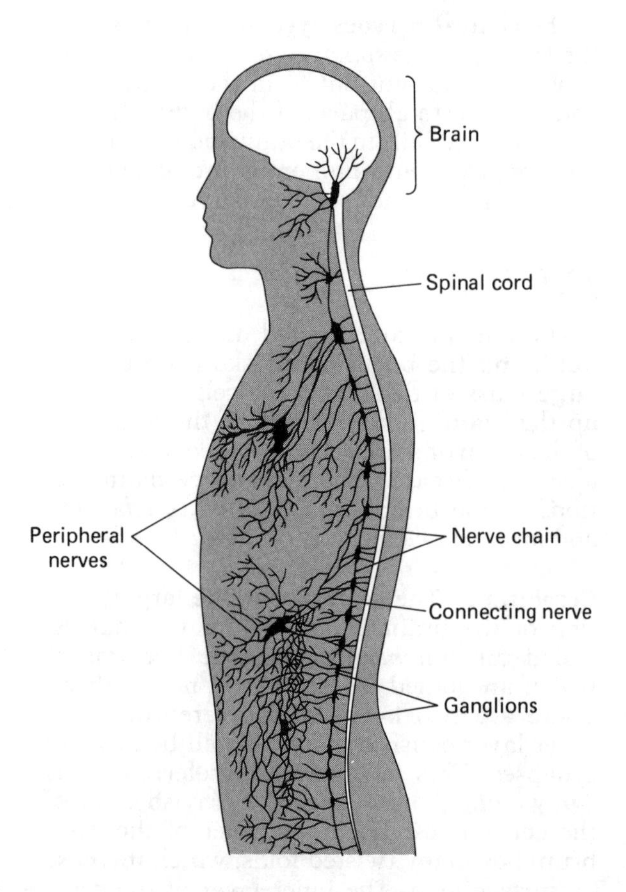

Fig. 20-6. Major parts of the nervous system.

The Peripheral Nervous System

The **peripheral nervous system** is made up of the numerous pairs of nerves that are located outside the brain and spinal cord. These nerves connect to all parts of the body, carrying impulses to and from the central nervous system. Some nerves connect the brain to the sense organs and internal organs. Other peripheral nerves connect the spinal cord to the skeletal muscles. (See Fig. 20-6.)

Behavior

The variety of responses that an organism makes to different stimuli is called its **behavior.** The stimuli may originate either outside or inside the organism's body. The way in which an organism behaves shows how the organism adjusts to changes in its environment. Behavior is controlled either by chemicals (hormones), a nervous system, or both.

Behavior in Simple Organisms

Tropism. Although they lack a nervous system, plants automatically respond to stimuli. This type of behavior is called a **tropism.** (See Fig. 2-5, page 19.) A tropism is the simplest type of behavior, involving only motion toward or away from a stimulus. In general, tropisms aid the survival of an organism. Tropisms are never learned. Instead, this type of behavior is inborn. Chemicals appear to be the controlling factor in most tropisms. You will learn more about tropisms in Chapter 22.

Taxis. Simple protozoans, like the paramecium, also respond to stimuli such as light, touch, and chemicals. The entire body of the unicellular organism moves toward or away from the stimulus in a response known as a **taxis.** Although lacking a nervous system, the organism has the inborn ability to respond in a way that aids its survival (Fig. 20-7).

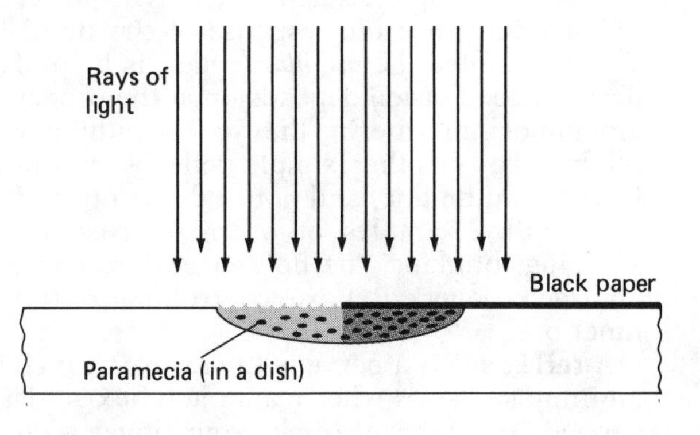

Fig. 20-7. An example of a taxis: paramecia move away from strong light.

Behavior in Complex Organisms and Humans

As you have learned, simple organisms without nervous systems respond to a stimulus only by moving toward or away from it. The simplest type of response that is controlled by a nervous system is known as a *reflex*. For example, the hydra, which has a very primitive nervous system, responds by reflex action to outside stimuli such as food and temperature changes. More complex animals, including humans, also have simple reflex responses. In addition, complex organisms exhibit other types of behavior that are more complicated than reflex responses. Human behavior varies from the relatively simple inborn type, carried out from birth, to complex types that require a great deal of time and effort to learn. Some types of human behavior are so complex, that they are still not well understood.

The major behavior types include the *simple reflex act*, the *conditioned reflex response*, *instinct*, and the *voluntary act*.

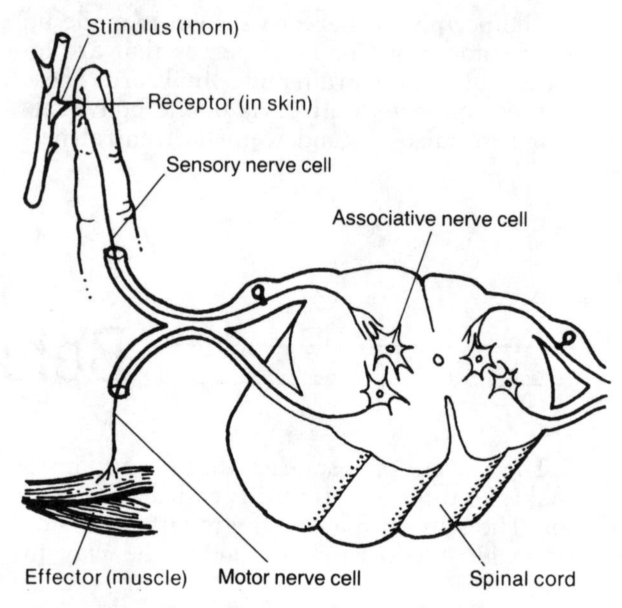

Fig. 20-8. The reflex arc of a simple reflex act.

Simple (or Unconditioned) Reflex Act. The **simple reflex act** is a beneficial, rapid, direct response to a stimulus. Reflex behavior is inborn. The response, which is either the contraction of a muscle or the secretion by a gland, never varies. It is automatically carried out without thought when a particular stimulus is present.

As you saw in your laboratory investigation, the response of your eye to bright light was a decrease in the size of your pupil. The response to removing the light was an increase in the size of your pupil. Exerting your will power did not influence the response of the pupil. This act, called the *pupillary reflex*, is helpful because good vision depends upon the proper amount of light entering the eye. The pupillary reflex, like all other simple reflexes, is unlearned, automatic, and not under control of your will. Examples of other reflexes are sneezing, blinking, the flow of gastric juice, and the knee jerk. Reflexes control many vital functions.

A **reflex arc** is a nerve pathway over which an impulse passes when a simple reflex act is carried out. If, for example, your finger accidentally touches a sharp pinpoint and, without thinking about it, you rapidly withdraw your hand, the impulse travels over a reflex arc, as shown in Fig. 20-8.

The sense organ (receptor), which in this case is a nerve ending in the skin, receives the stimulus and starts a nerve impulse.

A *sensory nerve cell* passes this impulse to the spinal cord. (Because the reflex arc travels through the spinal cord, rather than the brain, the response is extremely rapid.)

An *associative nerve cell* in the spinal cord relays the impulse to a motor nerve cell and also to the cerebrum. (The cerebrum is not part of the reflex arc, but it makes you aware of what is happening.)

The *motor nerve cell* passes the impulse to an effector, which may be either a muscle or a gland.

The *effector*, which in this case is a muscle of the arm, responds by contracting and pulling the hand away from the harmful stimulus.

Conditioned Reflex Response. This type of behavior, in which an inborn reflex is triggered by a new type of stimulus, is learned. When the nervous system links together two events that occur at nearly the same time, a **conditioned reflex response** is formed. The response that would ordinarily have been the reaction to one stimulus becomes the reaction to a second stimulus. The second stimulus becomes a substitute for the first one. Although forming a conditioned response usually requires repetition of the two stimuli, it does not necessarily require that one be conscious of what is going on.

Ivan Pavlov (1849–1936), a Russian scientist, experimented with the behavior of dogs. Pavlov showed that the dogs could be taught to change their behavior, that is, their behavior could be *conditioned.* He observed that when a hungry dog was shown food, its mouth "watered." This was a simple reflex act in which saliva was secreted in response to the presence of food, the stimulus. Just before presenting a hungry dog with food, Pavlov rang a bell. The dog would lift its head and ears in response to the sound of the bell. Pavlov repeated this procedure many times.

At first, the dog's mouth would water just when it was given food. In time, the dog's mouth watered shortly after the bell-ringing, even though it was not given any food. Pavlov concluded that the ringing of the bell had become linked in the dog's brain with the subsequent presence of food. In other words, the dog responded to the new stimulus (the bell-ringing) just as it had responded to the original stimulus (the presence of food). (See Fig. 20-9.)

The *conditioned-response pathway* is the nerve pathway that is established when conditioning occurs. Each reflex act has a separate reflex arc, or pathway, in the nervous system. From each reflex arc, associative nerve cells reach the cerebrum. Here an impulse can pass from one reflex arc to a second reflex arc. Then the impulse can pass along to the motor nerve cell of the second reflex arc. In the dogs of Pavlov's experiment, associative nerve cells in the cerebrum made it possible for the impulses from the sensory nerve cells of the ears to cross over to the motor nerve cells leading to the salivary glands.

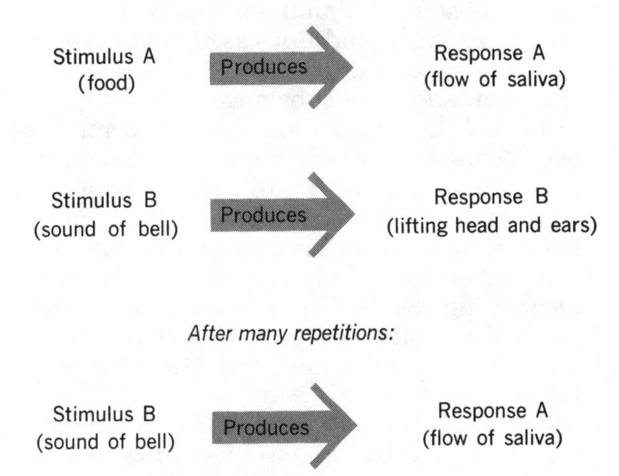

Fig. 20-9. Forming a conditioned reflex response.

Examples of conditioned responses in humans are: likes, dislikes, unexplainable fears, mouth-watering when thinking of a favorite food, and the linking of words with their meanings.

Instinct. An **instinct** is inborn behavior consisting of a chain of reflex acts. The completion of the first reflex act becomes the stimulus that starts the second reflex act. This in turn becomes the stimulus for the next reflex act in the chain. Eventually the act is completed. Though some learning of behavior occurs in animals, instincts help most animals get food, avoid danger, mate, rear young, and find shelter. Examples of instincts are nest-building by birds and web-spinning by spiders.

Voluntary Act. When an action or behavior is controlled by the will, it is called a **voluntary act.** Stimuli that produce voluntary responses often begin within the cerebrum. Voluntary behavior is conscious behavior and requires some kind of intelligence. This type of behavior is very complex, involving many nerve pathways and many connecting nerve cells in the cerebrum. Examples of voluntary acts include imagining, reasoning, making decisions, solving problems, and remembering. Such behavior makes it possible for humans to adjust to a greater variety of conditions than other organisms can.

Voluntary behavior can often be changed. Undesirable behavior patterns, such as eating with dirty hands, can be improved. Desirable behavior, such as good personal hygiene, can be acquired, or learned. Among the major methods of learning are trial and error, imitation, and forming habits.

Trial and Error. This type of learning takes place when a person tries an act, makes mistakes, tries again, and repeats the process until he or she learns the act correctly. A child who solves a jigsaw puzzle after several attempts has learned to fit the pieces together by trial and error.

Learning by Imitation. This type of learning takes place when a person attempts to do something by watching someone else and then trying to do it the same way. A child who is learning how to write does so by imitating the teacher.

Habit. Learned behavior that becomes an automatic response as a result of repetition is called a **habit.** In the early stages of acquiring a habit, the act is voluntary, conscious, and often difficult. After practicing, the act becomes involuntary, unconscious, easier, faster, and more skillful. Examples of habits are obeying traffic signals, proper penmanship, smoking cigarettes, and playing scales on a piano.

The *habit pathway,* like that of a conditioned response, must be established. When we decide to make an act habitual, the stimulus starts in the cerebrum. The nerve impulse then passes over pathways from the cerebrum to the proper muscles. With repeated use of the new pathway, the impulse crosses the synapses between nerve cells more easily. In time, voluntary control lessens, and the act is performed unconsciously and smoothly. Thus, for example, repetition of scales on a piano becomes an easy habit.

Habit Formation. Because habits are basically voluntary acts, good habits can be formed and bad ones broken. For example, the bad habit of cracking knuckles is initially voluntary, and can be stopped if a person consciously decides to do so.

Problem Behavior Associated With Drugs

Drugs that are helpful to sick people are extremely dangerous to healthy people when used without a doctor's advice. Especially dangerous are those drugs that act by speeding up or slowing down the nervous system. Such drugs often interfere with the nervous system to the extent that a person's behavior becomes abnormal and dangerous to himself and others. The most frequently *abused,* or improperly used, substances are listed below.

Marijuana and *hashish* are mild **hallucinogens** (drugs that affect sensory perceptions) that can cause a person to have a distorted sense of time, poor motivation, loss of short-term memory, and impaired reflexes. These drugs, which are usually smoked, can also cause lung damage.

LSD (or *lysergic acid diethylamide*) causes **hallucinations,** that is, the user sees things that do not exist. Hallucinations brought on by LSD can be frightening. Some users have been known to suffer long-term psychological problems.

Cocaine, which can be injected, smoked, or snorted, is a **stimulant.** It affects the central nervous system, causing it to speed up the body's activities. Repeated use of cocaine causes nervousness, sleeplessness, loss of appetite, and **addiction,** a state of drug dependency and regular use. Users of cocaine neglect eating, lose weight, and may experience irregular heartbeat, stomach problems, and mental problems. Abuse of cocaine can lead to death.

Crack is a powerful, smokable drug that is a purified form of cocaine. Crack is highly addictive and causes violent behavior in people who use it.

Speed and *amphetamines,* taken as pills or injected, stimulate the central nervous system, causing a decrease in appetite, and an increase in blood pressure and nervousness.

Nicotine interferes with normal heart action, digestion, and respiration. Nicotine, and other substances found in cigarette smoke, may cause such conditions as high blood pressure, ulcers, cancer, and heart disease. Nicotine is an addictive substance.

Alcohol decreases control of the cerebrum over behavior, thereby impairing speech, coordination, and reflex abilities. Alcohol is a type of **depressant** (slows down the central nervous system) that reduces inhibitions and can lead to violent behavior. In large amounts, alcohol injures the stomach, liver, and brain; alcohol can stop the activities of the brain, leading to death. Addiction to alcohol is known as *alcoholism.*

Opium and drugs derived from it (*heroin, morphine,* and *codeine*) are **narcotics** that relieve pain and cause drowsiness. Like cocaine, heroin is so addictive and dangerous that its manufacture has been prohibited by law. Heroin is usually injected and can lead to death by overdose.

Barbiturates and *tranquilizers* are addictive depressants (in pill form) that can cause depression, sleepiness, and impaired speech and coordination.

In large doses, many of these drugs can be fatal. The use of a drug in small, repeated doses becomes habitual, causing a need for more frequent and larger doses. In time, one becomes addicted to some drugs and then has great difficulty in overcoming the habit. The body cells become so dependent upon the drug that they cannot work normally. Lacking the drug, one suffers uncontrollable craving and great pain. Avoiding addiction is much easier than overcoming it.

Care of the Nervous System

Eat properly. For the nervous system to work normally, it requires a balanced, varied diet which includes such minerals as potassium and sodium, and such vitamins as thiamin.

Get plenty of rest. When the body is at rest, wastes are removed from nerve cells just as they are from all body cells.

Engage in some recreational activity. Switching activities from work to play or to a hobby relieves mental strains that interfere with normal nerve activities.

Avoid taking drugs without medical advice. Improperly used, drugs are damaging to the nervous system and a threat to your health and life.

Consult a physician regularly. The early discovery and treatment of disturbances of the nervous system can prevent physical and mental damage.

CHAPTER REVIEW

Science Terms

The following list contains all of the boldfaced scientific terms found in this chapter and the page on which each appears.

Matching Questions

On the blank line, write the letter of the item in column B which is most closely related to the item in column A.

Column A

____ 1. simplest behavior under chemical control

____ 2. controls voluntary, conscious acts

____ 3. a nerve pathway

____ 4. one stimulus substitutes for another

____ 5. simplest behavior under nerve control

____ 6. passes impulse to muscle or gland

____ 7. passes impulse to the spinal cord

____ 8. muscle or gland that responds to impulse

____ 9. regulates involuntary life activities

____ 10. learned behavior that becomes automatic

Column B

a. conditioned reflex response
b. effector
c. sensory nerve cell
d. a reflex
e. a tropism
f. medulla oblongata
g. cerebrum
h. a habit
i. cerebellum
j. reflex arc
k. motor nerve cell

Multiple-Choice Questions

On the blank line, write the letter preceding the word or expression that best completes the sentence or answers the question.

1. The nervous system controls responses to
a. chemical stimuli only
b. external and internal stimuli
c. external stimuli only
d. internal stimuli only — 1 ____

2. A change in the environment of a cat is called a
a. stimulus *b.* response *c.* food chain *d.* prey — 2 ____

3. The basic unit of the nervous system tissue is the
a. axon *b.* cell body *c.* nerve cell *d.* impulse — 3 ____

4. The nucleus of a nerve cell is located in the part of the cell called the
a. axon *b.* cell body *c.* neuron *d.* end brush — 4 ____

5. The axon of a nerve cell
a. receives impulses from nearby nerves
b. sends impulses to other nerve cells
c. sends impulses to the cell body
d. receives impulses from the end brush — 5 ____

6. Impulses pass over the gap of a synapse with the aid of
a. fibers *c.* ganglions
b. enzymes *d.* special chemicals — 6 ____

7. Nerve cells that carry impulses from sense organs to the brain are called
a. sensory *b.* motor *c.* receptors *d.* effectors — 7 ____

8. Which type of nerve cell causes glands to secrete?
a. sensory *b.* connecting *c.* motor *d.* receptor — 8 ____

9. Synapses occur between
a. cell bodies and dendrites
b. cell bodies and axons
c. end brushes and dendrites
d. cytons and end brushes — 9 ____

10. A fatty sheath covers the part of a nerve cell called the
a. end brush *b.* dendrites *c.* axon *d.* cyton — 10 ____

11. Clusters of cell bodies that are located along sensory nerves are called
a. ganglions *b.* cytons *c.* axons *d.* synapses — 11 ____

12. A nerve impulse that leaves the eye is eventually interpreted by the
a. optic nerve *b.* cerebrum *c.* cerebellum *d.* spinal cord — 12 ____

13. The part of the brain that controls the breathing rate of humans is called the
a. cerebrum *b.* cerebellum *c.* gray matter *d.* medulla oblongata — 13 ____

14. Fear of mice in humans is an example of a
a. voluntary response
b. voluntary act
c. conditioned response
d. habit — 14 ____

15. The nerve pathway carrying an impulse that causes a reflex act is called
a. voluntary arc *b.* synapse *c.* reflex arc *d.* instinct — 15 ____

16. Which is an example of instinctive behavior?
a. a robin brings a worm to its nest
b. the pupillary reflex
c. breathing faster when running
d. the solving of a maze by a mouse — 16 ____

17. A type of behavior that results from the linking in the brain of impulses from two different stimuli is called
a. instinct
b. conditioned response
c. stimulated response
d. simple response — 17 ____

18. The organ in which conditioned responses are established is the
a. spinal cord *b.* cerebrum *c.* medulla *d.* ganglion — 18 ____

19. Which group contains *only* voluntary acts?
a. mouth-watering, fear of snakes, blinking
b. iris movement, sneezing, coughing
c. breathing, coughing, clapping
d. reasoning, remembering, deciding — 19 ____

20. Although ants seem to act cleverly, their responses do not appear to involve
a. instincts *b.* reflexes *c.* nerve cells *d.* thought — 20 ____

21. Acts controlled by the will start with impulses in the
a. cerebellum *b.* cerebrum *c.* medulla *d.* spinal cord — 21 ____

22. When an activity is learned by watching and copying somebody else it is called
a. instinct
b. imitation
c. addiction
d. habit formation — 22 ____

23. Habits are considered to be
a. learned automatic acts
b. inborn reflexes
c. conditioned reflexes
d. acts of willpower only — 23 ____

24. The decision to form a good habit starts a nerve impulse in the
a. nerve cell of a ganglion
b. cerebellum
c. medulla oblongata
d. cerebrum — 24 ____

25. Cough medicine may contain drugs, such as codeine and alcohol, that can be addicting if taken for prolonged periods. When a cough persists, the best procedure is to

a. use any advertised remedy
b. consult your physician
c. ask your friends for advice
d. just stay in bed until the cough goes away

25 ____

Modified True-False Questions

In some of the following statements, the italicized term makes the statement incorrect. For each incorrect statement, write the term that must be substituted for the italicized term to make the statement correct. For each correct statement, write the word "true."

1. During exercise, an increase in your heartbeat is controlled by your *cerebellum*. 1 ____________

2. The flow of a glandular secretion is a type of *voluntary* act. 2 ____________

3. The pairs of nerves that lie outside the brain and spinal cord, carrying impulses to and from them, make up the *peripheral nervous system*. 3 ____________

4. When you go outside during the day and look at the sky, the size of your pupils *increases*. 4 ____________

5. A child born in China speaks Chinese well because it is a *reflex* behavior. 5 ____________

6. The spinal cord is composed of a tubelike mass of *bone* tissue. 6 ____________

7. The three membrane layers covering the brain are known as the *meninges*. 7 ____________

8. When an owl catches a mouse, it is an example of *conditioned* behavior. 8 ____________

9. When a paramecium turns toward a stimulus, its response is called a *taxis*. 9 ____________

10. Drugs that cause the central nervous system to speed up the body's activities are known as *depressants*. 10 ____________

Testing Your Knowledge

1. Explain the difference between the two terms in each of the following pairs:

a. sensory nerve cell and motor nerve cell ________________________________

____

b. receptor nerve cell and associative nerve cell

c. gray matter and white matter

d. reflex and instinct

e. habit and simple reflex

f. reflex arc and reflex act

g. stimulant and depressant

h. nerve and nerve cell

i. axon and dendrite

2. In what ways are simple reflex, habit, and conditioned response alike? Different?

3. *a.* Give three examples of human reflex acts.

b. Give at least three characteristics common to all simple reflex acts.

4. Name the three main divisions of the human brain and list the major uses of each.

a.

b.

c.

5. A person watching a television program is frightened by what he sees. His heart begins to beat faster. During this time, stimuli are received and nerve impulses start and move over a definite pathway. Trace the course of the stimulus and impulses by rearranging the following body parts in a proper sequence: cerebrum; eye; autonomic nervous system; optic nerve; medulla oblongata; cerebellum; heart muscle.

6. Describe the harmful effects of each of the following drugs on the body or the personality:

a. heroin ________________

b. speed ________________

c. alcohol ________________

d. nicotine ________________

e. marijuana ________________

f. cocaine/crack ________________

7. List the functions of the following:

a. skull ________________

____

b. meninges ________________

____

c. spinal column ________________

____

d. spinal fluid ________________

____

8. Describe the structure and function of the spinal cord. ________________

____

9. Identify the nerves that make up the peripheral nervous system. ________________

____

10. Explain the difference between innate behavior and acquired behavior. ________________

____

11. How does a reflex arc work (in the human nervous system)? ________________

____

21

The Sense Organs

LABORATORY INVESTIGATION

TESTING YOUR SENSE OF TASTE AND SENSE OF SMELL

A. Your teacher will supply you with slices of apple, pear, onion, and lemon. Work with a partner to perform this investigation.

1. Have your partner close his or her eyes. Place a piece of fruit in his/her mouth, while holding a different piece of fruit under the nose. Ask your partner to describe the smell and the taste. Record your partner's responses:

Smell ______________________________

Taste ______________________________

2. Repeat this process in reverse, holding the fruit that was tasted under the nose, and placing the other fruit in your partner's mouth. Again, record your partner's responses:

Smell ______________________________

Taste ______________________________

3. Repeat this process, this time placing a piece of fruit in your partner's mouth, while holding a piece of the same fruit under the nose. Do this procedure for each of the fruits used in Steps 1 and 2. Record your partner's responses:

Smell/Taste ______________________________

Smell/Taste ______________________________

B. Now vary this procedure, using different combinations of the pieces of apple, pear, onion, and lemon.

4. Record your partner's responses: ______________________________

C. Repeat Steps 1–4, with your partner testing and recording your responses to the different tastes and smells.

D. Answer the following questions, based on your findings.

5. Which was usually stronger—the sense of taste or smell? ______________________

6. Does the type of food item being smelled or tasted affect which sense is stronger? ______

Explain: ____

7. Does the sense of smell affect the sense of taste? ______________________.

Explain: ____

8. How many kinds of tastes could you identify? Describe them: ______________________

(**Caution:** Be sure to wash hands thoroughly before and after this lab investigation.)

The Sense Organs

The organs that are especially sensitive to stimuli are the **sense organs** (or *receptors*). Five well-known sense organs are the eyes, ears, nose, tongue, and skin. In addition to these, there are other organs that receive stimuli from inside the body. Among these internal receptors are the nerve endings in the stomach that make us aware of hunger and those in the throat that make us aware of thirst.

Each sense organ can detect only one type of stimulus, for example, light, or sound, or chemicals, and so on. When a sense organ detects a stimulus, a nerve impulse passes over the sensory nerve from the sense organ to the special region for sensations in the cerebrum. (Refer back to Fig. 20-5.) Then, the cerebrum sends impulses over motor nerve cells to muscles or glands. These organs carry out the responses that help the body adjust to changes in its environment. For example, extreme heat may be harmful to the body. When heat stimuli are received by the skin, impulses eventually reach the sweat glands, which then produce increased quantities of perspiration. The general effect of this reaction is to cool the body.

EYES

The eyes are nearly spherical in shape. Muscles are attached to the eyes and to bony sockets in the skull. These muscles enable the eyes to be moved in all directions. As shown in Fig. 21-1, the center of the front of the eye is clear and colorless. This part of the eye is called the **cornea.** A short distance behind the cornea is

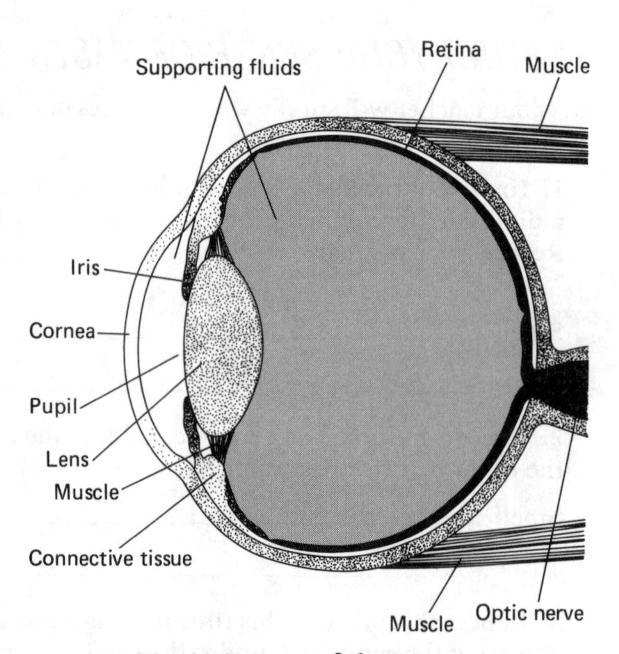

Fig. 21-1. Parts of the eye.

a circular, colored band, the **iris.** In the center of the iris is a dark circular opening, the **pupil.**

Immediately behind the iris and pupil is a clear, elastic structure, the **convex lens.** Muscle tissue and a tissue composed of fibers hold the lens in place. The **retina** is a thin layer of nerve cells that coats almost all the inner surface of the eyeball. The nerve cells of the retina are especially sensitive to light. There are two types of retinal nerve cells: **rods** and **cones.** Rods function to distinguish different shapes in dim light. Cones are sensitive to light of different colors. The retina is connected to the **optic nerve,** which in turn is connected to the cerebrum. Clear fluids fill the internal chambers of the eye and help maintain its shape.

Vision. Light rays coming from an object pass through the cornea, enter the pupil, and reach the lens. The lens bends the light rays, which then form an upside down image on the retina. This image is formed in the same way that a camera lens forms an image on film. The image on the retina produces impulses that reach the brain by way of the optic nerve. When the impulse arrives at the special area of the cerebrum that controls vision, it is reversed to appear right side up again, and you become aware that you see something.

The iris, which is composed of two sets of smooth muscles, regulates the amount of light that reaches the lens and retina. One set of muscles is arranged in a ring around the pupil. In bright light, this set contracts as the second set relaxes. These actions decrease the size of the pupil. The second set of muscles is arranged around the pupil like the spokes of a wheel. In dim light, this set of muscles contracts at the same time that the ringlike set relaxes. These actions enlarge the size of the pupil and let more light into the eye.

The muscle attached to the lens can change the shape of the lens. When you look at either a nearby or distant object, this muscle automatically changes the thickness of the lens and focuses a clear image on the retina.

Common Defects in Vision. When an image comes to a sharp focus on the retina (Fig. 21-2*a*), vision is clear. People who have certain defects in the shape of parts of the eye cannot see sharp images without the aid of eyeglass lenses or contact lenses that correct these defects.

Nearsightedness usually occurs when the eyeball is longer than normal. In such cases, a sharp image is produced in front of the retina instead of on its surface. Nearsighted people can see only nearby objects clearly. Concave lenses, which spread light, help correct this condition (Fig. 21-2*b*).

Farsightedness usually occurs when the eyeball is shorter than normal. In such instances, a sharp image cannot be produced on the retina. Farsighted people can see only distant objects clearly. Convex lenses, which gather light rays to a point, help correct this condition (Fig. 21-2*c*).

Astigmatism occurs when either the cornea or the lens is unevenly curved. In this condition, a hazy image is formed on the retina.

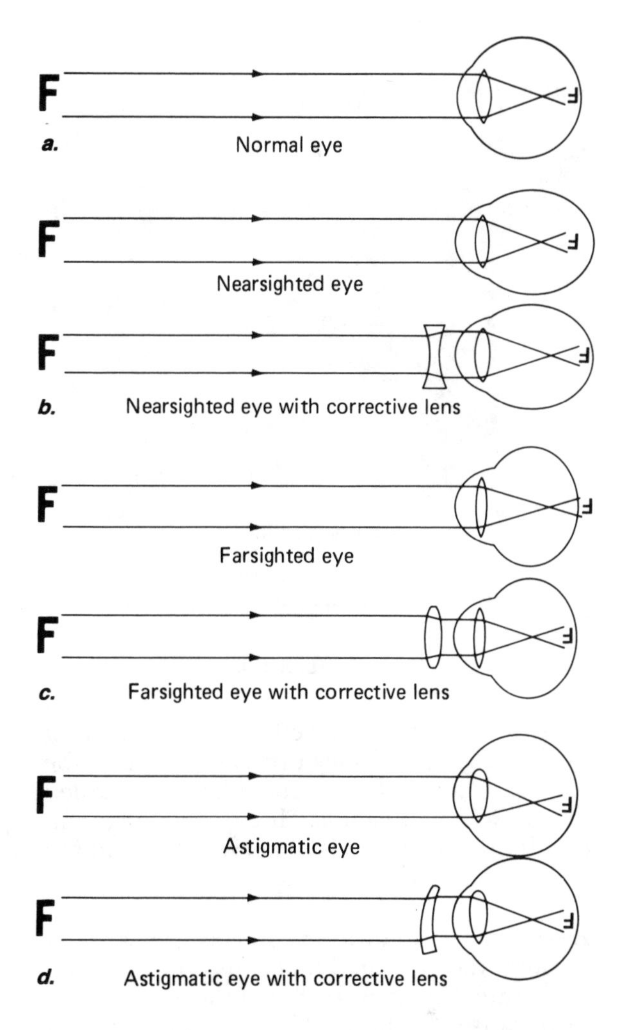

Fig. 21-2. Correcting poor vision.

Eyeglass lenses with curves that make up for the uneven curves of the cornea or the lens are used to correct astigmatism (Fig. 21-2*d*).

EARS

Each ear is divided into three regions: the **outer ear,** the **middle ear,** and the **inner ear.** Both the middle and the inner ear are located within a hollow portion of the skull bones (Fig. 21-3).

The outer ear consists of the fleshy projection from the side of the head and a hollow tube, called the **ear canal,** that leads inward.

The middle ear consists of the **eardrum,** which is a membrane; three little bones; and the **Eustachian tube,** which is a passageway leading to the throat.

The inner ear consists of a coiled, bony tube called the **cochlea.** It is lined with a membrane and filled with fluid. In the membrane are the

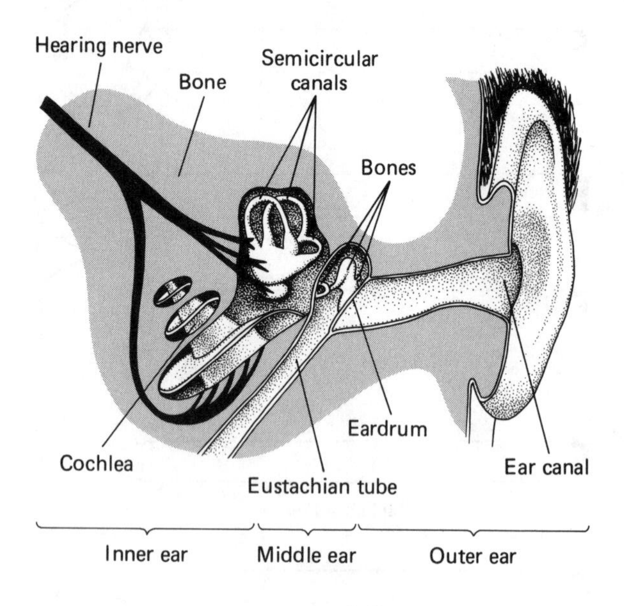

Fig. 21-3. Parts of the ear.

ends of sensory nerve cells from a branch of the hearing or **auditory nerve.** Attached to the cochlea is a set of three tubelike loops, called the **semicircular canals.** These canals also contain fluid and the ends of nerve cells from another branch of the auditory nerve.

Hearing. Sound waves entering the ear pass through the ear canal and cause the eardrum to vibrate. The movements of the eardrum, in turn, cause the three little bones to vibrate at the same speed as the eardrum. These bones then transfer the vibrations to a membrane in the cochlea and cause its fluid to vibrate. The motion of the fluid starts impulses in the nerve endings. These endings then pass the impulses to the hearing area of the cerebrum by way of the auditory nerve. When the impulse reaches the hearing area of the cerebrum, you become aware of the sound.

The Eustachian tube acts as a channel that equalizes the air pressure inside the middle ear cavity with the outside air pressure. Unequal air pressure interferes with hearing. When the eardrum is torn or the three little bones cannot move, hearing may be lost. This can occur even though there is no damage to the hearing nerve.

Maintaining Balance. Changes in the position of the head are detected by the semicircular canals. Bending or turning the trunk or head causes the fluid within these canals to move. This motion starts impulses in the nerve endings present in the canals. When these impulses reach the brain, return impulses may be sent to the muscles that can restore the body to its normal position, or help maintain its sense of balance.

SKIN

As you learned in Chapter 18, the dermis of the skin contains sensitive nerve endings. There are five different types of nerve endings, each of which can detect a different stimulus. Thus, one type of nerve ending is sensitive to heat, another to cold, and others to pressure, pain, and touch (Fig. 21-4). Although these nerve endings are present all over the body, there are more of them in the fingertips than anywhere else.

NOSE

Nerve endings located in the mucous membrane of the upper region of the *nasal passages* can detect odors. These nerve endings are connected to the cerebrum by way of the nerves

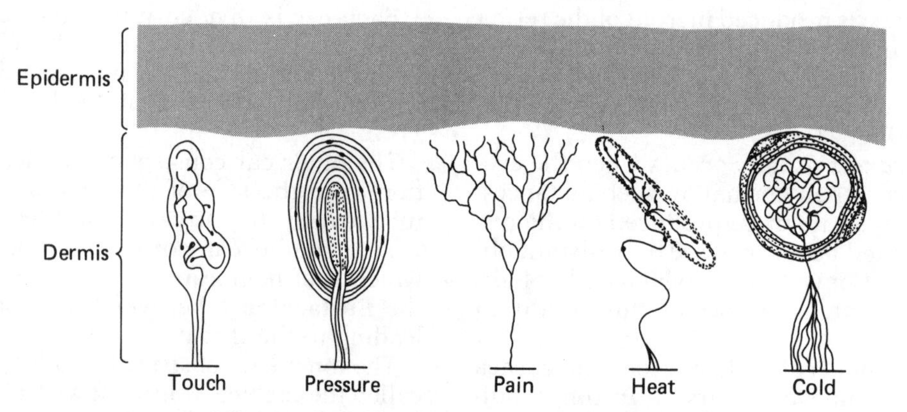

Fig. 21-4. The five kinds of nerve endings in the skin.

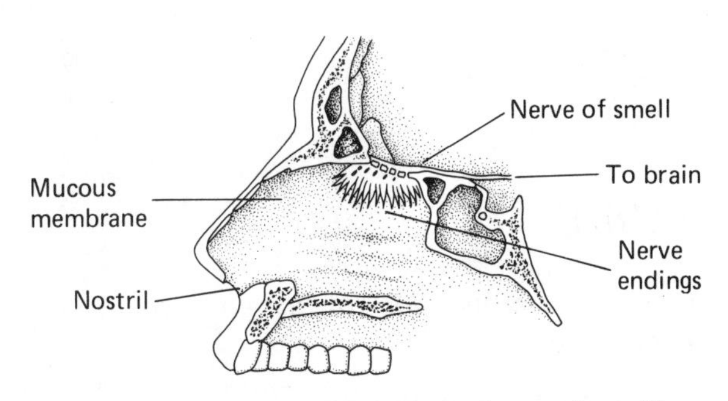

Fig. 21-5. Location of nerve endings of smell.

of smell (Fig. 21-5). Odors are actually chemical molecules in the air, which are breathed in by the nostrils. The senses of smell and taste are often combined while you eat.

TONGUE

The sense organ for taste is the tongue. Groups of special cells, called **taste buds,** are located in pores on the surface of the tongue. There are several types of taste buds, one of which is shown in Fig. 21-6. Each type of taste bud can detect a different taste.

Taste buds occur in groups in different areas of the tongue: for sweetness at the tip; sourness on the sides; bitterness at the back; and saltiness on the sides and tip.

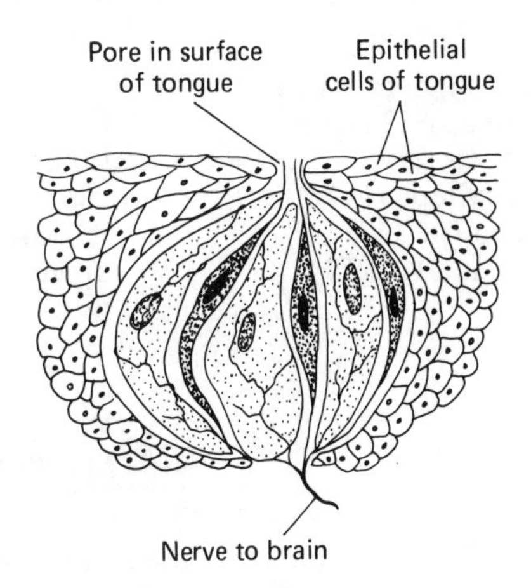

Fig. 21-6. A taste bud.

Care of the Sense Organs

Eyes

Prevent eyestrain by avoiding glare and by working and reading in well-lit places.

To rest your eyes, stare into the distance. This relaxes the lens muscle.

Avoid staring into the sun because ultraviolet rays destroy the delicate nerve cells in the retina.

Keep your eyes clean to avoid irritation and infection.

Have your vision checked regularly.

Prevent night blindness by including vitamin A in your diet.

Ears

Clean your ears with a washcloth. Avoid using sharp objects that can puncture the eardrum.

Keep both mouth and nostrils open when blowing your nose. This will prevent forcing mucus into the Eustachian tubes.

Avoid frequent exposure to extremely loud noises and loud music, since they can damage your hearing.

Tongue

Since extremes of temperature injure taste buds, avoid eating extremely hot or cold foods.

Avoid the use of alcohol and tobacco because these substances dull the sensitivity of the taste buds (in addition to harming other organs).

Nose

Nasal disorders should be promptly treated because obstruction of the nasal passages interferes with breathing and smelling.

Avoid the prolonged use of nose drops and similar medications because they injure the mucous membrane of the nose.

Skin

Wash regularly to remove dead skin cells and to keep the skin clear and free of infection.

Avoid excessive exposure to ultraviolet rays because they burn and destroy tissue, and may cause skin cancer.

CHAPTER REVIEW

Science Terms

The following list contains all of the boldfaced scientific terms found in this chapter and the page on which each appears.

auditory nerve (p. 252)
cochlea (p. 251)
cones (p. 250)
convex lens (p. 250)
cornea (p. 250)
ear canal (p. 251)
eardrum (p. 251)
Eustachian tube (p. 251)
inner ear (p. 251)
iris (p. 250)
middle ear (p. 251)
optic nerve (p. 250)
outer ear (p. 251)
pupil (p. 250)
retina (p. 250)
rods (p. 250)
semicircular canals (p. 252)
sense organs (p. 250)
taste buds (p. 253)

Matching Questions

On the blank line, write the letter of the item in column B which is most closely related to the item in column A.

	Column A	*Column B*
____	1. clear, colorless front of eye	*a.* retina
____	2. colored, circular band of eye	*b.* ear canal
____	3. dark opening in center of eye	*c.* Eustachian tube
____	4. layer of nerve cells inside the eye	*d.* iris
____	5. connects retina with the cerebrum	*e.* cochlea
____	6. hollow tube in outer ear	*f.* cornea
____	7. vibrating membrane in middle ear	*g.* taste buds
____	8. coiled bony canal in inner ear	*h.* nasal passages
____	9. special cells found on the tongue	*i.* pupil
____	10. nerve endings detect odors here	*j.* eardrum
		k. optic nerve

Multiple-Choice Questions

On the blank line, write the letter preceding the word or expression that best completes the sentence or answers the question.

1. When light enters the eye, the first part of the eye that the light passes through is the
a. retina *b.* pupil *c.* cornea *d.* lens 1 ____

2. The eye structure that regulates the amount of light that reaches the retina is the
a. iris *b.* pupil *c.* cornea *d.* lens 2 ____

3. As you read this sentence, an upside-down image of the print is formed in the eyes on the
a. optic nerve *b.* retina *c.* cerebrum *d.* lens
3 ____

4. Your sense of balance is maintained by impulses from nerve cells in your
a. bones *b.* tongue *c.* skin *d.* ears
4 ____

5. The Eustachian tubes are passageways that
a. connect to the semicircular canals
b. connect the nasal passages to the throat
c. connect the middle ear to the throat
d. connect the cochlea to the auditory nerve
5 ____

6. The vision defect that occurs when a hazy image is formed is known as
a. farsightedness *c.* astigmatism
b. nearsightedness *d.* shortsightedness
6 ____

7. Vibrations from sound waves are passed to a membrane in the cochlea by the
a. auditory nerve *c.* semicircular canals
b. three little bones *d.* eardrum only
7 ____

8. Impulses are started in the sensory nerve cells of the cochlea as a *direct* result of vibrations in the cochlea's
a. canals *b.* fluid *c.* membrane *d.* bones
8 ____

9. Air pressure is equalized between the outer ear and middle ear by the
a. Eustachian tube *c.* auditory nerve
b. semicircular canals *d.* cochlea
9 ____

10. Sound impulses are passed to the hearing area of the brain by the
a. optic nerve *c.* nerves of smell
b. auditory nerve *d.* dermis nerve endings
10 ____

11. How many types of nerve endings are found in the skin?
a. two *b.* three *c.* four *d.* five
11 ____

12. The nerve endings in the skin are sensitive to the following:
a. pressure, touch, temperatures, pain
b. pressure, pain, taste, touch
c. pain, temperatures, smell, touch
d. touch, taste, temperatures, sound
12 ____

13. The nerve endings of the skin are found in its
a. epidermis *b.* dermis *c.* oil glands *d.* follicles
13 ____

14. The nerve endings that detect odor are located in the
a. cartilage tip of the nose
b. openings of the nostrils
c. mucous membrane in nasal passages
d. nerves of smell in the cerebrum
14 ____

15. Taste buds are located in the
a. lower dermis of the tongue
b. pores on the surface of the tongue
c. roof of the mouth and gums
d. nasal passages' membrane
15 ____

Modified True-False Questions

In some of the following statements, the italicized term makes the statement incorrect. For each incorrect statement, write the term that must be substituted for the italicized term to make the statement correct. For each correct statement, write the word "true."

1. The retina is connected to the brain by means of the *auditory* nerve. 1 ______________
2. In dim light the size of the *cornea* of the eye increases. 2 ______________
3. The inner end of the ear canal is covered by a vibrating membrane called the *Eustachian tube*. 3 ______________
4. The hearing nerve endings are found inside the *cochlea*. 4 ______________
5. You know that you have heard a sound when impulses from your ear reach the *cerebellum*. 5 ______________
6. Hearing depends, in part, on the ability of three little bones in the middle ear to *vibrate*. 6 ______________
7. The ability to sense heat, cold, and pressure on the surface of the body depends upon different nerve endings in the *ears*. 7 ______________
8. Nerve endings that enable you to taste sugar are located at the *tip* of the tongue. 8 ______________
9. Taste buds that enable you to taste sour lemon are located at the *back* of the tongue. 9 ______________
10. The body's balance is maintained by impulses from the *Eustachian* canals. 10 ______________

Testing Your Knowledge

1. Explain the difference between the two terms in each of the following pairs:

   a. rods and cones ____

   ____

   b. convex lenses and concave lenses ____

   ____

   c. optic nerve and auditory nerve ____

   ____

   d. ear canal and semicircular canals ____

   ____

2. Tell the cause and correction for each of the following eye defects:

   a. astigmatism ____

   ____

b. farsightedness ______

______

c. nearsightedness ______

______

3. When a person listens to music, the sound waves are first received by the outer ear; then the vibrations and nerve impulses start and move over a definite pathway. Trace the course of the stimulus, vibrations, and impulses by rearranging the following body parts in a proper sequence: auditory nerve; three little bones; cochlea membrane and fluid; eardrum; cerebrum hearing area; ear canal; cochlea sensory nerve endings.

______

______

4. Each sense organ detects one type of stimulus, such as sound or chemicals. Based on what you know about the tongue and the nose, explain why you think the senses of smell and taste often combine.

______

______

5. Why would eyeglasses be necessary to correct the vision of a farsighted person?

______

______

6. If a hearing aid could not help someone with hearing problems, what might be another choice?

______

______

7. When a surgeon reduces the size of a person's nose (cosmetic surgery) would the sense of smell be altered?

______

______

Hormones and Chemical Control

LABORATORY INVESTIGATION

RESPONSE AND GROWTH IN PLANTS

A. In each of four paper cups, or other containers, plant three radish seeds. Place all of these plantings in a dark closet and water them regularly. When the seedlings have reached a height of about 2.5 cm, remove them from the closet.

1. Describe the appearance of the seedlings immediately after removal from the closet. ____

____

B. Place one container near a one-sided source of light. Set another container in a place where the seedlings will be lit, or *illuminated,* evenly from all sides. Place the third container on its side, making sure that these seedlings also will be lit evenly from all sides. Place the fourth container on its side in a dark closet. (Fig. 22-1.) Continue to water the plants as necessary and observe them for one week to note any changes in the direction of their growth.

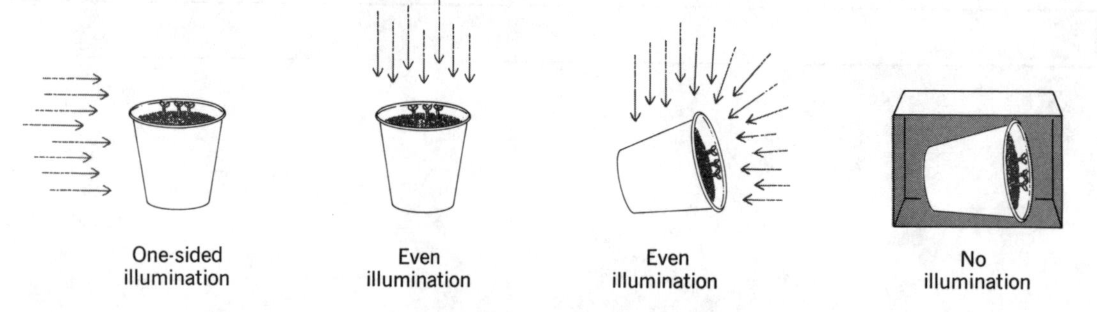

Fig. 22-1

2. In the following spaces, describe the growth of the plants in each of the four conditions, naming the stimulus (or stimuli) to which the plants responded:

a. After exposure to one-sided illumination. ______________________________

____

b. After exposure to uniform (even) illumination. ______

______

c. After having been placed on a side and exposed to uniform illumination. ______

______

d. After having been placed on a side in a dark closet. ______

______

3. To what stimuli did the four groups of radish stems respond? ______

______

Responses Controlled by Chemicals

Your observations of the radish plants show that these plants respond (as animals do) to such stimuli as light and gravity. However, plants respond much more slowly than animals do. Although plants and many simple organisms lack a nervous system, they do detect stimuli and respond to them. Since the responses of these organisms are automatic and usually beneficial, it is apparent that some control system exists.

The control system of plants depends upon certain chemicals, called **hormones,** which are produced within the cells. Some hormones are made near the top of the stem. Plant hormones control such activities as the production of flowers, ripening of fruits, dropping of leaves, and other responses to stimuli.

Chemical Control in Plants: Auxins and Tropisms

Your laboratory investigation showed that the stem of a plant turns toward light and away from the pull of gravity. The root turns toward the pull of gravity. As you learned in Chapter 20, this type of behavior, in which part of a plant moves in response to some stimulus, is called a *tropism.* These tropisms, or movements, are controlled by growth hormones, called **auxins.** Although no special gland is visible, these hormones are produced near the tip of each branch.

Auxins stimulate cell growth and thereby control the movement of some plant structures. Uneven distribution of auxin causes uneven growth. For example, when one side of a plant has more auxin than the other side, the cells grow faster on the side with more auxin. The resulting rapid, one-sided growth forces the stem tip to curve. Since light somehow reduces the quantity of auxin in cells directly exposed to it, cells in the light grow more slowly than cells in the shade. As a result, the more rapid growth and elongation of cells on the shaded side forces the stem tip to bend toward the source of light (Fig. 22-2). The radish seedlings in your laboratory work showed

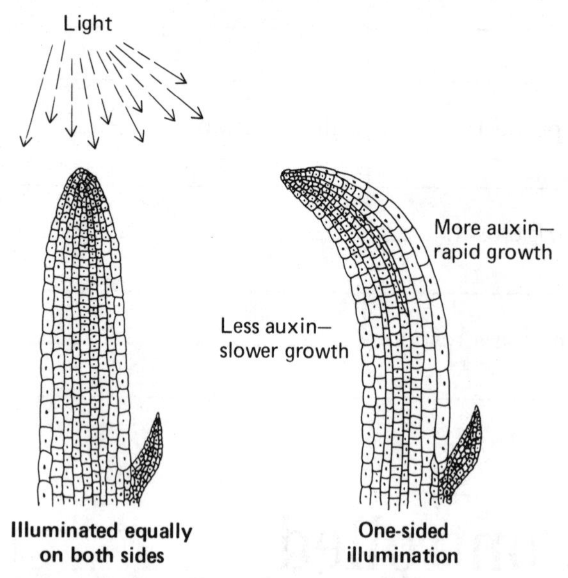

Fig. 22-2. The effect of auxin and light on stem growth.

this effect. The tropisms (growth and turning) of roots and other plant parts are also under chemical control similar to that in the stem.

Kinds of Tropisms. Each tropism is named according to the type of stimulus that causes the response. If the response movement is in the direction toward the stimulus, the tropism is said to be *positive;* if it is in the direction away from the stimulus, *negative.* Some of the different kinds of tropisms are as follows:

Phototropism is the response to light. The leaves of an ivy plant turn toward light, showing positive phototropism.

Geotropism is the response to the pull of gravity. The roots of most plants grow downward, showing positive geotropism. Most stems grow upward, showing negative geotropism (Fig. 22-3).

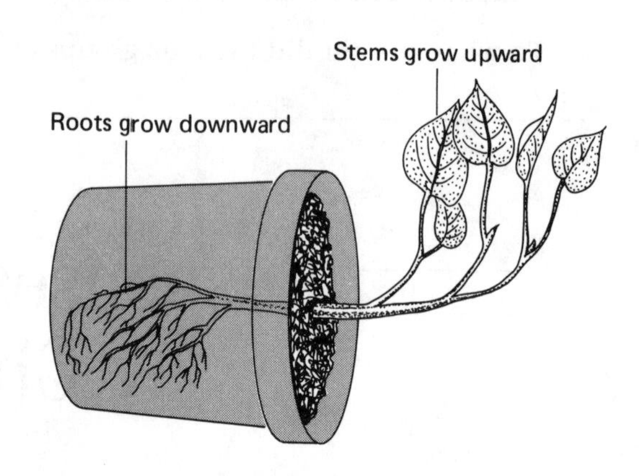

Fig. 22-3. Two tropisms in response to the pull of gravity.

Other tropisms include *thermotropism,* the response to heat; *chemotropism,* the response to chemicals; *hydrotropism,* the response to water; and *thigmotropism,* the response to contact with a solid object.

Chemical Control in Humans: The Endocrine System

In humans, some activities are controlled by hormones, and some by both hormones and the nervous system working together. The body's hormones and other chemicals are secreted by special glands.

A gland, such as the liver, that secretes a juice into a tube leading to another organ is called a **duct gland.** In addition to this type of gland, the body has numerous glands that have no tubes (Fig. 22-4). These glands are called **ductless glands,** or **endocrine glands.** The major ductless glands are shown in Fig. 22-5. Each of these glands is composed of cells that secrete a special hormone. Each hormone diffuses from its ductless gland directly into the bloodstream, which transports the hormone throughout the body. The ductless glands and their hormones make up the body's **endocrine system.**

A hormone acts on parts of the body that are at some distance from the ductless gland that secretes it. For this reason, hormones are often called "chemical messengers." Along with the nervous system, hormones help control and

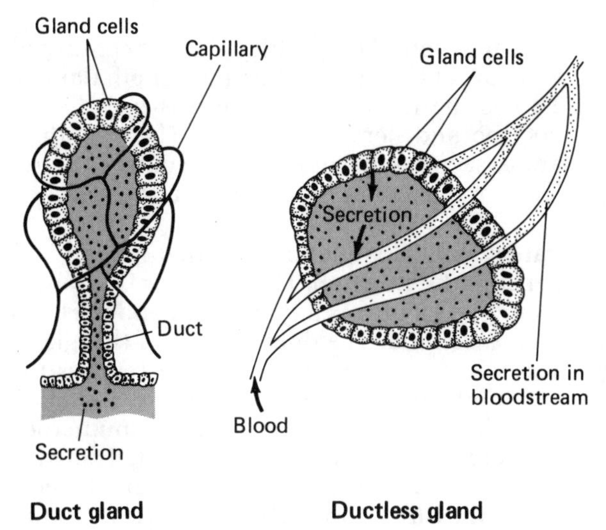

Fig. 22-4. A duct gland and a ductless gland.

coordinate the activities that go on inside the body. Hormones regulate growth and development, and help control metabolism. The hormones exert their control by either speeding up or slowing down the activities of different organs.

When a ductless gland releases too much or too little of a hormone, some organs fail to work normally. Such disturbances of these glands often result in disease.

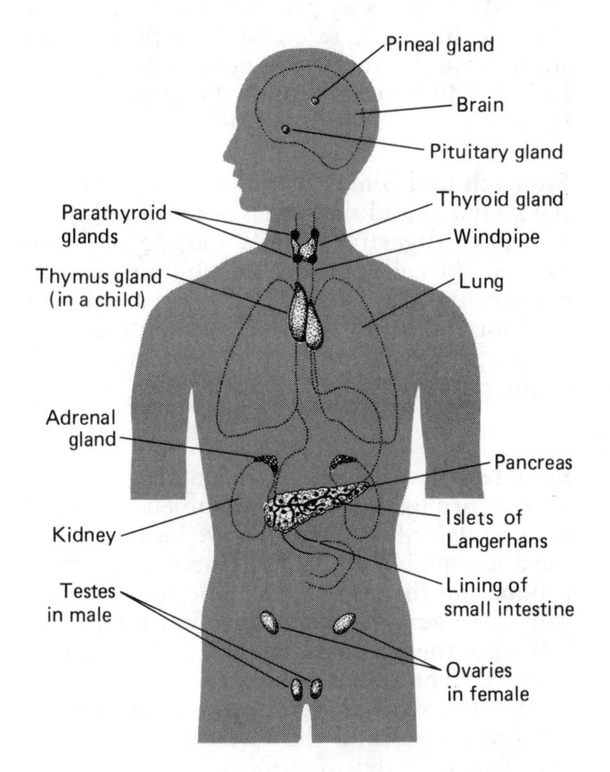

Fig. 22-5. Human ductless glands.

DUCTLESS GLANDS

Thyroid Gland. Located in front of the windpipe near the base of the neck is the **thyroid gland.** This gland is shaped like a bow tie. **Thyroxin** is the hormone secreted by the thyroid gland. Thyroxin speeds up the oxidation of nutrients in the cells, that is, it regulates the metabolic rate. In children, a lack of thyroxin can result in arrested growth and mental retardation. Thyroxin contains iodine, which is supplied to the thyroid gland by the blood. If the diet lacks iodine, or if the thyroid gland does not function properly for other reasons, a body disturbance results. Two important diseases of the thyroid gland are:

1. *Simple Goiter.* When the body does not receive an adequate supply of iodine, the thyroid gland fails to manufacture a usable form of thyroxin. In time, this results in an enlargement of the thyroid gland, called *simple goiter.* This disease is easily prevented by eating foods naturally rich in iodine. Among these are seafoods and vegetables that have been grown in seacoast regions. In the absence of such foods, iodized table salt can be added to food.
2. *Exophthalmic Goiter.* When the thyroid gland secretes excessive quantities of thyroxin, the rate of oxidation in the cells increases. This condition is usually accompanied by nervousness, restlessness, and loss of weight. In time, bulging eyes and an enlarged thyroid gland also develop. This type of thyroid enlargement is called *exophthalmic goiter.* Treatment of this disease involves partial destruction or partial removal of the thyroid gland by means of surgery or radiation.

Parathyroid Glands. The **parathyroid glands** are four small masses of similar tissue imbedded in the rear surface of the thyroid gland. The hormone of these glands, called **parathormone,** helps keep the amount of calcium and phosphorus in the blood at a constant level, which is important for proper nerve and muscle function, blood clotting, and proper growth of teeth and bones.

Adrenal Glands. These small glands are located one on top of each kidney. Each of the **adrenal glands** has an inner and an outer region.

1. The *inner region* of an adrenal gland—the *adrenal medulla*—secretes the hormone **adrenalin** when a person is angry, frightened, or excited. When adrenalin reaches the liver, the liver releases glucose into the bloodstream. As a result, extra energy is available for the "fight or flight" response that occurs in a stressful situation. Adrenalin also increases the breathing rate and causes the heart to beat faster and stronger. Other effects of adrenalin include a slowing down of peristalsis by causing some smooth muscles to stop contracting, and a faster clotting of blood when the body is cut. Since these changes make the body ready for emergencies, the adrenal glands are frequently called the "glands of combat."
2. The *outer region* of an adrenal gland—the *adrenal cortex*—secretes several hormones, the best known of which is **cortisone.** Among the activities of the hormones of the adrenal cortex are regulation of the body's balance of mineral salts and water, and of its use of sugars and proteins. Cortisone, which helps maintain healthy cartilage, is also given to people who have rheumatoid arthritis and certain allergies because it reduces inflammations.

Pituitary Gland. Attached to the base of the brain is a pea-sized gland, the pituitary. Since the **pituitary gland** secretes numerous hormones, many of which control the activities of other ductless glands, it is called the "master gland." Some of the major hormones of the pituitary gland are listed below.

1. **Growth hormone** regulates bone and muscle growth. A child who has too much of this hormone may grow so rapidly that he or she becomes a giant. Such a person may even reach a height of eight feet. A child who lacks this hormone may grow so slowly that he or she remains very short, becoming a dwarf who may never grow beyond a height of three feet.
2. *Thyroid-stimulating hormone* controls the activity of the thyroid gland.
3. *Sex-gland-stimulating hormones* control the development of the sex glands (the ovaries and the testes).
4. The hormone *ACTH* causes the adrenal glands to secrete cortisone.
5. **Prolactin** stimulates milk production.
6. One hormone helps control water balance by the kidneys.

In addition, the pituitary gland is connected to a part of the brain called the **hypothalamus.** This is the main connection between the endocrine and nervous systems. The hypothalamus secretes hormones that control the release of hormones by the pituitary gland.

Pancreas. This gland is located behind and beneath the stomach. The **pancreas** is a double gland—part of it is a duct gland and part is a ductless gland. The duct portion of this gland secretes the pancreatic juice that reaches the small intestine by way of the pancreatic duct. The ductless portion consists of numerous scattered groups of cells, called the **islets** (or *islands*) **of Langerhans,** which secrete the hormone **insulin** into nearby capillaries. Insulin lowers the body's blood sugar by controlling passage of glucose from the blood into the liver and into other cells. In this manner, insulin regulates the storage of glucose as glycogen in the liver and the use of glucose by all cells.

Normally, the secretion of insulin by the pancreas keeps the amount of glucose in the blood at a fairly constant level. When too little insulin is secreted, the glucose level of the blood becomes so high that life is endangered. This condition is called **diabetes.** In 1922, Canadian scientists *Frederick G. Banting* and *Charles A. Best* extracted insulin from the pancreas of animals. The researchers successfully treated diabetics by giving them injections of this insulin. Now scientists can produce insulin through genetic engineering techniques, rather than getting it from the glands of animals.

Stomach and Small Intestine. The lining of the stomach and the lining of the small intestine have **digestive glands** that secrete hormones. The cells of the stomach lining secrete **gastrin,** which stimulates the production of acid for digestion. The cells of the small intestine produce **secretin,** which affects the role of the pancreas, stomach, and liver in digestion.

Sex Glands. The female sex glands, called the *ovaries,* are located in the abdomen. The male sex glands, called the *testes,* are located below the abdomen. In addition to functioning in reproduction (by producing the egg and sperm cells), the **sex glands** produce hormones that influence the development of female and male body characteristics at **puberty,** or sexual maturity. The ovaries secrete **estrogens** (female sex hormones) and the testes secrete **androgens** (male sex hormones).

Thymus Gland. The **thymus gland** is located behind the breastbone, in the chest of young children. As a child matures, the thymus gland shrivels and almost disappears. The thymus produces hormones, called **thymosins,** that strengthen the immune system (see Ch. 23).

Pineal Gland. The **pineal gland** lies in a region of the brain between the left and right cerebral hemispheres. It produces a hormone, called **melatonin,** that is thought to affect natural cycles in the body's activities (called *biorhythms*) and the timing of puberty.

See Table 22-1 below to review the important endocrine glands and the hormones they secrete.

TABLE 22-1. IMPORTANT ENDOCRINE GLANDS AND THEIR HORMONES

Gland	*Hormone*	*Role of Hormone in the Body*
adrenal	adrenalin	Controls muscle reaction, blood pressure, and blood clotting
digestive glands	gastrin, secretin	Stimulate acid production for digestion
pancreas	insulin	Controls blood sugar level
parathyroid	parathormone	Controls calcium and phosphorus levels in blood
pineal	melatonin	Helps regulate pituitary gland and biorhythms
pituitary	growth hormone, ACTH, prolactin, etc.	Controls the growth of bones, etc.
sex glands	estrogens, androgens	Controls development of mature sexual traits
thymus	thymosin	Controls growth of certain white blood cells
thyroid	thyroxin	Controls rate of body growth

CHAPTER REVIEW

Science Terms

The following list contains all of the boldfaced scientific terms found in this chapter and the page on which each appears.

adrenal glands (p. 261)
adrenalin (p. 262)
androgens (p. 262)
auxins (p. 259)
cortisone (p. 262)
diabetes (p. 262)
digestive glands (p. 262)
duct gland (p. 260)
ductless glands (p. 260)
endocrine glands (p. 260)
endocrine system (p. 260)
estrogens (p. 262)
gastrin (p. 262)
growth hormone (p. 262)
hormones (p. 259)
hypothalamus (p. 262)
insulin (p. 262)
islets of Langerhans (p. 262)
melatonin (p. 263)
pancreas (p. 262)
parathormone (p. 261)
parathyroid glands (p. 261)
pineal gland (p. 263)
pituitary gland (p. 262)
prolactin (p. 262)
puberty (p. 262)
secretin (p. 262)
sex glands (p. 262)
thymosins (p. 263)
thymus gland (p. 263)
thyroid gland (p. 261)
thyroxin (p. 261)

Matching Questions

On the blank line, write the letter of the item in column B which is most closely related to the item in column A.

Column A

____ 1. plant growth hormones
____ 2. regulates calcium and phosphorus levels
____ 3. plant's response to light
____ 4. regulates muscle and bone growth
____ 5. speeds up oxidation of nutrients
____ 6. maintains healthy cartilage
____ 7. increases breathing and heartbeat rate
____ 8. secrete hormone insulin
____ 9. hormone from small intestine
____ 10. estrogens and androgens

Column B

a. adrenalin
b. secretin
c. thyroxin
d. gastrin
e. islets of Langerhans
f. auxins
g. sex gland hormones
h. pituitary gland
i. phototropism
j. cortisone
k. parathormone

Multiple-Choice Questions

On the blank line, write the letter preceding the word or expression that best completes the sentence or answers the question.

1. When the sun rises, most plants
a. respond in a positive way to light
b. show no response to light
c. start impulses in their nerve cells
d. stop oxidizing their food — 1 ____

2. Secretions from ductless glands pass
a. to duct glands
b. directly into the blood
c. out of the body
d. directly into the stomach — 2 ____

3. Another term for a hormone is
a. chemical messenger *b.* enzyme *c.* catalyst *d.* nerve control — 3 ____

4. A rapid rate of oxidation can be the result of increased secretion by the
a. lungs *b.* pineal gland *c.* thymus gland *d.* thyroid gland — 4 ____

5. A hormone noted for its value in treating some types of arthritis is
a. adrenin *b.* cortisone *c.* insulin *d.* thyroxin — 5 ____

6. Dwarfed people probably lack the hormone that controls
a. growth *b.* oxidation *c.* mental development *d.* respiration — 6 ____

7. Goiters are usually caused by the improper working of the
a. parathyroid glands
b. thyroid gland
c. sex glands
d. gastric glands — 7 ____

8. The hormone melatonin, which affects the body's natural cycles, is produced by which gland?
a. thyroid *b.* pancreas *c.* pineal *d.* thymus — 8 ____

9. Excess glucose in the blood is characteristic of the disease called
a. diarrhea *b.* anemia *c.* hemophilia *d.* diabetes — 9 ____

10. The secretion that plays an important part in normal mental development is
a. thyroxin *b.* adrenalin *c.* secretin *d.* insulin — 10 ____

11. Iodine in foods is needed by the body for forming the hormone
a. adrenalin *b.* insulin *c.* secretin *d.* thyroxin — 11 ____

12. The pituitary gland is attached to the base of the
a. kidneys *b.* brain *c.* thyroid *d.* lungs — 12 ____

13. Radiation and surgery have been found useful in treating some disturbances of the
a. adrenal glands *b.* cortex glands *c.* thyroid gland *d.* pituitary gland — 13 ____

14. The amount of glucose in the cells of the body and in the blood plasma is related to levels of the hormone
a. secretin *b.* glycogen *c.* insulin *d.* iodine — 14 ____

15. A hormone that a doctor would be likely to inject in a case of heart failure is
a. insulin *b.* ACTH *c.* adrenalin *d.* thyroxin — 15 ____

16. A hormone that regulates the amount of calcium in the blood is made by
a. the pituitary gland
b. the parathyroid glands
c. the pancreas
d. the adrenal glands — 16 ____

17. The small intestine secretes several enzymes and the hormone called
a. insulin *b.* cortisone *c.* glucose *d.* secretin 17 ____

18. Auxins are hormones that stimulate the growth of
a. bone tissue *b.* muscle tissue *c.* plant tissue *d.* gland tissue 18 ____

19. Secretin is the hormone that stimulates the actions of organs affecting
a. digestion *b.* secretion *c.* excretion *d.* respiration 19 ____

20. The turning of a sunflower as the sun moves is controlled by
a. the type of soil it is in
b. the muscular system of the plant
c. the receptors of the plant
d. the plant hormones 20 ____

Modified True-False Questions

In some of the following statements, the italicized term makes the statement incorrect. For each incorrect statement, write the term that must be substituted for the italicized term to make the statement correct. For each correct statement, write the word "true."

1. A plant's response to a stimulus is known as a (an) *auxin*. 1 ____________

2. The growth hormones of plants are called *adrenalins*. 2 ____________

3. Glands that secrete their juices directly into the bloodstream are called endocrine, or *duct* glands. 3 ____________

4. The thyroid gland needs the element *iodine* to function well. 4 ____________

5. Adrenalin causes the liver to release *iron* into the bloodstream. 5 ____________

6. The master gland that secretes many hormones that control the activities of other glands is called the *pituitary* gland. 6 ____________

7. The islets of Langerhans are found in the *adrenal* gland. 7 ____________

8. The scientists Banting and Best were the first to extract *insulin* from animals for the treatment of diabetes. 8 ____________

9. The digestive glands are found in the linings of the small intestine and the *liver*. 9 ____________

10. The hormones that influence the timing of sexual maturity are produced by the *puberty glands*. 10 ____________

Testing Your Knowledge

1. Explain the difference between the two terms in each of the following pairs:

a. positive tropism and negative tropism ____________________________________

____

b. phototropism and geotropism

c. duct gland and ductless gland

d. gastrin and secretin

e. ovaries and testes

f. pituitary gland and pineal gland

g. thyroid gland and parathyroid glands

2. Each of the following conditions is unusual or abnormal. For each, explain the probable cause and describe how the condition might be either prevented or treated.

a. simple goiter

b. diabetes

c. a child dwarf

d. The stem of a geranium plant in your classroom is growing sideways instead of upright.

3. Explain why you think it is beneficial for a plant to exhibit the various tropisms that it has. Give some examples.

4. Name the two glandular portions of the pancreas and the function of both parts.

Anabolic Steroids: Buildup for a Letdown?

News item: September 24, 1988. Canadian sprinter Ben Johnson runs fastest 100-meter dash in history during Olympic Games in Seoul, South Korea.

News item: September 27, 1988. International Olympic Committee accuses Johnson of cheating and takes away his gold medal.

What did Ben Johnson do to deserve such harsh punishment? His blood tested positive for a chemical similar to one that the body normally makes. The natural body chemical is a male hormone called testosterone. The chemical found in Johnson's blood was one of a family of artificial hormones called *anabolic steroids.*

Against Olympic rules, Johnson had taken the drug in an effort to improve his performance. Anabolic steroids have several effects on an athlete's body. Perhaps the most obvious effect is that they bulk up a person's muscles.

In only a few months, skinny would-be athletes can develop into powerful body builders. Weight lifters can improve their performance dramatically. Runners can run faster. Jumpers can jump higher and farther. Football linemen can more easily punch holes in the opposition's defenses. And they can do it more aggressively, since anabolic steroids pump up aggressiveness as well as muscles.

So what's wrong with taking steroids? For one thing, it gives the steroid-taker an unfair advantage over athletes who don't take drugs to improve their performance. It's also against the rules of most organized sports associations, so it is considered cheating.

Furthermore, in most cases the use of steroids is against the law. Anabolic steroids can be obtained only with a doctor's prescription. And most athletes who take the drugs get them without a prescription. So why do an estimated more than half-a-million high school, college, and professional male and female athletes cheat and break the law?

As one football star put it: "Our coaches know all about what we do. All they want to do is win. Winning is everything. If you want to be a starter, and all the glory that goes with it, you do what you have to do. Everyone's doing it anyway."

However, as Ben Johnson can testify, getting caught takes the shine off the glory. Johnson not only had to give up his Olympic gold medal, and the rewards that would have come with it, he also gave up a huge chunk of his pride and self-esteem. But even then, he was luckier than other athletes who have turned to steroids. Some actually have died from heart attacks. Others have committed suicide, driven to it by the effects that steroid abuse has on the brain. Still others have developed liver and kidney diseases or mental illnesses. And that is only a partial list of the devastating tragedies that can strike steroid abusers.

You may wonder how these ill effects can occur, since some steroids—like testosterone—are made by the body. How can they be harmful? First of all, a male's body produces about 2.5 to 10 mg of testosterone a day. (Women make much

less.) For any man, that is a healthy amount. But a steroid abuser takes 100 mg or more of steroids a day, or at least 10 to 40 times the amount that is normal for a male and thousands of times what is normal for a female. The body simply cannot adjust to such an overdose and it loses its ability to function normally.

1. *Anabolic steroids are*

a. natural male hormones.
b. natural female hormones.
c. artificial male hormones.
d. artificial female hormones.

2. *Anabolic steroids*

a. increase muscle bulk but decrease aggressiveness.
b. increase muscle bulk and increase aggressiveness.
c. decrease muscle bulk but increase aggressiveness.
d. decrease muscle bulk and decrease aggressiveness.

3. *Ben Johnson lost his Olympic gold medal because he*

a. lost the 100-meter race.
b. used anabolic steroids.
c. did not use anabolic steroids.
d. did not produce male hormones.

4. *In addition to the effects of steroid abuse described above, what other effects might it have on women? Give reasons for your answers. NOTE: You may have to do some library research to gather more information.*

____

____

____

5. *Carefully reread the full quote that begins "Our coaches know . . ." On a separate sheet of paper write an essay discussing who you think is responsible for the athlete's use of steroids—the school, the coach, the team, or the individual athlete. Explain the reason for your answer.*

23

The Immune System and Disease

LABORATORY INVESTIGATION

COMPARING THE EFFECTIVENESS OF ANTISEPTICS

Your team will be provided with four different *antiseptics,* such as hydrogen peroxide, tinctures of iodine and merthiolate, Mercurochrome, boric acid, etc. You will also receive two petri dishes containing nutrient agar mixed with sour milk. The sour milk contains millions of harmless bacteria. Nutrient agar is a convenient food-containing material on which bacteria grow well.

A. Out of blotting or filter paper, cut four small disks of equal size. With a glass-marking pencil, draw lines on the underside of one petri dish, dividing the dish into four equal sections. Number each section. Holding a disk with forceps, dip the disk into one of the antiseptics. After draining off the excess antiseptic from the disk, open the petri dish and set the disk on the agar in section 1. (In the table on the next page, record the name of the antiseptic for section 1.) Proceed similarly with each of the antiseptics. Promptly close the dish (Fig. 23-1).

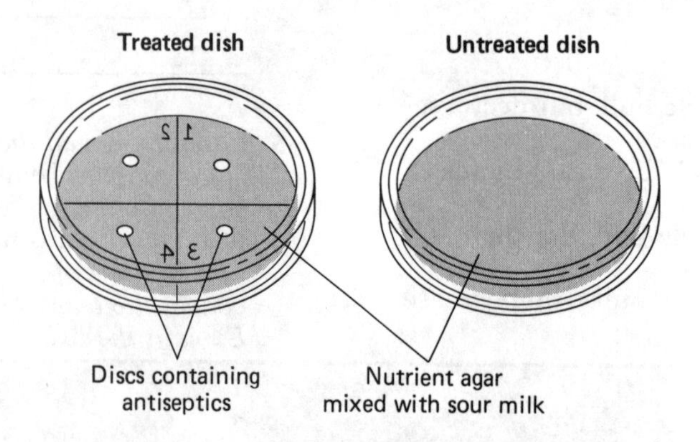

Fig. 23-1. Appearance of dishes before incubation.

Section Number	*Antiseptic*	*Change*
1		
2		
3		
4		
Untreated dish	none	

B. Do not open or put anything into the second petri dish. (Keep the dish closed.) Set both dishes into an incubator for 2 or 3 days. Using a pencil, shade the diagram (Fig. 23-2), showing the changes that occurred in both dishes after incubation.

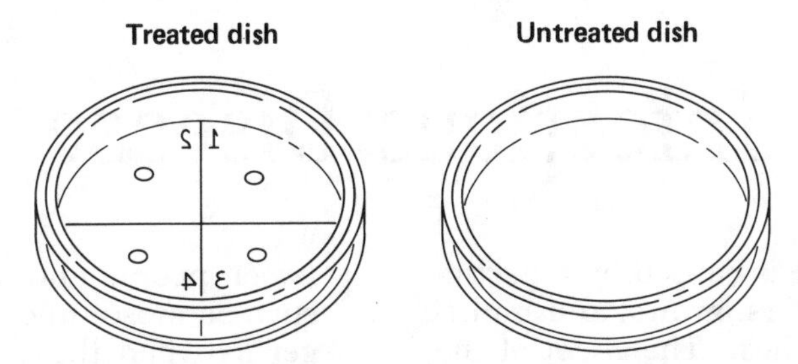

Fig. 23-2. Appearance of dishes after incubation.

1. Explain why it was necessary to keep the second dish closed, without adding an antiseptic to it. ____________________

____

2. After the incubation period, describe the changes in the untreated dish and in each section of your treated dish. (Record your description in the table.)

3. What are your conclusions about bacterial growth and the antiseptics tested? ____________________

____

Disease

The **antiseptics** you used in your experiment are chemicals that have been found to slow the growth of bacteria. (You probably observed, from your laboratory work, that all antiseptics do not work equally well.) Bacteria are among a group of tiny organisms, usually microscopic in size, that enter the bodies of other organisms and interfere with their normal activities. These various microorganisms, or **microbes,** that infect other organisms are also referred to as **pathogens.** They are more commonly known as *germs.* Any condition of the body in which there is a disturbance of normal body structure or activity is called a **disease.**

Through the research of scientists, we have learned the causes of many diseases and the ways to treat them. As a result, we have conquered a number of diseases that once caused many early deaths. The use of antiseptics by the surgeon *Joseph Lister* (1827–1912) was one advance in medicine that has helped to save many lives.

The chemist *Louis Pasteur* (1822–1895) proved that some diseases were caused by microbes that entered the body. Pasteur developed vaccines for *anthrax* and *rabies.* Anthrax is a disease of sheep, cattle, and humans. Rabies is a disease of dogs, raccoons, and other mammals, including humans.

Robert Koch (1843–1910), a physician, developed a reliable method for finding the pathogens that cause particular diseases. He discovered the bacterium that causes *tuberculosis.*

Diseases can be classified into two main groups—*infectious* and *noninfectious.*

Infectious Disease

An **infectious disease** is caused by a particular pathogen that enters, grows, and multiplies in the body of a host. The effect of the infection depends on the strength of the pathogen, how it enters the host, and how well the host resists infection. Most infectious diseases are *contagious* (or "catching") and are called **communicable diseases.** Examples are diphtheria and the common cold.

HOW PATHOGENS SPREAD

Pathogens from an infected (diseased) individual are spread to others in several ways:

1. Germs can be transferred from one person to another when the individuals touch—*direct contact.* For example, a cold can can be spread by kissing. Some very serious infectious diseases are spread by sexual contact.
2. A person can pick up germs by using objects that have been in contact with an infected individual—*indirect contact.* For example, athlete's foot can be caught by wearing the shoes of a person who has the infection.
3. By inhaling the *droplets* (tiny drops) that spray out of the nose of an infected person, another person can acquire the same disease. For example, the droplets discharged by the sneeze of an influenza patient can infect several people.
4. Certain animals—*animal carriers*—can transfer pathogens between people or between people and other animals. Animals such as mosquitoes, flies, and ticks spread germs when they bite other animals, but are themselves unaffected by the diseases the germs cause. Other animals that carry pathogens not only infect people when they bite them, but suffer from the disease as well, like a dog that has rabies.
5. Healthy people—*human carriers*—may carry pathogens in their bodies that do not harm them. However, these people can spread the germs to other people who might get sick from them. "Typhoid Mary," (*Mary Mallon,* 1870–1938), a cook who had recovered from typhoid fever, carried the germs of typhoid fever and spread the disease to hundreds of people.
6. Most pathogens can only live within an organism's body, and so must be transferred from one host to another to survive. However, some germs can live for a while in water, air, soil, and food. Infected people, insects, and other animal carriers, as well as **contaminated** (pathogen-containing) water, soil, food, air, and other objects, are known as disease vectors. **Vectors** are the objects and organisms that transmit infectious diseases. When germs that are in water or food are taken into someone's body, the germs multiply and make the person ill. For example, dysentery germs that have entered a water supply can cause the disease in people who drink that contaminated water.

HOW PATHOGENS ENTER AND AFFECT THE BODY

Microbes can enter the body by way of such body openings as the mouth, nose, and ears; through sexual contact; through breaks in the skin; and through contaminated needles and blood transfusions. Germs interfere with normal body activities by destroying cells and tissue or by releasing **toxins,** which are chemical compounds that poison the body. For example, the malaria microbe destroys red blood cells by entering and bursting them. Rabies virus produces toxins that destroy nerve tissue. The tetanus (lockjaw) bacterium produces a toxin that prevents muscles from relaxing, and that affects nerve tissue. And the bacterium that causes tuberculosis can destroy tissue in the lungs, bones, or glands.

THE IMMUNE SYSTEM: HOW THE BODY PROTECTS ITSELF AGAINST PATHOGENS

Various microbes are constantly invading or living in and on our skin and body openings. However, the bodies of humans and other organisms possess several natural defenses that guard against the diseases many pathogens may cause. These defenses make up the **immune system.**

1. The respiratory (nose and trachea), digestive, excretory, and reproductive tracts are lined by the *mucous membrane,* which secretes the slimy mucus to which microorganisms stick.
2. In addition to mucus, the linings of the nose and trachea have *cilia.* As the cilia move, they push the particles trapped in the mucus toward the nostrils and the throat. Sneezing and coughing eventually discharge the particles and mucus from the body.
3. If it is unbroken, the skin cannot be pierced by most microbes.
4. Tears, produced by the tear glands of the eyes, and hydrochloric acid, produced by the gastric glands of the stomach, are *body secretions* that are able to wash away and destroy some pathogens.
5. Microbes that enter the body are often destroyed by special white blood cells, called **phagocytes,** which surround and digest the microbes. Other special white blood cells, called **lymphocytes,** chemically attack harmful foreign microbes. There are also white blood cells that specifically destroy cells that have been infected.
6. Specific protein compounds, called **antibodies,** are produced by the lymph nodes, spleen, and thymus gland. Antibodies attack germs, viruses, and bits of foreign proteins, called **antigens.** The reaction of the antibodies against antigens and pathogens is called an **immune response.** Some antibodies dissolve germs; these antibodies are called *lysins.* Other antibodies neutralize the toxins produced by germs, and are called *antitoxins.* Another type of antibody helps the phagocytes engulf and destroy microbes.
7. **Interferon** is a protein produced by the body which interferes with the reproduction of viruses. It is also thought to fight viruses that cause some types of cancer.

BODY RESISTANCE TO DISEASE

Some individuals never get a particular disease, because their bodies are able to resist it. Others are not so fortunate. The ability of a body to resist the effects of pathogens is called **immunity.** There are two general types of immunity: natural immunity and acquired immunity.

Natural Immunity. Some people inherit the ability to resist a particular disease. Immunity that is inherited is called **natural immunity.** For example, some people are born with the ability to resist tuberculosis.

Acquired Immunity. The ability to resist a particular disease can be developed. Immunity that is developed is called **acquired immunity.** There are two kinds of acquired immunity: *active* and *passive.*

1. *Active Acquired Immunity.* In this kind of immunity, the body resists a disease because the body has already manufactured its own antibody to that disease. Active acquired immunity usually lasts for a long time—several years. This kind of acquired immunity often develops when a person naturally overcomes a disease. For example, when someone has recovered from chicken pox, he or she is usually immune

to this disease forever. The chicken pox antibody that has formed will destroy any chicken pox virus that may enter the body in the future.

Another method of making a person actively immune to disease is by injecting either a vaccine or a toxoid. A **vaccine** is made up of weakened or killed microbes. A **toxoid** is made up of the weakened toxin that is released by a microbe.

The injection, or **vaccination,** of a vaccine or toxoid causes the body to manufacture its own antibody against the particular microbe or toxin that was injected. Thus, the injection of smallpox vaccine—first used by the English doctor *Edward Jenner* (1749–1823) in the eighteenth century—produces immunity to smallpox. In fact, smallpox has been virtually wiped out by this method worldwide. Similarly, the injection of tetanus toxoid produces immunity to tetanus infection. These vaccinations usually do not cause any illness.

2. *Passive Acquired Immunity.* In this kind of immunity, the body resists or recovers from a disease because it has received an antibody that was produced by another person or animal that had the disease. Although this type of immunity is rapidly acquired and helps a sick person recover, the immunity lasts only a short time—a few weeks or months. For example, when measles antibody is injected into a measles patient, he or she usually recovers in a short while. However, this antibody injection may not safeguard the person against another measles attack.

   Injections of a special blood protein called **gamma globulin** are often used to give passive immunity to some diseases. Unborn babies have passive acquired immunity from the antibodies in their mother's blood; newborn babies get passive acquired immunity from the antibodies in their mother's milk.

DRUGS THAT HELP FIGHT DISEASE

Over the past few centuries, people have discovered and developed numerous drugs that fight disease-causing microbes. Some drugs prevent the growth of microbes, others slow their growth, and still others kill the microbes. With the assistance of these drugs, the body can rapidly fight off pathogens before they do much damage.

Antiseptics. Before the middle of the nineteenth century, deaths commonly occurred because surgical and other wounds became infected. When Joseph Lister introduced the use of carbolic acid as an antiseptic in surgery, the death rate from surgery dropped sharply. Use of antiseptics and sanitary procedures in hospitals is now routine.

Sulfa Drugs. When placed in wounds or taken orally, the sulfa drugs either slow the growth of many bacteria or kill them. Sulfa drugs are made from coal tar. *Sulfanilamide* and *sulfadiazine* are examples of these drugs.

Antibiotics. Like sulfa drugs, **antibiotics** slow the growth of certain bacteria or kill them. Antibiotics are chemicals produced by living things—usually by molds, which are fungi. Examples of antibiotics are *penicillin* (discovered by *Alexander Fleming* in 1929) and *streptomycin* (discovered by *Selman Waksman* in 1944). Antibiotics are not effective against viral infections.

Genetically-Engineered Medicines. Through modern scientific techniques, such as *DNA-splicing,* genetic engineers have been able to insert genes for special traits into microorganisms, such as bacteria, that reproduce rapidly. These microbes then make large amounts of beneficial chemicals, such as interferon and insulin, that can be injected into people to help fight infections and control diseases.

ORGANISMS THAT CAUSE DISEASE

There are six major types of organisms that can cause infectious diseases. See Fig. 23-3 for some examples of these organisms.

1. A *virus* is a very small particle that is considered to be on the borderline between living and nonliving things. A virus reproduces only when it is inside a living cell, which is destroyed in the process.

2. A *rickettsia* is an organism smaller than a bacterium but larger than a virus. Like viruses, rickettsias destroy cells. Diseases caused by rickettsias are spread by ticks, lice, and fleas.
3. Certain *bacteria* harm the body by destroying cells or by producing poisons. However, as you learned in a previous chapter, many bacteria do not cause disease and may even be helpful to other organisms.
4. Some *protozoa* invade the body and destroy cells. Malaria, African sleeping sickness, and amebic dysentery are caused by protozoans. But most species of protozoans do not cause diseases.
5. *Flatworms* and *roundworms* are parasites that harm the body when they attach themselves to tissues and absorb nourishment from them.
6. Some *fungi* harm the body by destroying the tissues from which they obtain nourishment. Many other types of fungi are harmless, or even beneficial.

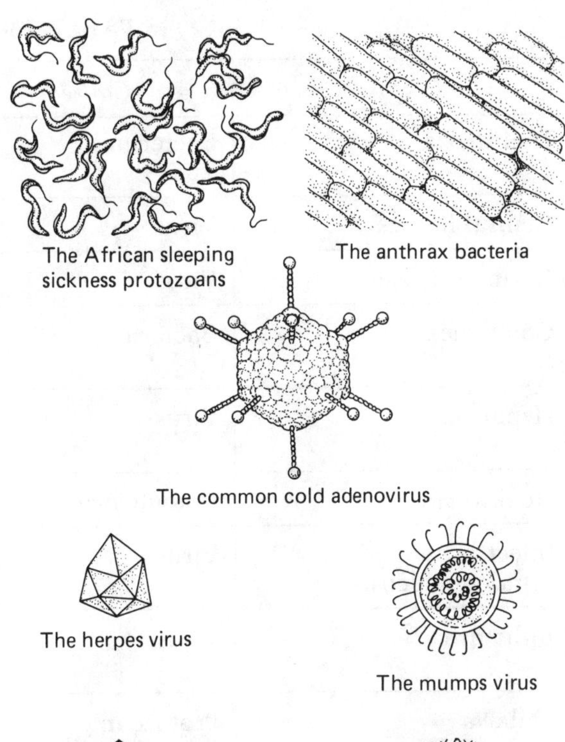
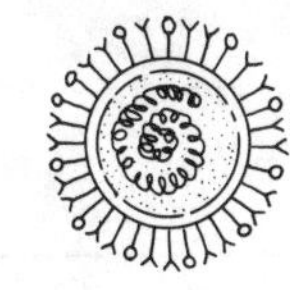

Fig. 23-3. Some disease-causing organisms.

IMPORTANT INFECTIOUS DISEASES

Table 23-1 lists, in alphabetical order, several important infectious diseases. The table also gives the cause of each disease and the method used to cure, treat, or prevent the illness. Parasitic diseases such as malaria, amebic dysentery, hookworm, and African sleeping sickness affect more than 1/4 of the people alive today; we still do not have very good drug treatments or cures for these illnesses.

TABLE 23-1. IMPORTANT INFECTIOUS DISEASES

Disease	*Cause*	*Cure, Treatment, or Prevention*
African sleeping sickness	Protozoan	Control of tsetse flies; some drugs.
AIDS	Virus	Avoid sexual contact or use condoms; do not share needles; AZT.
Amebic dysentery	Protozoan	Antibiotics; avoid infected water and people.
Athlete's foot	Fungus	Antifungal drugs; good sanitation.
Cholera	Bacterium	Vaccine of dead cholera bacteria; sanitation.
Common cold	Viruses	Avoid infected people; bed rest to recover.

(*continued*)

TABLE 23-1. (*Continued*)

Disease	*Cause*	*Cure, Treatment, or Prevention*
Diphtheria	Bacterium	Toxoid of diphtheria toxin; pasteurized milk.
Genital herpes	Virus	Avoid sexual contact or use condoms.
German measles	Virus	Vaccine of live German measles virus.
Gonorrhea	Bacterium	Antibiotics; avoid sexual contact or use condoms.
Hepatitis	Virus	Avoid infected food and needles; gamma globulin.
Hookworm	Roundworm	Antihelminthic drugs; wear shoes outside.
Infectious mononucleosis	Virus	Avoid infected people; bed rest to recover.
Influenza	Virus	Vaccine of dead flu virus; avoid sick people.
Malaria	Protozoan	Control of mosquitoes; quinine; chloroquine.
Measles	Virus	Vaccine of live measles virus; avoid contact.
Mumps	Virus	Vaccine of live mumps virus; avoid contact.
Pneumonia	Bacterium or virus	Antibiotics; avoid infected people.
Poliomyelitis	Virus	Vaccine of weakened live or dead polio virus.
Rabies	Virus	Vaccine of weakened live rabies virus.
Ringworm	Fungus	Antifungal drugs; avoid contact; sanitation.
Rocky Mountain spotted fever	Rickettsia	Spotted-fever vaccine; antibiotics; avoid ticks.
Scarlet fever	Bacterium	Antitoxin injections; antibiotics; avoid contact.
Smallpox	Virus	Vaccine of live cowpox virus; bed rest.
Staph infection	Bacterium	Antibiotics; avoid infected people.
Strep throat	Bacterium	Antibiotics; avoid infected people.
Syphilis	Bacterium	Penicillin; avoid sexual contact or use condoms.
Tapeworm	Flatworm	Antihelminthic drugs; thoroughly cook meats.
Tetanus	Bacterium	Toxoid of tetanus bacteria; cleanse wound.

TABLE 23-1. (*Continued*)

Disease	*Cause*	*Cure, Treatment, or Prevention*
Trichinosis	Roundworm	Thoroughly cook pork.
Tuberculosis	Bacterium	Antibiotics; tuberculosis vaccine; bed rest; avoid infected people.
Typhoid fever	Bacterium	Antibiotics; dead typhoid vaccine; sanitation.
Typhus fever	Rickettsia	Antibiotics; dead typhus vaccine; rat control.
Yeast infection	Fungus	Antifungal drugs; cleanliness.
Yellow fever	Virus	Yellow fever vaccine; mosquito control; rest.
Whooping cough	Bacterium	Vaccine of dead whooping cough bacteria.

Noninfectious Disease

A **noninfectious disease** is one that is usually caused by the presence or absence of some nonliving agent. Some noninfectious diseases are caused by dietary deficiencies, genetic defects, allergies to foreign substances, chemicals in the environment, and badly functioning organs and endocrine glands. Noninfectious diseases are not contagious. The most important types of noninfectious diseases include allergies, cancer, circulatory diseases, deficiency diseases, and hereditary diseases. Some diseases are caused by damage to the brain or nervous system. The cause of a number of diseases is still unknown.

ALLERGIES

An **allergy** is a harmful overreaction of the immune system to some foreign substance (i.e., antigen), which is usually a harmless protein or a low-level toxin, as in a bee sting. The reaction may be one of itching, swelling, or a burning sensation. The tendency to become allergic may be hereditary or may arise from emotional conditions. The substances responsible for allergies, which are referred to as **allergens,** may differ in different people. Among such substances are pollen, dust, foods, polluted air, proteins in animal fur, and various medical drugs. Hay fever and hives are examples of allergic reactions. Allergens cause a person's antibodies to produce **histamine,** the substance that causes allergic reactions like sneezing. Drugs used to fight such effects are called **antihistamines.**

Allergy tests help determine to what substances a person's body reacts abnormally. Once the cause is known, the reactions can be prevented by avoiding contact with the offending substance or by building body resistance to the substance. Resistance can sometimes be built by means of repeated injections of the offending allergen.

CANCER

Ordinarily, the division and growth of cells in the body is controlled by the genes. When this control is lost or disrupted, cancer may develop. **Cancer** is an abnormal growth of tissue (which can form tumors) occurring as the result of rapid, uncontrolled division of cells. Cancer cells overpower normal cells and prevent their proper functioning. One factor that can cause cancer is long-term or frequent exposure to strong radiation such as ultraviolet light, X rays, and atomic radiation. Another factor is contact with chemicals like coal tars, benzene, arsenic compounds, asbestos, and cigarette smoke. Research has shown a strong connection between lung cancer and cigarette smoking. Air that is polluted with exhaust fumes from automobiles and chimneys may also be a factor in lung cancer. And poor dietary habits, such as eating very fatty meats, may lead to cancer in various organs.

Although cancers are not contagious in people, it is thought that some cancers may be caused by a virus. In addition, certain genes may play a role in turning on and off the production of cancerous cells. And the immune system also may be responsible, if it fails to recognize and destroy cancerous cells.

Tests that can detect cancer include X ray, biopsy, and smear tests. In a **biopsy,** a bit of living tissue is examined under the microscope and searched for cancer cells. In a smear test, fluid from the suspected organ is spread on a microscope slide and is similarly examined. If detected early enough, most cancers are usually curable by means of radiation, surgery, or chemicals (chemotherapy). The best way to prevent cancer is to avoid exposure to the substances known to cause the disease.

Seven signs are considered to be warning signals. Each may be only the result of a minor condition that can be readily treated by a physician. However, if any of these signs arise, a physician should be consulted so that, if necessary, proper steps for a cure can be started at once. The seven danger signals of cancer are:

1. Sores that do not seem to heal in a reasonable time.
2. A wart or similar condition that starts to grow very rapidly.
3. A permanent lump anywhere on the body.
4. Unexplained, irregular bleeding from any body opening.
5. Coughing or hoarseness that continues without apparent cause.
6. Frequent digestive disturbances.
7. A change in normal bowel or bladder habits that continues for a long period of time.

CIRCULATORY DISEASES

Some diseases of the heart (coronary disease) and blood vessels are the result of damage caused by the process of aging. The tendency to develop these diseases also may be hereditary in some people. In some cases, the walls of the arteries harden as a result of the deposit of calcium. In other cases, a fatty deposit called **cholesterol** accumulates on the inner surface of the arteries, narrowing the passageways and causing a disease known as *atherosclerosis.* Both of these conditions can lead to high blood pressure and heart damage, because the heart has to work harder to pump blood through these vessels. Such damage can lead to heart attacks and strokes.

There are no definite preventive measures for many circulatory diseases. However, it is advisable to have a life-style that includes relaxation and rest; daily moderate exercise; a diet low in saturated fats (solid fats like butter and lard); avoidance of tobacco and addicting drugs; and regular medical examinations. Treatments include bypass surgery for blocked arteries; heart transplants and artificial valves; and medicine for high blood pressure.

Rheumatic heart disease occurs when heart valves are damaged and cannot function properly. Sometimes surgery can help this condition. However, if *streptococcus* infections like *strep throat* are treated with penicillin, this disease can be prevented.

DEFICIENCY DISEASES

The lack, or *deficiency,* of particular nutrients or hormones can lead to some diseases. For example, the lack of iron can result in *anemia;* lack of vitamin C can result in *scurvy.* Other conditions caused by dietary deficiencies include beriberi (vitamin B_1), rickets (vitamin D), osteoporosis (calcium), and kwashiorkor (protein). Supplying the body with the substance it lacks usually prevents or controls the condition. For example, a lack of the hormone insulin—due to its insufficient secretion by the pancreas—causes *diabetes.* Injections of insulin are used to control diabetes.

Another type of deficiency disease is **immune deficiency.** In this condition, the immune system lacks or loses the ability to attack pathogens and to destroy diseased cells. This condition is usually fatal because the person cannot fight off infections of any kind. Sometimes this condition is genetic, and a person is born with it. In other cases, a person may get it later in life.

AIDS (*acquired immune deficiency syndrome*) is caused by the contagious HIV virus (human immunodeficiency virus). The virus causes an immune deficiency because it destroys the special white blood cells that fight infections. As a result, the body cannot protect itself against invasions of other pathogens, such as those that cause cancers and pneumonia. It is these secondary infections that usually kill a person who has been infected with HIV virus. There is no cure for AIDS yet, but there are some medicines that help the condition.

HEREDITARY DISEASES

Diseases that "run in the family" are called **hereditary diseases.** Hemophilia, the "bleeder's disease," is an example. It is possible for unaffected people to carry a recessive gene for such a condition without being aware of it. Offspring who inherit two such recessive genes will have the disease. Other hereditary diseases include cystic fibrosis, muscular dystrophy, sickle-cell anemia, Huntington's disease, and Tay-Sachs disease. Recent advances in genetic studies are helping to pinpoint the genes that cause these diseases, and may someday lead to cures for some of them.

AUTOIMMUNE DISEASES

An **autoimmune disease** is one in which the immune system fails to recognize normal cells of the body as "self" and, as a result, attacks those very cells that it would normally defend. There are about eighty disorders of this type, some of which may actually have a genetic basis. A damaging autoimmune response can occur when some antigens resemble one of the body's own normal cell types or tissues. Then the antibodies destroy these normal tissues along with the antigens.

An example of an autoimmune disease is *multiple sclerosis,* in which special white blood cells attack the protein that coats nerve cells, leading to a disruption of nerve impulses and loss of control in parts of the body. Although this condition is not yet treatable, scientists are working on a possible cure that would use antibodies to fight off the antibodies that attack the nerve cells. Other conditions of this type include juvenile diabetes, in which the pancreas secretes too little or no insulin, and rheumatoid arthritis, a crippling condition in which the tissues of the joints become inflamed and swollen. An autoimmune condition is similar to the immune response that occurs when the body rejects transplanted organs.

CHAPTER REVIEW

Science Terms

The following list contains all of the boldfaced scientific terms found in this chapter and the page on which each appears.

Matching Questions

On the blank line, write the letter of the item in column B which is most closely related to the item in column A.

	Column A	Column B
____	1. drugs that kill bacteria	*a.* phagocytes
____	2. organisms that cause diseases	*b.* acquired immunity
____	3. blood cells that digest microbes	*c.* antibodies
____	4. protein compounds that attack germs	*d.* allergy
____	5. inherited ability to resist disease	*e.* antibiotics
____	6. developed ability to resist disease	*f.* antihistamines
____	7. made up of weakened or killed microbes	*g.* atherosclerosis
____	8. overreaction of the immune system	*h.* natural immunity
____	9. drugs that fight allergic reactions	*i.* pathogens
____	10. a type of circulatory disease	*j.* toxoid
		k. vaccine

Multiple-Choice Questions

On the blank line, write the letter preceding the word or expression that best completes the sentence or answers the question.

1. The effectiveness of antiseptics can be measured by
a. the rate at which they are oxidized
b. the clear zones that occur in agar plates containing bacteria
c. the strength of their odors
d. the burning sensation they produce in a wound — 1 ____

2. A child born today can expect to live about 75 years; one born in 1900 had a life expectancy of 50 years. The most important reason for this increase in life expectancy is the
a. improvement in food production *c.* conquest of heart disease
b. conquest of many diseases *d.* improvement in housing conditions — 2 ____

3. Who introduced the use of antiseptics in surgery?
a. Lister *b.* Jenner *c.* Pasteur *d.* Salk — 3 ____

4. Louis Pasteur proved that
a. polio was caused by a germ *c.* penicillin came from a mold
b. tuberculosis could be fatal *d.* germs were often responsible for disease — 4 ____

5. The bacterium that causes tuberculosis was discovered by
a. Joseph Lister *b.* Edward Jenner *c.* Louis Pasteur *d.* Robert Koch — 5 ____

6. The disease called African sleeping sickness is caused by
a. fungi *b.* tapeworms *c.* protozoans *d.* a dog — 6 ____

7. Poisons produced by bacteria are called
a. toxins *b.* toxoids *c.* vaccines *d.* spores — 7 ____

8. The body tissue that is injured by the malaria protozoan is
a. muscle *b.* blood *c.* bone *d.* nerve — 8 ____

9. Body cells that can destroy bacteria are the
a. red blood cells *b.* bone cells *c.* nerve cells *d.* white blood cells — 9 ____

10. A secretion of the stomach that can kill some bacteria is
a. rennin *b.* thyroxin *c.* hydrochloric acid *d.* adrenin — 10 ____

11. Smallpox is rare in the world now because most people have received
a. antibiotics *b.* smallpox antitoxin *c.* smallpox vaccine *d.* sulfa drugs — 11 ____

12. Antibodies are produced in the body when
a. germs enter the body
b. germs leave the body
c. antiseptics enter the body
d. antibiotics enter the body — 12 ____

13. Natural immunity depends on
a. vaccines *b.* toxoids *c.* heredity *d.* antibiotics — 13 ____

14. The smallpox virus vaccine was first used by
a. Albert Sabin *b.* Jonas Salk *c.* Edward Jenner *d.* Joseph Lister — 14 ____

15. A toxoid injection stimulates the body to make its own
a. antibodies *b.* toxins *c.* red blood cells *d.* lysins — 15 ____

16. The immune deficiency disease AIDS is caused by
a. passive immunity *b.* a virus *c.* an allergy *d.* a bacterium — 16 ____

17. The type of immunity that is most rapidly developed in the body of a patient is
a. passive immunity
b. active immunity
c. toxoid immunity
d. natural immunity — 17 ____

18. Because of the development of an effective vaccine, no child need suffer from
a. colds *b.* malaria *c.* athlete's foot *d.* measles — 18 ____

19. Sulfa drugs are effective against diseases caused by
a. bacteria *b.* worms *c.* deficiencies *d.* heredity — 19 ____

20. Who discovered the antibiotic penicillin?
a. Selman Waksman
b. Jonas Salk
c. Edward Jenner
d. Alexander Fleming — 20 ____

21. The organisms that are disease agents are
a. green algae
b. viruses
c. bread mold
d. bacteria in yogurt — 21 ____

22. Rickettsias are disease agents that are
a. smaller than viruses
b. smaller than bacteria but larger than viruses
c. larger than bacteria
d. larger than an ameba — 22 ____

23. African sleeping sickness can be prevented by destroying
a. tsetse flies *b.* houseflies *c.* mosquitoes *d.* rabid dogs — 23 ____

24. Measles is caused by a
a. yeast *b.* virus *c.* mold *d.* bacterium — 24 ____

25. Hereditary diseases are
a. contagious *b.* not contagious *c.* spread by insects *d.* spread by birds — 25 ____

Modified True-False Questions

In some of the following statements, the italicized term makes the statement incorrect. For each incorrect statement, write the term that must be substituted for the italicized term to make the statement correct. For each correct statement, write the word "true."

1. Diseases that can be caught from other people are called *infectious* diseases. 1 ____________

2. Dysentery germs can be spread by contaminated *air.* 2 ____________

3. A clean, unbroken *skin* is good protection against germs. 3 ____________

4. Cilia and mucus in the *digestive* system help to protect you against bacteria that enter your body. 4 ____________

5. The ability of the body to resist disease is called *allergy.* 5 ____________

6. A person who has had a disease and recovered from it, has acquired a(an) *passive* immunity to that disease. 6 ____________

7. A disease in which cells divide abnormally is *cancer.* 7 ____________

8. Disease-fighting chemicals that are produced by certain molds are called *antitoxins.* 8 ____________

9. A tissue biopsy can be used to detect *heart disease.* 9 ____________

10. The lack of a nutrient or hormone can lead to development of a *hereditary* disease. 10 ____________

Testing Your Knowledge

1. Explain the difference between the two terms in each of the following pairs:

a. toxoid and vaccine ____________________

____

b. infectious and noninfectious ____________________

____

c. natural immunity and acquired immunity ____________________

____

d. active acquired immunity and passive acquired immunity ____________________

____

____

e. antibody and antibiotic ____________________

____

f. antibiotic and antitoxin ______________________________

____

2. List the six major types of organisms that cause infectious diseases and briefly describe the damage they do to the body.

a. ______________________ *d.* ______________________

b. ______________________ *e.* ______________________

c. ______________________ *f.* ______________________

3. Describe how each of the following structures helps keep your body healthy and free of disease.

a. cilia ______________________________

b. skin ______________________________

c. white blood cells ______________________________

d. tear glands ______________________________

e. gastric glands ______________________________

4. Complete the table.

Disease	*Cause*	*Cure or Treatment*
Anemia	*a.* ____________	*b.* ____________
Beriberi	*c.* ____________	*d.* ____________
Diabetes	*e.* ____________	*f.* ____________
Measles	*g.* ____________	*h.* ____________
Rabies	*i.* ____________	*j.* ____________

5. What are two ways in which bacteria may cause harm to the human body?

____

____

6. What structures would form the first line of defense of the human body?

____

____

7. How can the spread of AIDS be contained?

____

____

Ecology

LABORATORY INVESTIGATION

TESTING ABILITY OF SOILS TO HOLD WATER

A. Set up three funnels of the same size on tripods. Place a small, loose wad of absorbent cotton in the neck of each funnel. Fill each funnel to within 2 cm. from the top with one of the following dry soils: beach soil (sand); clay soil from an eroded slope or well-worn path; and garden soil (loam). Place a graduated beaker or similar receptacle under each funnel. Use a glass-marking pencil to label each beaker as shown in Figure 24-1.

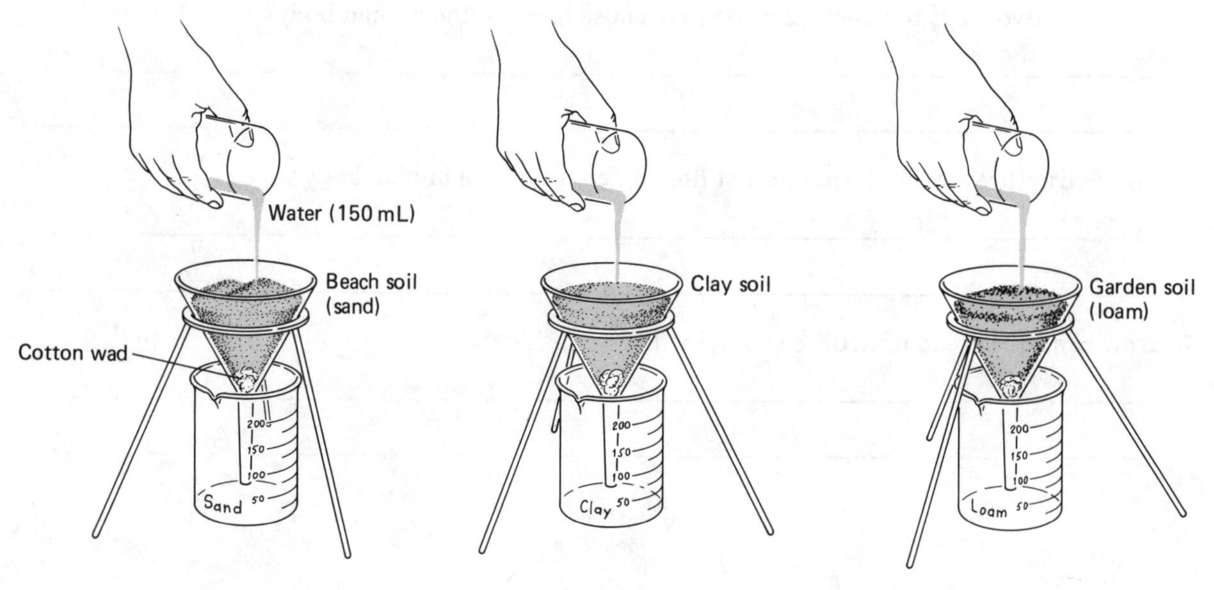

Fig. 24-1

B. Pour 150 milliliters of water into a container. (More or less water may be used according to the capacity of the funnels.) Slowly pour all of the water into the funnel containing beach soil. Wait five minutes and measure the amount of water collected in the beaker. Repeat this procedure with each of the other two soils. Record the measurements below. Also note whether the water drained through each soil slowly or rapidly.

Amount of Water Passing Through Soil in 5 Minutes	
Beach soil	______ mL
Clay soil	______ mL
Garden soil	______ mL

C. 1. *a.* Which type of soil allows water to drain most rapidly? ______

______

b. least rapidly? ______

2. *a.* Which type of soil holds the most water? ______

b. the least? ______

3. What do you think accounts for the capacity of soils to hold water? ______

______

4. Which type of soil would best support plant life?

______

Explain. ______

______

Living and Nonliving Parts of the Environment

An organism's natural **environment** consists of all the living and nonliving things in the area where it lives. *Living* parts of the environment include all organisms. *Nonliving* parts include such things as air, water, soil, light, and heat. As you learned in Chapter 2, all living things carry out certain life activities to survive. An organism's environment supplies all the energy and materials that the organism needs to perform its life processes. For example, in your laboratory investigation you tested the ability of different soils to hold water. The amount of water that soil can hold, and the various minerals in the soil, affect the types and number of plants that can grow in a particular environment. In turn, the type of plant life determines the kinds of animals that the environment will support. Ultimately, the life activities of living things have an effect on the nonliving parts of their environment, such as the soil. The study of the complex relationships between living things and their environment is known as **ecology**.

Populations, Communities, and Ecosystems

You have probably heard that the human population on Earth is over 5 billion people. On a smaller scale, some nations have a human population of only a few million people, while other nations have a population of more than 100 million people. In nature, a **population** is made up of all the organisms of the same species living in a particular place.

For example, all the alligators living in Everglades National Park in Florida make up the total population of alligators for that area. In addition to the alligators, there are many populations of different species of animals, plants, and other organisms. All these populations together make up the natural community of the Everglades. A **community** is composed of all the organisms living in the same place. All organisms in a community interact with, and affect the lives of, other organisms in that community. For example, in the Everglades community, various species of birds, fish, reptiles, plants, insects, and fungi come into contact with each other in the course of carrying out their life activities.

There are different types of communities. In general, communities that are in or around a body of water are called **aquatic;** those that are on land are called **terrestrial.** Table 24-1 on page 287 shows some common aquatic and terrestrial communities and the types of organisms that live in them.

Living things in a community interact with the nonliving parts of their environment. Together, all these living and nonliving things, and their interactions, make up an **ecosystem.** The primary source of energy in an ecosystem is the sun. (See Fig. 24-2.)

Fig. 24-2. An example of an ecosystem.

TABLE 24-1. AQUATIC AND TERRESTRIAL COMMUNITIES

Type of Commuity	*Common Organisms*
Pond (Aquatic)	Algae, water lilies, dragonflies, frogs, bass, kingfisher, beaver
Desert (Terrestrial)	Barrel cactus, creosote bush, rattlesnake, kangaroo rat, road runner
Seashore (Aquatic)	Seaweed, beach grass, snail, crab, seagull, mussels
Evergreen forest (Terrestrial)	Hemlock, birch, moose, squirrel, owl, gray wolf

Adaptations in Ecosystems

As you learned in Chapter 11, organisms have special adaptations that enable them to survive in their particular environments. For example, a whale, which inhabits cold seawater, has thick layers of fat that help keep its body warm. The kind of environment that an organism lives in is called its **habitat.** A whale's habitat would be cold seawater, or the ocean. The specific type of habitat an organism lives in depends on its needs. These needs include proper temperature, moisture, shelter, sunlight, and a source of food. Many types of organisms can live together in the same habitat, as part of a community. However, the specific needs and activities of the different organisms will vary. For example, seals, whales, and sharks may all live in the same area, or habitat; but they will each eat a different type of fish as their main source of food, and they will most likely get their food at different levels in the water.

The particular way an organism lives within its habitat is called its **niche.** An organism's niche may include such factors as what it eats, where it gets the food, and where it finds shelter in its habitat.

Adaptations to Living in Water. Organisms that live in aquatic habitats have special adaptations for living in the water. For example, plants like water lilies have air chambers that allow gases to diffuse inside the plant and enable the leaves to float on the water's surface. Because they float, the leaves can get the air and light that they need for photosynthesis. Fully aquatic animals, such as fish and lobsters, have *gills,* which absorb the dissolved oxygen that is used in respiration. Aquatic air-breathing animals, like ducks and beavers, have webbed feet to help them swim, and waterproof feathers and fur.

Adaptations to Living on Land. Plants that live on land usually have well-developed root systems that anchor the plant in the soil while obtaining nutrients from it. Most land plants also have strong supporting tissue that enables the plant to grow upright and get more air and sunlight. Plants in drier habitats, like cacti in deserts, have thick stems that can store water from the infrequent rains. Plants in cold, windy areas tend to grow lower to the ground.

Terrestrial animals have lungs (or in the case of insects, air tubes) that take in oxygen from the air for respiration. In cold regions, animals tend to have more body fat and thicker fur. In every ecosystem, the plants, animals, and other organisms have a variety of adaptations that help them adjust to the other living and nonliving parts of the environment.

FOOD CHAINS AND FOOD WEBS

In any particular type of ecosystem, many different organisms can be found living together. One kind of animal may feed upon a specific plant. Another type of animal may feed upon the plant-eating animal. For example, a zebra may eat grass and may, in turn, be eaten by a lion. Such a group of organisms (grass, zebra, lion) in which one organism depends upon another for food is called a **food chain.** The energy and matter in a natural community flow through the organisms in a food chain.

Food chains often overlap one another. An animal that is part of one food chain may eat an animal that is also part of another food chain. The criss-crossing of many food chains is called a **food web.** For example, plant-eating animals, or **herbivores,** often eat more than one type of plant. Flesh-eating animals, or **carnivores,** usually eat more than one type of herbivore. Some carnivores may even eat one another. Animals that eat both plants and other animals are called **omnivores.** When plants, plant-eaters, and flesh-eaters die, as a result of aging or other causes, the remains of their dead bodies are consumed by the organisms that bring about decay. Fig. 24-3 illustrates a food web.

PRODUCERS, CONSUMERS, AND DECOMPOSERS

Every food chain begins with green plants or, in an aquatic habitat, algae. The plant or alga absorbs energy from the sun and makes its own food by photosynthesis. These plants—grasses, for example—are called **producers.** Animals such as grasshoppers that feed upon green plants are called **primary consumers.** Other animals, like frogs, that feed upon grasshoppers are called **secondary consumers.** And animals such as raccoons that feed upon frogs are called **tertiary consumers,** and so on (see Fig. 24-4).

When producers and consumers die, special types of bacteria and fungi feed upon their bodies and decay them. These organisms are

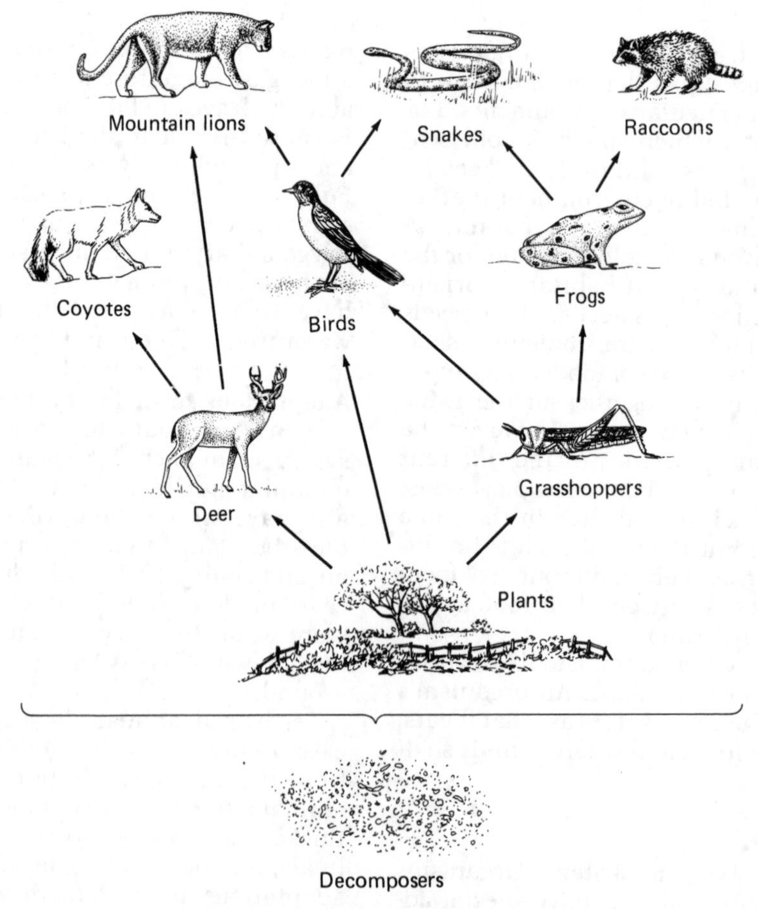

Fig. 24-3. An example of a food web.

called **decomposers,** or **saprophytes.** The decomposers break down the complex organic compounds of the dead organisms to simple inorganic compounds. The inorganic compounds become part of the soil and the atmosphere. Eventually, these inorganic compounds are taken in by producers, which use them again as raw materials when they manufacture food. The abundance or scarcity of producers in an ecosystem determines the abundance of animal life, since the producers are at the base of every food chain.

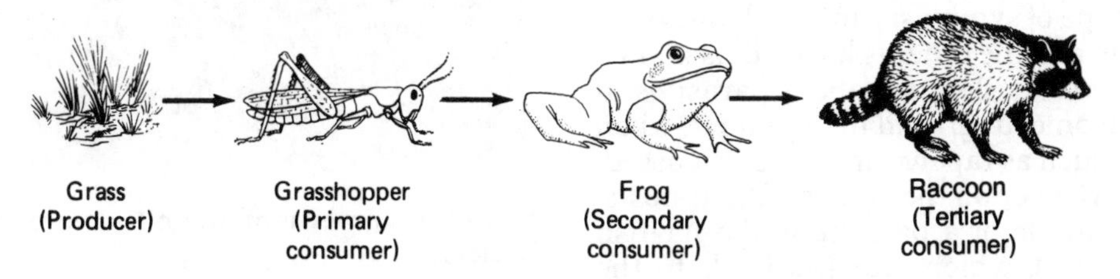

Fig. 24-4. Feeding levels in a food chain.

Competition and Cooperation

Competition. Although every organism has its own specific niche in its habitat, **competition** exists within and between species for certain resources. Plants compete with each other for access to sunlight and water. As a result, tall plants like trees often crowd out smaller plants like bushes.

Among the herbivores, many animals compete with one another for access to a variety of vegetation. Members of the same monkey species, for example, will compete with each other to feed at a favorite fruit tree, and they will also compete with other monkey species that eat at the same fruit tree. This competition is reduced when different parts of the vegetation are eaten, such as when one species eats the leaves, and the other species eats the fruit on the same tree.

Among the carnivores, there is also competition within and between species. Carnivores that feed on other living animals (by killing them first) are called **predators.** The animals that they kill and eat are called **prey.** Animals that eat the leftover parts of dead animals are called **scavengers.** Owls are predators that feed on field mice, which are the prey. Hawks are also predators that feed on field mice. Hawks and owls are in competition for the same prey, but this competition is reduced because owls feed mainly at night and hawks feed during the day. Vultures are scavengers; they eat the remains of mice and other prey that the owls and hawks may not finish. Sometimes scavengers also compete with predators by stealing their prey from them.

Cooperation. Frequently, different organisms live in **cooperation,** that is, they cooperate in their life activities and aid each other's survival. **Social animals** are members of the same species that live together in colonies and help one another. For example, in a bee colony (beehive), the workers and queen (females) and the drones (males) all carry out different activities that benefit the hive. Worker bees gather food for the other hive members and care for the young that are produced by the queen and drones. Some small mammals also live in colonies and exhibit cooperative behavior—a few members of the colony stay alert to warn the others of predators.

When two different types of organisms live together, they are referred to as **symbionts,** and the relationship is known as **symbiosis.** You have already learned about the alga and fungus that make up a lichen (see Chapter 5). In this type of symbiosis, both partners benefit by getting a place to live in exchange for nutrients from each other, so it is known as *mutualism.*

Another type of symbiosis, known as *commensalism,* occurs when two organisms live together and one benefits while the other neither benefits nor suffers. An example of this would be sharks and remoras—the remoras attach by sucker disks to the underside of sharks, thus getting a free ride and scraps of food left over from the shark's feeding.

A third type of symbiosis, in which one partner benefits and the other is harmed, is known as *parasitism.* You learned about parasites in Chapter 5 (some fungi) and in Chapter 7, when organisms such as tapeworms were discussed. Parasites live in or on the bodies of their hosts. Another example of a parasite is the hagfish (see Chapter 8), which attaches itself to the body of another fish, draining blood and nutrients from the host until it dies.

Microorganisms are also found in various symbiotic relationships. Plants such as the soybean, clover, and alfalfa have beneficial bacteria that grow in small lumps, or *nodules,* on their root branches, as shown in Fig. 24-5.

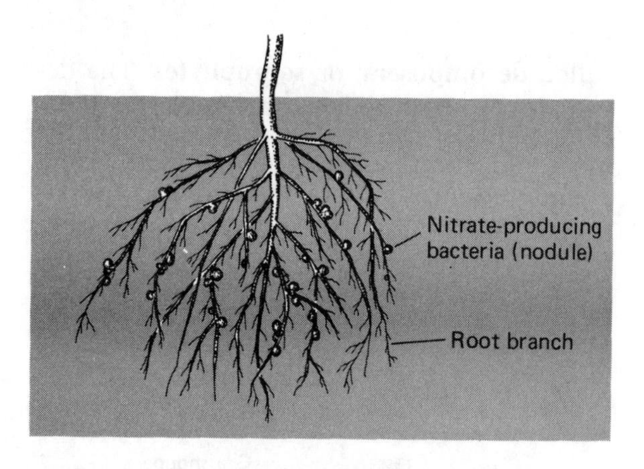

Fig. 24-5. Clover root and nitrate-producing bacteria.

These bacteria produce nitrates, which are mineral salts needed by the plants. In turn, the plants provide the bacteria with food, moisture, and a place to live. (You will learn more about the relationship between bacteria and plants in Chapter 25.)

Effect of Living Things on Their Environment

As a result of their life activities, living things gradually cause changes in their environment. Three important examples follow:

1. *Soil Formation.* **Soil** is made up of tiny rock particles, water, air, and **humus** (decayed organic matter). Lichens, which can live on bare rock, break down some of the rock into small particles. These particles and dead lichens provide a thin layer of soil in which such plants as moss can grow. Eventually, more rock is broken down and dead moss plants accumulate, forming a thicker layer of soil that can anchor and support other plants. In time, the rock becomes covered with a thick layer of soil in which many types of plants grow.

   Recall from your laboratory investigation that you tested three types of soil: clay, sand, and garden (loam). *Clay* consists of very fine rock particles and some humus. Clay is packed so firmly that little water is absorbed and few plants can grow. *Sand* has rock grains and a little clay and humus. Water drains through sand rapidly and some plants can grow in it. *Loam,* or garden soil, has sand, clay, and more humus. Loam absorbs water well and holds more air, water, and minerals than the other two soil types can. Many kinds of plants can grow well in garden soil.
2. *Changing the Mineral Content of Soil.* Most plants, like wheat and corn, require nitrates and other mineral compounds, which they absorb from the soil. Other organisms, such as the nitrate-producing bacteria discussed above, put nitrates back into the soil and plant life.
3. *Decay of Dead Organisms.* The wastes that animals produce and the remains of dead organisms are broken down in the process called *decay.* This process is carried out by the decomposers—bacteria and fungi. These organisms feed upon dead material and break it down to simple substances. Many of these substances fertilize the soil by seeping down and becoming part of it.

Succession of Communities

The organisms in a community are often responsible for an accumulation of soil, wastes, dead branches, and other natural objects. As these materials accumulate, the character of the environment changes. Natural fires, changes in climate, and the activities of people also change the character of the environment.

Organisms that are not suited to the changed conditions either die out or go elsewhere. Then organisms from other communities may invade the changed community. If these organisms are suited to this new environment, they settle there and form a community that is different from the original one. Over a long period of time, the new community may be followed by still another type of community, and so on. This gradual change of one community into another is called **biological succession.** For example, a pond community eventually will be replaced by a grassland, or meadow, community. In turn, the grassland community will be succeeded by a forest community. The various stages of succession usually appear next to each other in an area, with transition zones between them. (See Fig. 24-6.)

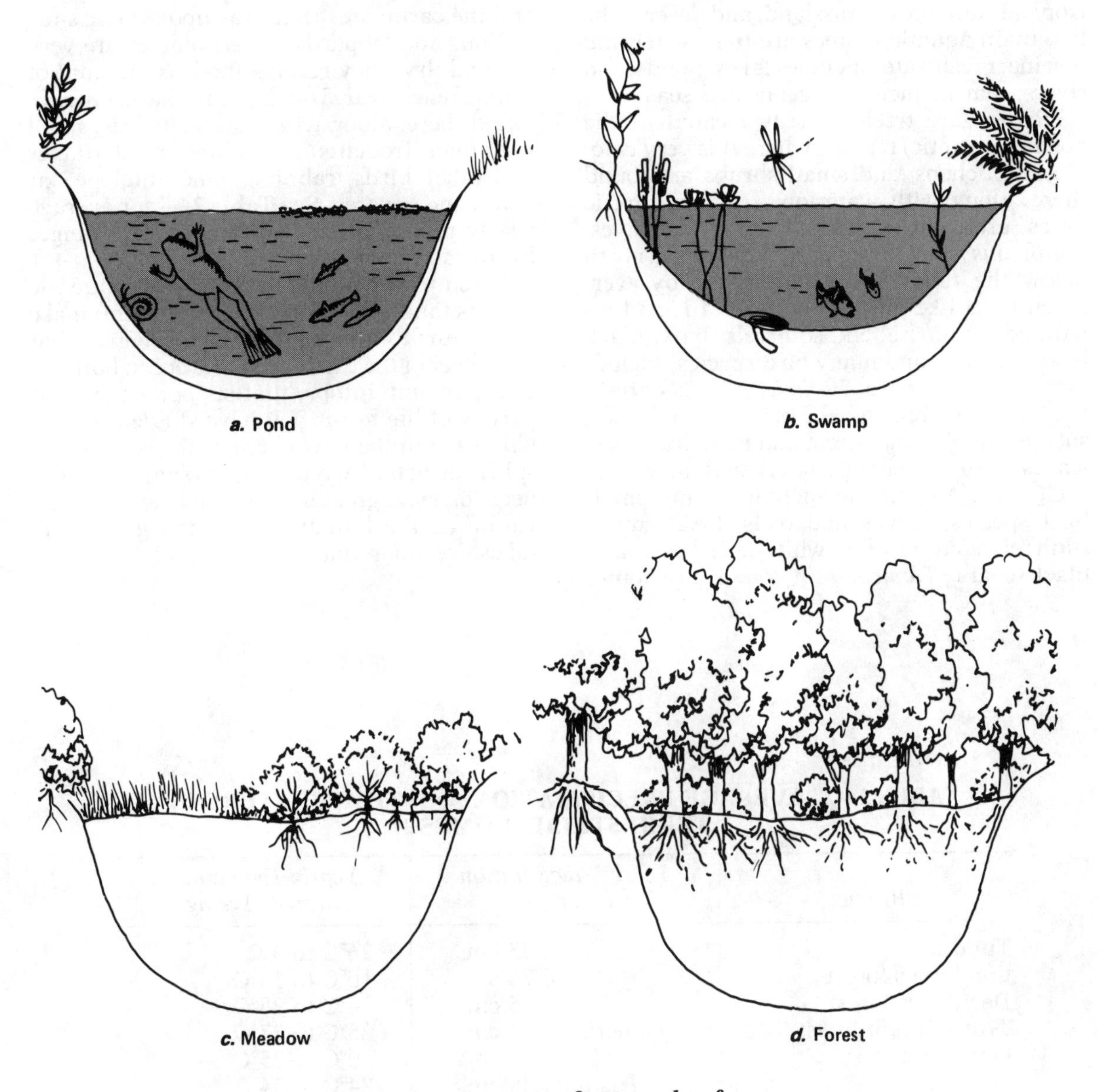

Fig. 24-6. Succession from pond to forest.

Biomes and the Biosphere

The final community to appear in a biological succession is called the **climax community.** A climax community is often dominated by several large, long-lived tree species and the animals that live in and around them. The specific climax community of an area depends upon the climate there. Major climax communities that cover broad regions are known as **biomes.**

The main types of biomes found on land are the tundra, coniferous forest, deciduous forest, tropical rain forest, grassland, and desert. The two main aquatic biomes are freshwater and marine. Freshwater includes lakes, ponds, and rivers; marine includes oceans and seas.

Tundras are treeless areas located in the northern (Arctic) regions where it is very cold. Mosses, lichens, and small shrubs are found there, along with waterfowl, caribou, arctic hares, arctic foxes, musk oxen, and wolves. *Coniferous forests,* or *taigas,* found in the north below the tundras, are dominated by evergreen trees like pine, spruce, and fir, and inhabited by moose, elk, squirrels, lynx, black bears, wolves, and many bird species, including hawks and owls. *Deciduous forests* are in the temperate regions; the broad-leaved trees, such as maple, oak, birch, and elm, drop their leaves with the change of seasons to winter each year. Animal life includes many small bird species, hawks and owls, foxes, mice, squirrels, some reptiles, white-tailed deer, and black bears. *Tropical rain forests* are found around the equator of the earth. This type of biome is hot, receives much rainfall, and has the greatest variety of plant and animal life, including hundreds of broad-leaved tree species, countless insect species, numerous colorful birds, reptiles, amphibians, and many monkeys, bats, sloths, anteaters, tapirs, armadillos, and jaguars. *Grasslands* receive some rainfall, but not enough to support forests. They are inhabited by herds of antelope, zebra, bison, and other types of herbivores, and the carnivores that prey upon them, such as lions and leopards. *Desert* biomes are very hot and dry; they receive the least amount of rainfall each year. Tough plants like cactus are found there, along with snakes, lizards, small mammals (rodents), scorpions, road runner and other birds, rabbits, some antelope, kit foxes, and coyotes. See Table 24-2 for average yearly precipitation and temperature ranges for the six main terrestrial biomes.

Taken together, the terrestrial and aquatic biomes (and all the organisms in them) make up the earth's **biosphere.** The biosphere is that thin layer of the earth, from the ocean bottoms to the mountaintops, that supports the great variety of life forms you have studied. As you will learn in the next chapter, the earth's biosphere is actually a delicate system that is in need of ever greater protection and preservation as the human population grows and takes over more habitats.

TABLE 24-2. AVERAGE PRECIPITATION AND TEMPERATURE OF TERRESTRIAL BIOMES

Biome	*Yearly Precipitation (Average)*	*Yearly Temperature Range (Average)*
Tundra	*less than* 25 cm	−26°C to 4°C
Coniferous forest	35 to 75 cm	−10°C to 14°C
Deciduous forest	75 to 125 cm	6°C to 28°C
Tropical rain forest	*more than* 200 cm	25°C to 27°C
Grassland	25 to 75 cm	0°C to 25°C
Desert	*less than* 25 cm	24°C to 34°C

CHAPTER REVIEW

Science Terms

The following list contains all of the boldfaced scientific terms found in this chapter and the page on which each appears.

aquatic (p. 286)
biological succession (p. 291)
biomes (p. 292)
biosphere (p. 292)
carnivores (p. 288)
climax community (p. 292)
community (p. 286)
competition (p. 289)
cooperation (p. 289)
decomposers (p. 289)
ecology (p. 285)
ecosystem (p. 286)
environment (p. 285)
food chain (p. 288)
food web (p. 288)
habitat (p. 287)
herbivores (p. 288)
humus (p. 290)
niche (p. 287)
omnivores (p. 288)
population (p. 286)
predators (p. 289)
prey (p. 289)
primary consumers (p. 288)
producers (p. 288)
saprophytes (p. 289)
scavengers (p. 289)
secondary consumers (p. 288)
social animals (p. 289)
soil (p. 290)
symbionts (p. 289)
symbiosis (p. 289)
terrestrial (p. 286)
tertiary consumers (p. 288)

Matching Questions

On the blank line, write the letter of the item in column B which is most closely related to the item in column A.

	Column A	Column B
____	1. study of living things and their environment	*a.* scavengers
____	2. all members of same species in one area	*b.* community
____	3. way an organism lives within its habitat	*c.* omnivores
____	4. each organism feeds upon another one	*d.* humus
____	5. all the different organisms in one area	*e.* ecology
____	6. animals that eat both plants and animals	*f.* symbiosis
____	7. decayed organic matter found in the soil	*g.* herbivores
____	8. all the biomes in the world together	*h.* population
____	9. animals that eat dead animal "leftovers"	*i.* food chain
____	10. two different organisms living together	*j.* biosphere
		k. niche

Multiple-Choice Questions

On the blank line, write the letter preceding the word or expression that best completes the sentence or answers the question.

1. Included in the nonliving environment are
a. air, soil minerals, water, light
b. plants, air, food, water
c. bears, oaks, water, minerals
d. minerals, light, snakes, birds
1 ____

2. Plants that possess floating leaves are suited to life
a. on land *b.* in water *c.* on land or in water *d.* as parasites 2 ____

3. When their food is scarce, robins in a community must either starve to death or
a. make their own food
b. compete with hawks
c. change their diet to fish
d. move to another similar community 3 ____

4. Plants are considered to be
a. mainly predators
b. food producers
c. decomposers
d. primary consumers 4 ____

5. For breathing, water-dwelling animals usually possess
a. lungs *b.* gills *c.* air channels *d.* long snouts 5 ____

6. Cactus plants can withstand conditions that are
a. very moist *b.* hot and humid *c.* cold *d.* very dry 6 ____

7. A whale can live in a cold-water environment because it
a. breathes air
b. has thick fur
c. has thick layers of fat
d. has a large tail fin 7 ____

8. Most plants grow best in a soil made up of
a. only sand
b. only clay
c. only humus
d. sand, clay, and humus 8 ____

9. Grass plants and clover plants growing in the same field may compete with one another for
a. proteins *b.* fat *c.* warmth *d.* light 9 ____

10. Two animals that may compete with each other for food are the
a. bear and grasshopper
b. horse and mountain lion
c. horse and cow
d. mouse and owl 10 ____

11. Mice are the prey of cats. This means that
a. cats eat mice
b. mice eat cats
c. dogs chase mice
d. owls eat mice 11 ____

12. A lichen consists of two organisms; one is a fungus and the other is
a. a bacterium *b.* an alga *c.* a virus *d.* a flower 12 ____

13. The relationship between the tapeworm and the human is that of
a. two independent animals
b. saprophytes
c. parasite and host
d. beneficial symbiosis 13 ____

14. A cooperative relationship exists between
a. clover and certain bacteria
b. cats and clover
c. sparrows and sparrow hawks
d. cats and mice 14 ____

15. A predator-prey relationship normally exists between
a. man and wheat
b. sheep and dogs
c. mountain lions and deer
d. mice and grasshoppers 15 ____

16. The outstanding characteristic of symbionts is their ability to
a. cooperate
b. compete
c. be completely independent
d. cause disease 16 ____

17. Certain bacteria put back into the soil substances that many plants take out of the soil. The most important of these substances is
a. iron sulfide *b.* water *c.* nitrates *d.* carbon dioxide 17 ____

18. Producer organisms, consumer organisms, and decomposer organisms make up
a. food webs *b.* parasites *c.* a succession *d.* decay 18 ____

19. Which are the first organisms in a food chain?
a. fungi
b. bacteria
c. algae and plants
d. carnivores
19 ____

20. Bacteria of decay help
a. form oxygen
b. enrich soil
c. form water
d. deplete soil
20 ____

21. In a schoolroom aquarium consisting of goldfish and elodea plants,
a. the fish and plants have a favorable effect on each other
b. the fish and plants have no effect on each other
c. the fish and plants have a harmful effect on each other
d. the fish help the plants but the plants harm the fish
21 ____

22. The feature of a polar bear that enables it to withstand the Arctic cold is
a. heavy fur *b.* white color *c.* sharp teeth *d.* sharp claws
22 ____

23. If some grass seeds started to grow in a forest, we could expect the trees to
a. help the grass plants to grow
b. keep the grass plants warm
c. compete with the grass plants for light
d. provide the grass plants with carbon dioxide
23 ____

24. In a hive, the different types of bees
a. compete *b.* cooperate *c.* eat one another *d.* are predators
24 ____

25. Organisms that help make soil from rock are
a. bacteria *b.* spirogyra *c.* mushrooms *d.* lichens
25 ____

Modified True-False Questions

In some of the following statements, the italicized term makes the statement incorrect. For each incorrect statement, write the term that must be substituted for the italicized term to make the statement correct. For each correct statement, write the word "true."

1. Special body features that help an organism fit into its environment are called *adaptations*.
1 ________

2. Ducks have webbed feet and breathe by means of *gills*.
2 ________

3. Bees are social animals that *compete* with other members of the hive.
3 ________

4. Green plants are *consumer* organisms because they make food from inorganic substances.
4 ________

5. *Decomposer* organisms can feed upon either dead plants or dead animals.
5 ________

6. The green frog is an example of an animal that is often part of a *pond* community.
6 ________

7. Cactus plants are usually found in *forest* communities. 7 ____________

8. A *grassland* community can eventually change into a forest community. 8 ____________

9. The gradual replacement of a pond community by a grassland community is called biological *invasion*. 9 ____________

10. Major communities that cover broad regions of land or water are called *biospheres*. 10 ____________

Testing Your Knowledge

1. Explain the difference between the two terms in each of the following pairs:

   a. aquatic and terrestrial ____________

   b. predator and prey ____________

   c. competition and cooperation ____________

   d. producer and consumer ____________

   e. herbivore and carnivore ____________

   f. habitat and niche ____________

   g. ecology and ecosystem ____________

   h. biome and biosphere ____________

2. Explain why lichens can become established in areas where it is impossible for other organisms to live (especially plants).

____

____

3. Although hawks may sometimes carry off chickens, explain why the hawks may still be beneficial to farmers. ______________________________

____

____

4. Explain why a balanced aquarium containing water plants, goldfish, and a catfish may require little attention except for the addition of some fish food and water (to replace evaporated water).

____

____

____

5. Many plants and animals die every day. Explain why, after a month, the earth is not cluttered with the dead bodies of these organisms.

____

____

____

6. Describe how light, temperature, and precipitation vary with position on the earth's surface.

____

____

7. Describe the flow of energy in an ecosystem in terms of a pyramid of energy.

____

____

8. What conditions are necessary for a stable, self-sustaining ecosystem?

____

____

Forest Fires: How Best to Reduce the Risk?

Yellowstone National Park sprawls across more than two million acres of forests, lakes, rivers, and streams in the northwest corner of Wyoming. It is home to countless species of animals and plants, including grizzly bears and brown bears, elk and bison, bighorn sheep and mountain lions, mice, and moles, owls and hawks, and spruce, fir, and lodgepole pine trees. It is a preserved wilderness visited by two million tourists a year. In short, it is a national treasure.

Although almost no one is against protecting this natural treasure, many people differ on how to do the job. And among the most controversial protection issues is that of forest fires.

In 2002, the administration of President George W. Bush decided to ask Congress to ease various environmental laws so 10,000,000 acres of overgrown federal forest land could be thinned more rapidly in the hope to reduce widespread forest fires. The idea behind this philosophy, as outlined by the Agriculture Undersecretary (Mark Rey), "is to reduce the long-term effects of catastrophic fire." Thinning the federal lands now would provide a favorable environmental impact in the future.

The goal of this legislative and administrative plan (as put forth by the President) is a more "common sense" approach to the protection of vast amounts of forest land. Many of these lands have become choked with brush, twigs, dead leaves and small trees that act as kindling. Fires burned about 6.5 million acres of forest land in 2002, about double the 10-year average.

Forest fires are viewed as economic disasters by some and have the potential for the disruption of tourism. For example, what would be the consequences to parks and local economies if future fires were allowed to burn naturally? What if a huge fire destroyed an entire link in a food web? On the other hand, it is only during fires that some trees, like the cones of the lodgepole pines, open like popcorn kernels and the wind distributes their seeds far and wide. It's only the heat of the forest that can make the cones pop. So, in spite of the loss of many mature trees, hundreds of new trees are able to sprout in the rich soil after a fire.

However, many conservationists say that changing the environmental laws would actually be catering to the timber industry and result in widespread abuses, i.e., commercial logging of valuable trees, excessive thinning deep in the woods, and overabundance of road building in sensitive areas. In addition, the legislative program would by-pass the environmental impact the new laws were set up to create, and there would be fewer studies of the potential harm to endangered species.

The legislative changes and the administrative changes would apply to all 190 million acres the government classifies as at risk of fire. Most of that land, a combination of federal, state, and private acreage, is in the western United States. The park officials and ecologists are at odds with the local politicians and business owners over these new changes and the thinning of the forests. The debate is not a simple one and will not be resolved easily.

1. *Why do you think occasionally setting small fires, or letting small fires burn out naturally, might prevent large fires later?*

____

____

____

____

2. *On a separate sheet of paper, write a brief essay contrasting the arguments for and against the proposed new federal regulations to protect our forests. NOTE: You will first have to do some Internet research to gather more information on the topic.*

25

Conservation of Natural Resources

LABORATORY INVESTIGATION

TESTING HOW POLLUTED WATER AFFECTS PLANTS

A. Your teacher will supply you with 3 jars, some aquarium sand or gravel, a bowl, soap, some automobile engine oil, and 3 elodea plant cuttings. Work with a partner to perform this investigation.

1. Put about 2 cm of clean sand or gravel into each of the 3 jars. Label the jars A, B, and C.

2. Set one elodea plant cutting into each of the jars as shown in Fig. 25-1. (Be sure the plants are all about equal in length.)

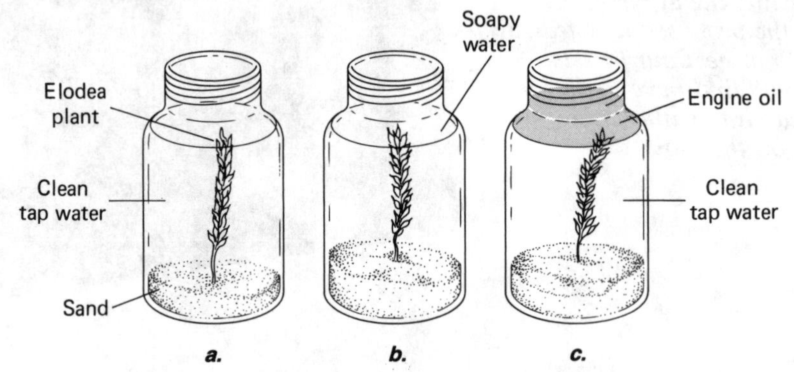

Fig. 25-1. Setup for polluted water investigation.

B. Pour enough water into the bowl to cover your hands. Wash your hands in the water with the soap. Save the soapy water. (Rinse your hands off in a sink.)

3. Pour enough clean tap water into Jar A to cover the plant.

4. Pour enough soapy water from the bowl into Jar B to cover the plant.

5. Pour enough clean tap water into Jar C to reach just below the top of the plant (about 3 mm). Then add some automobile engine oil to cover the plant and make the level of liquid in Jar C equal to that of Jars A and B. (Take care not to spill any engine oil on your hands.)

C. Set the three jars next to each other in a place where all will receive the same amount of sunlight. If an aquarium lamp is available, place it over the jars and turn it on each day. Otherwise, place the jars on a sunny windowsill.

6. Why did you use the same type of plant and the same amount of sand in all jars?

____

7. Why did you use three plants that were all the same size? ______________

____

8. Why must the level of the liquid and the amount of light be the same for all three jars?

____

D. Check the jars every day for about two weeks, keeping a record of changes you observe in the plants. At the end of the two weeks, answer the following questions:

9. How does the plant in each jar look? ______________

____

10. Which of the plants seems to be the most healthy? Which, if any, of the plants grew?

____

11. Give some possible explanations for your findings. ______________

____

The Balance of Nature

Before America was colonized by Europeans, its soil was fertile and its wilderness was vast. It seemed as though America's natural resources could never be used up. However, as a result of overfarming, overcutting of timber, and overhunting, these resources began to dwindle. Much of the fertile soil, like the loam in your previous lab experiment, became poor, like the sand and clay soil. Forests were clear-cut and destroyed. (*Clear-cutting* is the removal of all the trees in an area.) Lakes and rivers were polluted. And many species of wildlife became extinct.

NATURAL BALANCE

The **natural balance** (or *balance of nature*) is the tendency for the numbers of organisms in a given community to remain relatively unchanged from year to year. However, since natural conditions always change, this balance may be temporarily upset. However, the normal feeding habits of the organisms in the community may then restore the natural balance in the area. For example, after a good growing season when food is plentiful, an increase in the number of rabbits in an area is often followed by an increase in the number of hawks, which eat rabbits. As well-fed hawks reproduce in greater numbers, they eat more rabbits. This causes the number of rabbits to decrease. As a result, there is less food for the many hawks, and some of them may die of starvation or move away. Soon, as the number of hawks decreases, the number of rabbits begins to increase. Eventually, both the number of rabbits and the number of hawks return to approximately their original level in the area.

For many years this traditional view—that there is a balance of nature—has given scientists a model for studying ecosystems. Now, more and more biologists are beginning to think that what was considered a natural balance is actually a constant state of change, in which the numbers of organisms are always fluctuating. According to this new view, the numbers of hawks and rabbits will continue to vary (increase and decrease) rather than go back to any stable "original level" in their community.

NATURAL CYCLES OF ENERGY AND MATTER

As organisms carry out their life activities, they take in certain materials from the environment and they return other materials to it. The repeated shifting of particular materials between organisms and the environment forms **natural cycles.** Among the most important natural cycles are those dealing with *energy*, *carbon dioxide* and *oxygen*, *water*, and *nitrogen*.

Energy Cycle. In all food chains and food webs, energy from the sun is transferred to plants, and from them to animals. The transfer of the sun's energy to all types of organisms in an ecosystem is called the **energy cycle** (Fig. 25-2). As energy is transferred from one organism to the next, some is lost as heat in life activities. The energy lost is continually replaced in nature by the sun.

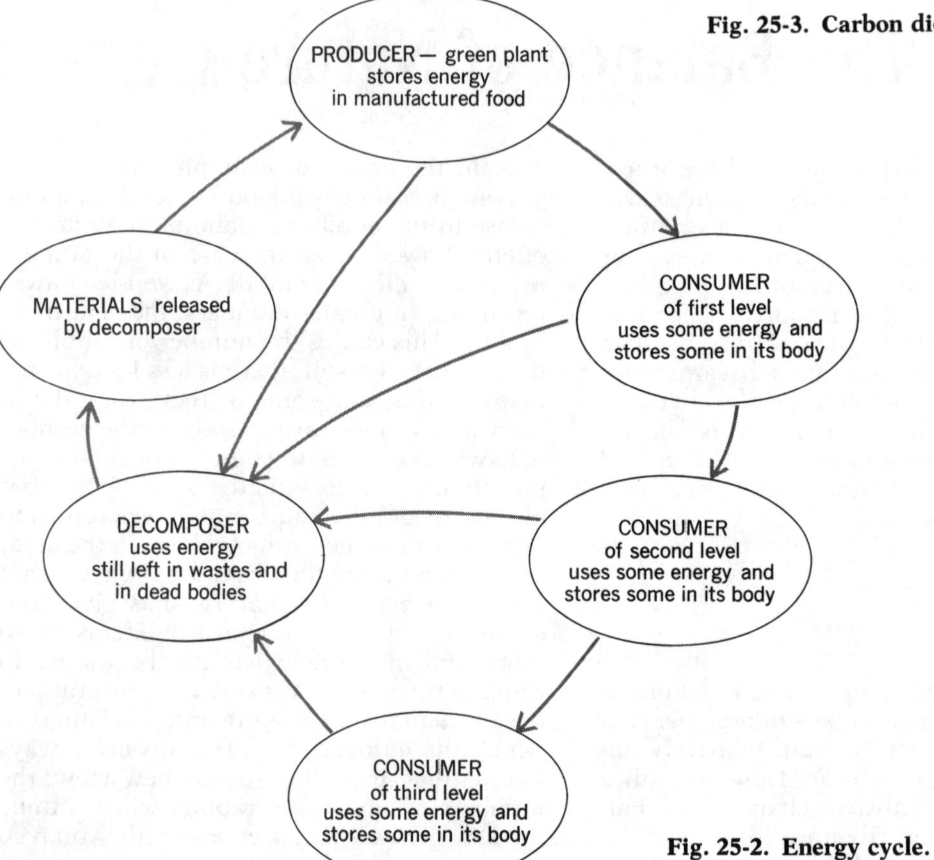

Fig. 25-2. Energy cycle.

Carbon Dioxide-Oxygen Cycle. The life processes of respiration and photosynthesis cause carbon dioxide and oxygen to circulate between plants and animals. In this cycle, called the **carbon dioxide-oxygen cycle,** plants and animals supply each other with the gases needed for life activities (Fig. 25-3).

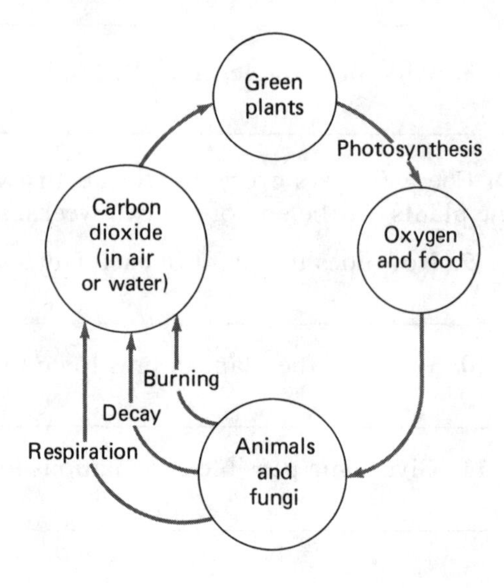

Fig. 25-3. Carbon dioxide-oxygen cycle.

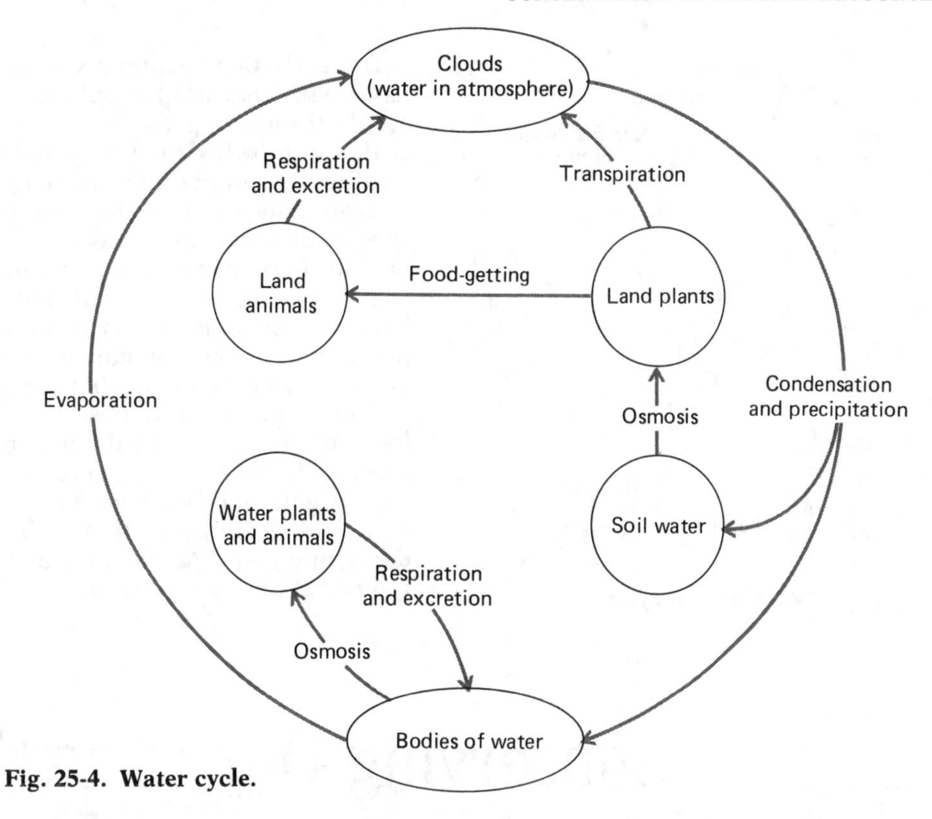

Fig. 25-4. Water cycle.

Water Cycle. As water evaporates, it enters the atmosphere. Then, as a result of condensation, water vapor leaves the atmosphere and enters the soil or returns to bodies of water. Living things take in this water and eventually release it. The movement of water through ecosystems is called the **water cycle** (Fig. 25-4). See Fig. 25-5 for an illustration of this cycle.

Nitrogen Cycle. In making the proteins they need, plants require the element *nitrogen.* Although nitrogen gas makes up four-fifths of the air, plants cannot use nitrogen in this form.

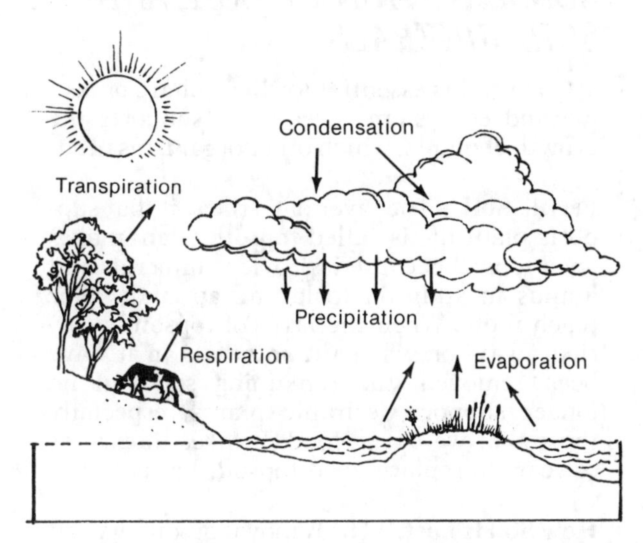

Fig. 25-5. The water cycle.

Only after nitrogen has been combined with other elements, forming *nitrates* in the soil, can it be used by plants. This natural process of combining nitrogen with other elements is called **nitrogen fixation.** Certain bacteria, such as those that live in the roots of clover, carry out the process of nitrogen fixation. (Refer back to Fig. 24-5.) After a plant has used the nitrogen to make its own protein, the plant may be eaten by an animal. After digestion, the animal's body changes the plant protein to animal protein. Dead plants and animals are decomposed by decay bacteria and fungi. Decay bacteria, which live in soil, break down the proteins of dead plants and animals and release the nitrogen present in them.

This complex cycle in which nitrogen circulates from the air to bacteria and plants, from plants to animals, and then back to bacteria and the air is called the **nitrogen cycle** (Fig. 25-6).

CHANGES IN THE NATURAL BALANCE

There are many natural events that upset the balance of nature. Among these are fires, floods, hurricanes, tornadoes, droughts, earthquakes, volcanic activity, and disease. It often takes years for a community to return to normal after such events occur.

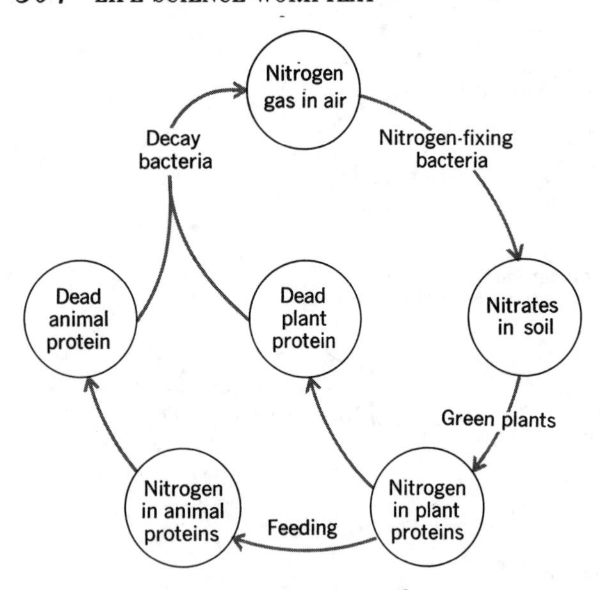

Fig. 25-6. Nitrogen cycle.

The activities of humans have often interfered with the natural balance, too. For example, by cutting down forests, mining for coal, and plowing the soil, people have made it difficult or impossible for many wild plants and animals to exist. By importing foreign species of organisms, people have upset the balance in many natural communities. A species that is not harmful in its native area may become destructive when it is released into a new region. This often happens when there are no natural predators of that species present.

Other human activities that have had a harmful effect on wildlife and their habitats include illegal hunting and overfishing, mining for gold and other minerals, paving over land and draining wetlands, building large dams, and polluting the air, water, and soil by industrial and other means.

Conserving Our Natural Resources

The effects of human disturbances on ecosystems are often more severe and longer lasting than those of natural upsets. It has therefore become necessary to practice conservation of **natural resources,** which include soil, water, air, forests, fuels, minerals, and wildlife. **Conservation** means using natural resources wisely so that they may be replenished, preventing further damage and waste, and preserving the natural balance of organisms.

Natural resources are classified as either renewable or nonrenewable. **Renewable resources,** which eventually can be replaced by natural processes, include air, water, soil and soil minerals, forests, and wildlife. **Nonrenewable resources** cannot be replaced after they have been used up or destroyed. These include fossil fuels (oil, natural gas, coal) and mineral ores. However, even though fresh water, trees, and good soil are renewable, the renewal process may take so long that these really may be considered *depletable resources*. The ever-increasing human population is using up more and more of the earth's renewable and nonrenewable resources.

Another grave threat to our resources is the pollution humans produce. **Pollution** is the addition of harmful substances into the air, land, or water. It has been said that pollution is actually a resource in the wrong place. An example of this occurs when good soil is washed away and ends up clogging a stream; the soil is a resource on the land, but a pollutant in the water.

CONSERVATION OF SOIL AND SOIL MINERALS

Good soil is essential for the welfare of people and ecosystems because it supports the growth of plants, which other organisms need.

Fertile Soil. The layer of fertile soil that supports plant life is called **topsoil.** Loam makes good topsoil because it provides minerals and humus in addition to letting air and water reach roots. When the layer of topsoil is carried away, or when the minerals in it have been removed, the remaining soil can no longer support desirable plants, especially those that we use as food. It takes thousands of years to replace good topsoil that is lost.

How Soil is Lost. The removal of soil by wind or moving water is called **erosion.** More than

30% of America's original topsoil has been lost in the past 500 years. The major causes of soil erosion are wind, water, and human activities such as strip mining, clear-cutting of trees, and poor farming practices.

In dry regions where vegetation is scarce, soil is uncovered and unprotected. Such exposed soil is often loose and easily removed by winds. Swiftly moving streams, runoff rainfall, and floods also remove large quantities of soil. The removal of trees and other vegetation by people exposes the underlying bare soil. Such unprotected soil then can be eroded by wind and water. And when crops like corn are planted in straight rows, the exposed soil between the rows is easily eroded.

Soil is also lost as a result of land development projects. Potentially good farmland is often paved over and built upon. Soil also may be harmed by chemical pollutants from industry and agriculture.

Conserving Soil. The erosion of soil can be prevented in several ways. After harvesting a crop, farmers should not leave the soil uncovered. Instead, a *cover crop,* like alfalfa, should be planted. Alfalfa plants spread and grow close together, thus covering the soil. And nitrogen-fixing bacteria in alfalfa roots help to enrich the soil. Adding humus to the soil also helps put nutrients back into the soil and prevent its erosion.

Strip cropping—planting alternate strips of two different crops—is another soil-saving technique. Strips of clover are planted between strips of corn, for example (Fig. 25-7). The corn provides no ground cover, but the clover does, thus preventing erosion. Nitrate-producing bacteria in clover roots also help to enrich the soil.

Fig. 25-7. Crops planted in strips.

In **contour plowing** (Fig. 25-8), grooves (furrows) are made across slopes along their curves, or *contours*. Unlike up-and-down furrows, contour furrows slow the process of topsoil erosion from rain runoff. **Terraces** are

Fig. 25-8. Contour plowing on a slope.

steplike, flat areas that are dug on steeply sloping farmland (Fig. 25-9). The terraces also slow the rainfall runoff that could carry away topsoil.

Fig. 25-9. Crops planted on terraces.

In addition, rows of trees and hedges reduce the force of strong winds. Such plantings around fields are called **windbreaks** (Fig. 25-10). When the force of the wind is broken, loose soil is less likely to be blown away.

Fig. 25-10. Windbreak planting.

How Soil Minerals are Lost. The continuous year-to-year growing of only one type of crop in a field removes large amounts of soil minerals. This practice does not allow time for natural cycles to restore the minerals. Eventually, such soil becomes incapable of supporting plant life. The soil then lies bare and can be easily eroded.

Conserving Soil Minerals. The permanent removal of essential minerals from the soil can be prevented by the rotation of crops, by the addition of fertilizers, and by letting the land lie unused, or **fallow.**

Repeatedly changing one crop with another in the same field is called **rotation of crops.** In a rotation scheme, cotton may be planted first for a few years. After gathering the cotton, a pod-bearing plant like peanuts is grown in the same field for two years. Then, after harvesting the pod-bearing crop, cotton is planted again, and so on. Nitrogen-fixing bacteria that live in the roots of pod-bearing plants restore nitrates that the cotton plants have absorbed from the soil.

Fertilizers are substances that are added to soil to enrich its mineral content. Most fertilizers include inorganic chemicals, such as sodium nitrate, and organic materials, like manure or decaying plant matter (*compost*). Lime is also added to soil to reduce its acidity and aid plant growth. And, by letting land lie fallow for a year, farmers allow natural processes to replace minerals in the soil.

CONSERVATION OF WATER

Water is essential for the life activities of all living things. People use water for home, agricultural, industrial, hydroelectric, recreational, and waste disposal purposes. As the human population increases, so does its need for fresh water. But the supply of fresh water is limited. Waste water often has to be purified to be used again.

How Water is Wasted or Made Unfit for Use. Water is often wasted when a large amount of rainfall runs into streams (that empty into a lake or sea) and is unavailable for the soil. This *excessive runoff* may be the result of the removal of vegetation from hillsides, improper plowing, or failure to install dams in streams that are likely to flood. A lot of fresh water is also lost as a result of agricultural and industrial use.

Water containing impurities that are harmful to living things is called *polluted water.* Water is often polluted by sewage from homes and by chemical wastes from factories and farms. Impurities in water can kill fish and aquatic plants and cause serious diseases in humans. As you saw in your laboratory investigation, water containing pollutants, like soap and oil, does not support plant life as well as clean water does. In nature, detergents that enter streams and oil that is spilled into oceans cause great damage to aquatic ecosystems and loss of wildlife.

Pollution of the oceans harms marine life and people who eat polluted seafood. Insecticides and fertilizers that the rain washes off farmlands pollute nearby streams and lakes. Another threat to the purity of aquatic ecosystems is acid rain. **Acid rain** is formed when the air pollutants sulfur dioxide and nitrogen oxides (from industry) dissolve in falling rain. The acidic rainfall kills off life in ponds and lakes. Putting lime into such bodies of water helps to restore the natural chemical balance and aquatic life forms.

Conserving and Protecting Water. Excessive runoff of rainfall can be prevented by planting grass or trees on hills, by contour plowing and terracing, and by building dams. These practices not only conserve water but also conserve soil. In fact, water conservation aids conservation of forests and wildlife as well. Proper management of reservoirs, watersheds, and reforestation projects (replanting trees) all help to conserve fresh water.

Water can be protected against pollution by using sewage-disposal plants and purification processes instead of allowing sewage and other wastes to empty directly into lakes, streams, and oceans. Other protective measures include the construction of safer oil tankers (to prevent oil spills) and the prohibition of ocean dumping of sewage, medical, nuclear, and chemical wastes.

CONSERVATION OF FORESTS

Forests are valuable for numerous reasons. Since the forest floor contains humus and rotting vegetation, it absorbs water and thereby prevents erosion and floods. Forests help regulate the climate, remove excess carbon dioxide from the air, and release oxygen into the air. They provide the wood necessary for producing building materials, furniture, paper, and chemicals, as well as providing sources of medicines and foods. Many wild animals depend on the forests for their natural habitat. In addition, many people enjoy using the forests for viewing wildlife, for camping, and for other outdoor activities.

How Forests are Destroyed. Forests have been damaged or destroyed by numerous causes. Parasitic fungi are responsible for the disease and death of many trees. Examples of fungus diseases are white-pine blister rust and Dutch elm disease. Many insects damage and even kill trees. Examples are the tent caterpillar, which eats leaves, and the wood-boring beetle, which tunnels into branches and trunks.

Acid rain damages forests by acidifying the soil and killing off trees. This is a serious problem for forests in many industrialized regions. (See Fig. 25-11.)

Fig. 25-11. A forest damaged by acid rain.

Forest fires also destroy or damage thousands of acres of forests every year. Some fires are caused naturally by lightning and volcanic eruptions, and may even be necessary for forest regrowth and succession. However, the greatest damage is often done by humans. People destroy forests by carelessly starting fires, by improper lumbering practices that do not promote regrowth of trees, by cutting trees down for fuel, and by practicing "slash-and-burn" farming, which destroys tropical rain forests around the world.

Conserving Forests. Several practices are necessary for good *forest management.* As with other crops, only mature trees should be harvested. These should be replaced with plantings of young trees. However, there is also a need to protect the "old growth" forests of truly ancient trees, because they are unique habitats in which rare wildlife species may live.

Selective cutting, rather than clear-cutting of forests, should be practiced by loggers. Diseased trees should be removed. All imported trees and other plants should be inspected to prevent the introduction of new fungus and insect pests. Harmful insects should be controlled by introducing organisms that feed upon the insects, without upsetting the local ecology. Careful chemical control of tree diseases also may be used.

Lookout towers, organized fire fighters, and the use of modern firefighting equipment can help limit harmful forest fires. Educating the public to be careful when tending campfires can also help prevent forest fires.

Millions of acres of forests have been set aside by the government as forest reserves and national parks. These areas are protected from logging and other damaging practices. This helps to conserve soil, water, and wildlife, as well as trees.

CONSERVATION OF WILDLIFE

Although no longer the primary source of food or clothing for most people, wildlife is valuable to us in many ways. Tourism based on wildlife viewing is an important business in many nations. Commercial fishing provides employment and food throughout the world. When left undisturbed by humans, wildlife tends to maintain its own natural balance. Unfortunately, there are few undisturbed areas of wildlife today. It is important to protect wildlife, since the health of ecosystems, and of the earth itself, depends on the preservation of the many complex webs of life.

How Wildlife is Destroyed. Many types of wild animals and plants have disappeared in recent decades as the result of human activities. People have overhunted animals, like the passenger pigeon, to the point that they have become extinct. Numerous other animals, like the bison and whooping crane, are recovering from near extinction. The world's catch of commercially valuable fish is even becoming smaller each year due to overfishing.

Many animals are illegally killed for their skins, meat, fur, horns, or other parts, while thousands more are trapped for the pet trade and scientific research. But by far the greatest threat to wildlife is the destruction of habitat due to human activities, such as **deforestation** (extensive tree cutting) and land development projects. In addition, our chemical pollutants harm wildlife, causing problems in reproduction, loss of food sources, and death.

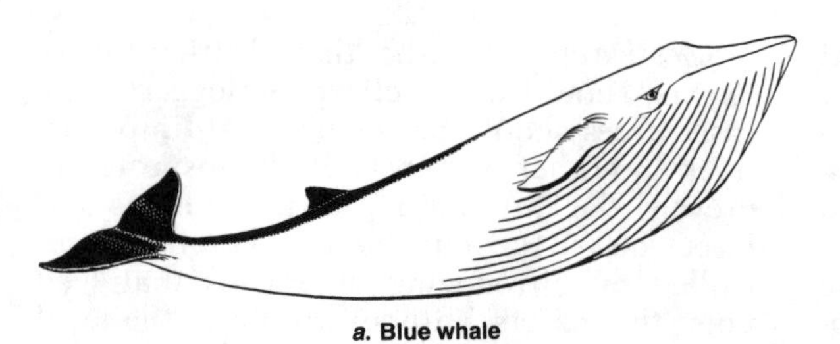

a. Blue whale

b. Bald eagle

Fig. 25-12. Two endangered species that are protected by law.

Conserving Wildlife. Governments do several things to help wildlife. For instance, government agencies build *fish hatcheries* where young fish, like trout, are raised and protected. When old enough, the trout are released into lakes and streams.

By setting up *feeding stations* during the winter months, wildlife agencies help many birds survive. Special parks, called *bird sanctuaries,* provide undisturbed breeding places for wild birds. In many states, laws limit or even prohibit the killing of certain birds.

Game laws, which control sports hunting, are designed to insure that there will be enough animals left alive to breed. Among such laws are those that forbid hunting during the breeding season or taking of females and young, and that limit the number of animals that one person may hunt. Wildlife is usually protected from hunting in *national parks, state parks,* and *wildlife sanctuaries,* where the animals are provided with a safe, well-managed habitat. In addition, laws such as the *Endangered Species Act* and the *Marine Mammal Protection Act* have been passed to protect rare species from illegal hunting and trade. (See Fig. 25-12.)

Private conservation organizations and scientific agencies also protect wildlife by researching the animals' needs, keeping track of population sizes, protecting critical habitats, breeding rare species in captivity, and reintroducing some rare species back into the wild when it can be safely done.

OTHER CONSERVATION AND POLLUTION PROBLEMS

Conserving Minerals. Industrialized nations use thousands of tons of mineral ores each year, most of which are mined in the poorer, developing nations. Not only is the supply of such resources limited, but the processes of mining and refining different minerals can cause habitat loss and pollution of other resources. Individuals and nations both have to be more responsible in the way they use these limited resources. The current movement to reuse, or **recycle,** more solid wastes like aluminum cans, glass bottles, and plastic packaging (chemically derived from petroleum deposits), is an important step toward reducing wastes and conserving resources. The search for cleaner, cheaper energy sources, like solar energy, is also important to reduce our dependence on oil, gas, coal, and uranium deposits.

Protecting Our Atmosphere. Air pollution is a threat to human health and to the living resources of ecosystems. As you have learned, acid rain, which forms from pollutants in the air, threatens the health of lakes and forests. Carbon monoxide, carbon dioxide, sulfur dioxide, and nitrous oxide gases are also harmful, irritating the eyes and lungs. Carbon dioxide (from the burning of forests and fossil fuels) and methane gas (from agricultural activities) trap heat from the sun close to earth's surface. Many scientists have warned that this heat build-up may cause a "**greenhouse effect,**" which is a warming up of the atmosphere. It is feared that this increase in heat could lead to a **global warming** (that is, an increase in average temperatures worldwide), which could have a drastic effect on our lives and on ecosystems. Scientists are researching the possible consequences, and governments and industries are trying to limit carbon dioxide emissions to avoid any negative effects on the environment.

Another compound that is a resource in one place and a pollutant in another is **ozone** (O_3). In the lower atmosphere, ozone is an irritating gas that forms as a result of human activities and which makes up a large part of **smog** (visible, hazy air pollution). In the upper atmosphere, ozone forms naturally, creating a layer of gas that shields the lower atmosphere from the harmful effects of the sun's ultraviolet rays.

Unfortunately, certain chemical compounds in air pollution from aircraft exhaust and aerosol sprays break down these ozone molecules in the upper atmosphere, thinning the protective layer and letting more ultraviolet rays reach the earth. It is thought that this increase in radiation could cause a rise in skin cancer incidences and may also cause more genetic mutations. Governments, industries, and scientists are working together to reduce the pollutants that harm the earth, and to preserve those parts of the environment that protect the earth's atmosphere and the living systems that depend upon it.

CHAPTER REVIEW

Science Terms

The following list contains all of the boldfaced scientific terms found in this chapter and the page on which each appears.

acid rain (p. 306)
carbon dioxide-oxygen cycle (p. 302)
conservation (p. 304)
contour plowing (p. 305)
deforestation (p. 307)
energy cycle (p. 302)
erosion (p. 304)
fallow (p. 306)
fertilizers (p. 306)
global warming (p. 308)
greenhouse effect (p. 308)
natural balance (p. 301)
natural cycles (p. 302)
natural resources (p. 304)
nitrogen cycle (p. 303)
nitrogen fixation (p. 303)
nonrenewable resources (p. 304)
ozone (p. 308)
pollution (p. 304)
recycle (p. 308)
renewable resources (p. 304)
rotation of crops (p. 306)
smog (p. 309)
strip cropping (p. 305)
terraces (p. 305)
topsoil (p. 304)
water cycle (p. 303)
windbreaks (p. 305)

Matching Questions

On the blank line, write the letter of the item in column B which is most closely related to the item in column A.

	Column A	*Column B*
____	1. transfer of sun's energy to organisms	*a.* pollution
____	2. wise use of natural resources	*b.* renewable resources
____	3. extensive overcutting of trees	*c.* erosion
____	4. can be replaced by natural processes	*d.* conservation
____	5. cannot be replaced after used up	*e.* sanctuaries
____	6. harmful substances in the environment	*f.* deforestation
____	7. increase in average world temperatures	*g.* water cycle
____	8. removal of soil by wind and/or water	*h.* energy cycle
____	9. enrich mineral content of the soil	*i.* nonrenewable resources
____	10. parks where wildlife is protected	*j.* fertilizers
		k. global warming

Multiple-Choice Questions

On the blank line, write the letter preceding the word or expression that best completes the sentence or answers the question.

1. If the plants and animals of a natural community are left alone for several years, their numbers generally
a. increase slowly
b. remain the same
c. decrease slowly
d. decrease rapidly
1 ____

2. The exchange of carbon dioxide and oxygen between the plants and animals in a forest is best described as the
a. respiration cycle
b. nitrogen cycle
c. water cycle
d. carbon dioxide-oxygen cycle
2 ____

3. The nitrogen cycle is important for plants because it enables them to get
a. water from the soil
b. nitrogen gas from the air
c. nitrates from the soil
d. protein from animals
3 ____

4. Nitrogen-fixing bacteria grow naturally on the roots of the
a. geranium plant
b. celery plant
c. clover plant
d. potato plant
4 ____

5. Human activities that have upset the balance of nature include all of the following, except
a. building large dams
b. volcanoes and earthquakes
c. mining for gold
d. paving over land
5 ____

6. The phrase that most closely describes the conservation of natural resources is
a. no longer using freshwater reserves
b. closing farms to preserve topsoil
c. preventing pollution of the air
d. using water, soil, and forests wisely to preserve the balance of nature 6 ____

7. Renewable resources include all of the following except
a. air and water *c.* soil and forests
b. mineral ores *d.* wildlife 7 ____

8. Substances that may be helpful in one area, but are pollutants when washed into another area include all of the following except
a. parasitic fungi *c.* insecticides
b. fertilizers *d.* topsoil 8 ____

9. Forests, grasses, and farm plants are supported by the soil layer called
a. sand *b.* topsoil *c.* subsoil *d.* eroded soil 9 ____

10. The destructive process that removes topsoil is called
a. erosion *c.* contour plowing
b. the water cycle *d.* terracing 10 ____

11. Two natural causes of soil erosion are
a. wind and runoff water *c.* farming and irrigation
b. wind and lightning *d.* runoff water and fungi 11 ____

12. Plowing across a hill instead of up and down is called
a. strip plowing *c.* contour plowing
b. terrace plowing *d.* cover plowing 12 ____

13. Peanuts are a good crop to use in a scheme of crop rotation because their roots bear
a. humus *b.* root hairs *c.* nitrogen-fixing bacteria *d.* decay bacteria 13 ____

14. How can soil erosion on very hilly farmland best be reduced?
a. frequent plowing *c.* removing dead trees
b. growing cotton *d.* terracing 14 ____

15. A belt of trees growing all around a field of corn is an example of
a. a cover crop *c.* strip cropping
b. a forest succession *d.* a windbreak 15 ____

16. Contour plowing is most likely to help in the
a. evaporation of water *c.* making of topsoil
b. reduction of erosion *d.* making of nitrates 16 ____

17. Mineral content of soil can be restored by the addition of
a. manure *b.* lime *c.* saltpeter *d.* cotton plants 17 ____

18. The reforestation of a burned-over area eventually results in increased
a. soil water *b.* floods *c.* aridity *d.* runoff 18 ____

19. After an entire forest has been cut down for lumber, we can expect the soil to
a. improve *b.* form layers *c.* remain the same *d.* be eroded 19 ____

20. Special parks in which birds can breed without being disturbed by people are called
a. bird hatcheries *c.* licensed hunting grounds
b. bird sanctuaries *d.* game preserves 20 ____

21. Robins often eat earthworms which live in moist soil. If the soil should become dry, we could expect the number of
a. robins to increase
b. robins to decrease
c. robins to remain the same
d. earthworms to remain the same
21 ____

22. The energy cycle on earth begins with the energy that comes from
a. soil *b.* sunlight *c.* respiration *d.* gasoline
22 ____

23. In which cycle do bacteria of decay play an important part?
a. nitrogen *b.* water *c.* oxygen *d.* energy
23 ____

24. When an organism is freed of its natural enemies and has plenty of food, the individuals of that species should
a. decrease in number
b. remain the same in number
c. increase in number
d. migrate
24 ____

25. The extinction of the passenger pigeon was caused by
a. a poultry disease
b. overhunting by people
c. drought and erosion
d. increases in the hawk population
25 ____

Modified True-False Questions

In some of the following statements, the italicized term makes the statement incorrect. For each incorrect statement, write the term that must be substituted for the italicized term to make the statement correct. For each correct statement, write the word "true."

1. Evaporation and condensation are two steps in the *nitrogen* cycle. 1 ____________

2. The depth of fertile *topsoil* in the United States is now less than it was 500 years ago. 2 ____________

3. The planting of alternate rows of wheat and alfalfa is an example of *hillside terracing*. 3 ____________

4. Wheat that is planted in soil having plenty of *nitrates* can be expected to produce a crop that is rich in plant proteins. 4 ____________

5. Factories that are located near streams often contribute to the *conservation* of fresh water. 5 ____________

6. A disease that is responsible for killing thousands of elm trees is caused by a type of *rabbit*. 6 ____________

7. A species of bird that nearly became extinct in the United States is the *whooping crane*. 7 ____________

8. Fungi and *nitrogen-fixing* bacteria help break down the bodies of dead plants and animals on a forest floor. 8 ____________

9. The warming up of the earth's atmosphere, as a result of gases trapping heat from the sun, is known as the *recycle* effect. 9 ____________

10. Proper disposal and purification of sewage helps to control the problem of *air* pollution. 10 ____________

Testing Your Knowledge

1. Explain the difference between the two terms in each of the following pairs:

   a. renewable resource and nonrenewable resource

   b. terracing and contour plowing

   c. strip cropping and crop rotation

   d. organic fertilizer and inorganic fertilizer

   e. decay bacteria and nitrogen-fixing bacteria

   f. selective tree cutting and forest clear-cutting

2. Explain how a sudden increase in the number of snakes in an area may affect the number of frogs in that area.

3. If you live in a large city, such as Chicago or New York, why might you be interested in a proposal that Congress establish a national park in an undeveloped area in Oklahoma?

4. If you live in the countryside, why might you be concerned about the water or air pollution problems of a large city 500 kilometers away?

5. Define, give an example, and explain the value or function of each of the following:

   a. cover crop

   b. fish hatchery

   c. natural balance

6. Explain why the sun is actually the indirect, primary energy source for all carnivores.

Oil Spills: Must We Trade the Environment for Energy?

"Fill it up," you'll say one day to a service station attendant. And then, with your hand clutching a $20 bill and your eye on the gas pump, you'll wait to see just what price you have to pay for your full tank of gasoline.

Unfortunately, you and everyone else who uses products made from crude oil—the name for oil fresh out of the ground—may actually be paying a much higher price than the one shown on the gas pump. The higher price is not measured in dollars but in birds, fish, otters, shellfish, polluted beaches, lost jobs, and lost vacation spots.

The fact is that crude oil is not always conveniently located near our homes, businesses, factories, and farms. A lot of it lies deep beneath the ground in the Middle East or in the most northern part of Alaska. It is literally oceans away from some of the major oil-consuming nations in the world. This means that billions of barrels of oil must be transported across the oceans every year. Unfortunately, not all of that oil gets where it's supposed to be going. Some of it ends up in oceans and other large bodies of water. And that's when the environmental price of oil soars.

To get an idea of what this price can come to, imagine that you are sailing on the beautiful, clear body of water called Prince William Sound, off the southern coast of Alaska. There, on an early spring day in 1990, the huge oil tanker *Exxon Valdez*, loaded with 200 million liters of crude oil, set sail from the town of Valdez. Valdez is where the pipeline from the northern Alaska oil fields ends.

But after traveling only 40 kilometers, the gigantic vessel—which was longer than three football fields laid end to end—smashed into a hidden reef. The reef sliced gashes through the ship's hull and oil came pouring out—42 million liters of it!

Why does crude oil kill living things? There are a number of reasons. For one thing, crude oil contains chemicals that are poisonous to many animals and plants. In addition, such oil sticks to the feathers of birds and the fur of otters. This robs the feathers and fur of their insulating, or heat-holding, ability. So the warmblooded birds and otters froze to death in the icy Alaskan waters. Many animals that were rescued and cleaned off still died later as a result of poisoning from the oil's chemical fumes.

The oil spill of the *Exxon Valdez* was only one of many that occur each year around the world. And as big as it was, it set no record. An even greater spill occurred on November 19, 2002. The Prestige, a 26-year-old tanker carrying 77,000 tons of crude oil, sank some 270 kilometers from the northwestern Spanish coast. This tragedy was one of the world's worst oil spills and the worst environmental disaster for Spain.

The Prestige, with its cargo of 77,000 tons of crude oil, sprang a leak on November 13, then snapped in two and sank six days later. A tide of foul-smelling sludge was unleashed on Spain's richest fishing grounds and the oil was seen floating to the northwest coast of Galicia.

The Greek captain (Apostolos Mangouras) of the tanker was arrested and charged with disobeying authorities and harming the environment. He was jailed in Teixeriro, Spain for 80 days until his bail was posted by a London insurance company.

Days later the oil was seen moving toward the region of Landes near Bordeaux in France. The oil patches, numbering around 100, were seen heading east into French waters. The slicks ranged in size but were seen to be about 30 meters in diameter.

As ugly as this sight was, the effects were even uglier. The oil spill claimed the lives of birds, countless fish, and many other animals and plants that lived in the water and along the shorelines. In addition, the cost of the cleanup ran into the millions of dollars.

1. *Two major sources of crude oil lie under*

a. Italy and Alaska.
b. Valdez and Seattle.
c. Alaska and the Middle East.
d. The Middle East and Valdez.

2. *The oil spill of the* Exxon Valdez *equaled*

a. 42 million barrels.
b. 42 million quarts.
c. 42 million gallons.
d. 42 million liters.

3. *Crude oil kills living things for the following two reasons:*

a. It is poisonous and reduces insulation.
b. It is poisonous and increases insulation.
c. It reduces insulation and forms ice.
d. It increases insulation and gets stickier.

In what ways might the possibilities of oil spills from ships be reduced?

____

____

5. *On a separate sheet of paper, write one or two sentences describing the benefits of oil to society. Then write one or two sentences describing how the use of oil may harm society. Finally, write one short paragraph giving your opinion of what can be done to improve the benefits while reducing the harm.*

Glossary

abdomen in insects, the hind region of the body behind the thorax, or midregion
absorption the passage of simple substances into the internal parts of a plant or animal; in higher animals, passage of nutrients into bloodstream
acid rain pollutants formed when sulfur dioxide and nitrogen oxide gases from industry dissolve in falling rain
acquired immunity immunity to a disease that is developed, either by the body making its own antibodies or by receiving a vaccination
acquired traits traits that develop during an individual's lifetime and which are not inherited
adaptations special traits that enable an organism to survive in its environment
addiction an unhealthy state of drug dependency and regular use
adrenal glands small ductless glands situated on each kidney; secrete adrenalin and cortisone
adrenalin hormone secreted by the adrenal glands when an individual is angry or frightened
AIDS the disease, or acquired immune deficiency syndrome, caused by the contagious HIV virus
air sacs (alveoli) microscopic pockets at the ends of bronchioles in the lungs; alveoli exchange carbon dioxide in the blood for oxygen
algae simple plantlike organisms that live in water or moist places; they have chlorophyll and make their own food by photosynthesis
alimentary canal the food tube; consists of the mouth, throat, esophagus, stomach, small intestine, and large intestine
allergens substances, such as pollen, dust, and foods, that cause an allergic reaction
allergy a harmful overreaction of the body's immune system to some foreign substance
amino acids organic compounds composed mainly of carbon, oxygen, hydrogen, and nitrogen; the end products of the digestion of proteins
amphibians class of coldblooded vertebrates that have a slimy skin; spend juvenile stage in water (use gills) and adult stage on land (use lungs)
androgens the male sex hormones secreted by the testes, or male sex glands, in animals
angiosperms the flowering seed plants, which produce seeds enclosed in a fruit
annelids segmented worms having three cell layers and complex organ systems
antibiotics drugs that slow the growth of certain bacteria or kill them; usually produced by molds
antibodies protein compounds made by the body, which attack germs, viruses, and antigens
antigens bits of foreign protein, which the body's immune system attacks
antihistamines drugs used to fight the effects of the body's allergic reactions, like sneezing
antiseptics chemicals that slow bacterial growth
anus opening of the rectum through which undigested solid wastes pass out of the body
aorta largest artery in the body; leads oxygen-rich blood out of left ventricle of the heart
aquatic a type of natural community that is found in or around a body of water; living in water
arachnids class of arthropods having two body sections and four pairs of walking legs attached to the front section
arteries blood vessels that lead blood away from the heart to capillaries throughout the body

arthropods phylum of invertebrates having two or three body sections, hingelike joints between the sections, and an outer body shell

artificially inseminated procedure in which the male's sperm is placed in the female's oviduct or uterus

artificial respiration procedure in which pressure on the lungs is rhythmically increased and decreased by an outside force to aid breathing

artificial selection intentional breeding of organisms by people for specific traits

asexual reproduction offspring are produced by only one parent

assimilation change of digested substances into new living material in the body's cells

atom smallest unit of an element that still has the properties of that element

ATP (adenosine triphosphate) compound in cells that stores the energy released by oxidation; direct source of energy for life activities of cells

atria (auricles) two upper chambers of the heart; receive blood from veins and pump it down into the ventricles

auditory nerve passes sound vibration impulses from nerve endings in the ear to the brain

autoimmune disease immune system fails to recognize normal cells as "self" and attacks own body cells

auxins plant hormones that control cell growth and movement of some plant structures

axon longest branch of a nerve cell; carries nerve impulse from the cell body to the end brush

bacteria one-celled, microscopic organisms, lacking nuclei; some cause diseases

balanced diet daily food selection that provides sufficient energy and nutrients needed by the body

behavior responses of an organism to changes in its environment

biceps muscles located on the upper side of the arm between the elbow and shoulder

bile juice secreted by the liver; breaks fat into tiny particles during process of digestion

binary fission asexual method of reproduction in which offspring are produced by the splitting of an organism into two equal parts

biological succession gradual change of one type of natural community into another

biomes major climax communities that cover broad regions of the earth

biopsy procedure in which a bit of living tissue is examined under the microscope for cancer cells

biosphere all the biomes of the earth together

blood a tissue that flows through the body, taking nutrients to and wastes from cells

blood tissue flows through body; composed of red blood cells, white blood cells, platelets, and plasma

blood transfusion transference of blood from a healthy person to another person in need

blood type determined by special proteins on red blood cells; people inherit one of four major types

blood vessels tubes that transport blood throughout the body

body cells cells that play no direct part in fertilization

bone tissue strong connective tissue having deposits of hard minerals between its living cells

bony fishes class of fish having a bony skeleton, overlapping scales, and a swim bladder

brain large mass of delicate nerve cells in the head

breastbone located in front of body, where ribs meet; along with ribs, encases the chest cavity

breathing inhalation and exhalation of air

breeds groups of similar organisms within a species

bronchi two large hollow branches, from the trachea to the lungs

bronchial tubes smaller branches of the bronchi

bronchioles microscopic tubes that branch from the bronchial tubes

budding asexual method of reproduction by which young organisms form from projections growing out of parent; unequal splitting of cytoplasm

bulb enlarged plant bud that grows underground and can develop into a complete plant

calorie amount of heat needed to raise the temperature of 1 gm of water 1° Celsius

cancer an abnormal growth of tissue due to uncontrolled division of cells

capillaries microscopic blood vessels having very thin walls; connect small arteries and veins

carbohydrates organic compounds composed of carbon, oxygen, and hydrogen

carbon dioxide-oxygen cycle circulation of carbon dioxide and oxygen between the atmosphere and living things

carnivores flesh-eating animals

cartilage makes up the skeleton of jawless fishes and sharks and rays

cartilage tissue strong, slippery, flexible supporting tissue of animals; is softer than bone

cartilaginous fishes class of fish having skeletons of cartilage; the sharks and rays

catalysts substances that speed up reactions by lowering the energy required; they are not chemically changed by the reaction

cell body control center of a nerve cell; it includes the nucleus and surrounding cytoplasm

cell division the division of one cell into two, two cells into four, and so on

cell membrane outermost living structure of all cells; encloses the cytoplasm, and controls passage of materials into and out of the cell

cell wall outermost structure of a plant, fungus, or algal cell; usually consists of cellulose

cells microscopic, boxlike units of which all living things are constructed

central nervous system the brain and spinal cord; receives and sends out impulses throughout body

cerebellum part of brain located between cerebrum and medulla oblongata; controls balance and coordination
cerebrum top, largest part of brain; center for thinking and control of voluntary activities
chemical digestion breakdown of food molecules by enzymes; occurs in mouth, stomach, and small intestine
chloroplasts small green bodies containing chlorophyll, found in plant and algae cells; where photosynthesis occurs
chlorophyll green compound that traps light energy for food-making in plants; acts as a catalyst in photosynthesis
cholesterol fatty deposit that accumulates on inner surface of arteries
chordate phylum of animals having gill slits, a notochord, and a nerve cord at some point in their development; includes all vertebrates
chromosomes composed of genes and proteins; dense rods that appear when cell divides
cilia tiny, vibrating hairs that line the surface of some epithelial cells
circulation transport of oxygen, digested nutrients, and waste materials throughout the body
class classification division for related organisms, above the order level and below the phylum level
climax community the final natural community to appear in a biological succession
cochlea coiled, bony fluid-filled tube in inner ear
coelenterates phylum of invertebrates composed of two main cell layers with a jellylike material between the layers
coldblooded animals organisms whose body temperature changes with that of the environment
communicable diseases (contagious, or catching diseases) diseases caused by pathogens that can be passed from one individual to another
community the entire population of living things existing in a given area
competition the need for, or use of, the same resources by different organisms
compound a substance composed of two or more elements united chemically in a definite proportion by mass
compound eyes in insects, large eyes consisting of thousands of tiny eyes; can detect moving objects
compound microscope scientific tool for observing tiny objects and organisms; has two lenses
conditioned reflex response learned response that results from the linking of two stimuli that repeatedly occur at the same time
conebearers (conifers) nonflowering seed plants that bear cones; also called gymnosperms
cones retinal nerve cells that are sensitive to light of different colors
conifers see *conebearers*
connective tissue makes up bone and cartilage; connects skin to tissue below it
conservation using natural resources in ways that restore them, prevent their waste, and preserve the natural balance of organisms
contaminated organisms or objects that contain pathogens and can spread disease
contour plowing farming technique in which grooves are made across slopes along curves
control group in an experiment, the subject group that is not exposed to the variable being tested
controlled breeding a type of artificial selection carried out to produce best traits of a breed
controlled experiment scientific experiment performed to test a hypothesis; includes a control group and an experimental group
convex lens clear, elastic structure in the eye, located behind the iris and pupil
cooperation different organisms working together to perform life activities and aid survival
cornea clear, colorless structure in the center of the front of the eye
cortisone hormone secreted by the adrenal glands
cotyledons the embryonic seed leaves of a young plant
crossbreeding (hybridization) method by which organisms of different breeds are mated to combine their desirable traits
crustaceans class of arthropods having two body sections, two pairs of antennae, and five pairs of legs; mainly aquatic
cysts small sacs formed in an animal's muscles, in which young parasitic worms lie dormant
cytoplasm all the material in a cell between the cell membrane and the nucleus
data in an experiment, the record of all scientific observations
daughter cells in cell division, the two complete new cells that are formed after mitosis and division of the cytoplasm
decomposers (saprophytes) organisms that feed upon and decay the bodies of dead organisms
deficiency disease illness resulting from the lack of a special substance or nutrient the body needs
deforestation extensive tree cutting and destruction of forests
dendrites short, branching fibers that project from a nerve cell body; receive stimuli and send messages to the cell body
depressant drug that slows down the central nervous system
dermis inner, thicker layer of the skin; contains oil glands, sweat glands, nerve endings, capillaries
diabetes condition in which too little insulin is secreted, so the glucose level of the blood is too high
diaphragm in mammals, a sheet of muscle below the lungs that separates the chest and abdomen regions to aid in breathing
diffusion spreading and mixing of molecules of two or more substances

digestion breaking down of food into simpler compounds that the cells of the body can use
digestive glands organs that secrete juices consisting of water, enzymes, and other compounds; found in stomach and small intestine
disease any body condition in which there is a disturbance of normal body activity or structure
DNA (deoxyribonucleic acid) compound in nuclei that makes up the genes (hereditary traits); only substance known that can make a copy of itself
dominant character inherited trait that overshadows the recessive trait and keeps it from showing in a hybrid
dominant genes genetic material that controls the dominant character traits
duct gland gland that secretes its juice into a tube leading directly into some organ
ductless gland (endocrine gland) gland that secretes its juice (hormone) directly into the blood, which carries the hormone throughout the body
ducts tubes that connect the digestive glands to the food tube
ear canal hollow tube of the outer ear that leads inward
eardrum membrane located in the middle ear
echinoderms phylum of marine invertebrates that have a round or star-shaped body covered with many spines
ecology study of the complex relationships between living things and their environment
ecosystem interactions of all the living and nonliving things in a natural community
effectors the muscles and glands that respond to motor nerve cell impulses, causing an effect
egg the female reproductive cell
egg-laying mammals the mammals that lay eggs (as reptiles do) and feed young on milk
electron microscope special microscope that allows scientists to see tiny structures within cells
element substance that cannot be changed to any other substance by ordinary chemical means
elimination movement of solid undigested wastes out of the body
embryo undeveloped organism formed by divisions of a fertilized egg
end brush many tiny branches at the end of an axon of a nerve cell; sends nerve impulse from the axon to the dendrites of an adjoining nerve cell
endocrine glands see *ductless gland*
endocrine system the ductless (endocrine) glands and their hormones
end products nutrients after they have been digested into simpler compounds in solution
energy cycle transfer of the sun's energy to all types of organisms
English system system of measurement that uses the foot for length, pound for weight, and quart for volume
enriched food a food to which vitamins have been added
environment all the living and nonliving things in the area in which an organism lives
enzymes organic catalysts that enable chemical reactions in the body to take place rapidly at body temperature
epidermal tissue outermost covering of a plant
epidermis in plants, outermost layer of cells covering the plant body; in animals, thin, outer layer of the skin
epithelial tissue in animals, tissue composed of cells that lie close together, covering the body and lining the internal organs
erosion removal of soil by the forces of wind and moving water
esophagus (gullet) part of the food tube that connects the throat with the stomach
estrogens the female sex hormones secreted by the ovaries, or female sex glands, in animals
Eustachian tube passageway leading from middle ear to the throat
evolution process of change that occurs in living things over time
excretion ridding the body of wastes that result from the other life activities of cells
exhalation the act of letting air out of the lungs
exoskeleton the outer body shell that covers all arthropods
experimental factor the variable that is tested in the experimental group
experimental group subject group that is exposed to the variable being tested
external fertilization in some aquatic species, the male's sperm swims to, and unites with, the female's eggs outside the female's body
extinct any species that no longer exists today
fallow in agriculture, land that is left to lie unused for some time
family classification subdivision above the genus level and below the order level
fats organic compounds composed of carbon, hydrogen, and oxygen; solid at room temperature
fatty sheath layer of fatlike material around the axon of a nerve cell
ferns vascular spore plants
fertilization union of a sperm cell and an egg cell in sexual reproduction
fertilizers substances that are added to soil to enrich its mineral content
fiber the hard, cellulose part of food plants that is not digested but is used by the body as roughage
five-kingdom system classification scheme in which all organisms are grouped into one of five major kingdoms
flagellum the whiplike tail that forms on a sperm cell, enabling it to move
flatworms flat, ribbon-shaped worms consisting of three cell layers and simple organ systems

fleshy root enlarged root containing stored foods; can grow into a complete plant by vegetative reproduction
food chain a group of organisms, starting with algae or plants, in which each organism feeds upon the one before it in the group
food web the criss-crossing of many food chains in an area
fraternal twins twin offspring that arise from two separate fertilized eggs; can be of opposite sexes
fronds the leaves of a fern plant; carry out photosynthesis and form spores on their undersides
fruit ripened ovary of a flower; contains seeds and remains of flower parts attached to it
fulcrum the point about which a lever moves
fungus microscopic and plantlike organisms that obtain their nutrients from other living and dead organisms
gall bladder sac in which bile from the liver is temporarily stored
gamma globulin a special blood protein, often injected to give immunity to a disease
ganglions enlarged regions of nerves composed mainly of sensory nerve cell bodies
gastric glands digestive glands that secrete gastric juices, located in the inner lining of the stomach
gastrin hormone secreted by the digestive glands of the stomach; stimulates production of acid
genes areas on the chromosomes, made up of DNA; control all cell activities and determine heredity
genetic engineering scientific development of new types of plants and animals by inserting genes for desirable traits from different species
genetics the science that studies heredity
genus classification subdivision above the species level and below the family level
geographic isolation physical separation of populations
germinate sprouting of a plant embryo in a seed
global warming an increase in average temperatures worldwide due to atmospheric warming
glucose a sugar (simple carbohydrate) made in a leaf by photosynthesis
glycogen starch made by the liver from excess glucose; stored in the liver
grafting artificial method of vegetative reproduction in which a scion (twig or bud) is attached to a stock (rooted part of plant)
greenhouse effect a warming up of the atmosphere due to heat trapped close to earth's surface
growth hormone secreted by the pituitary gland, it regulates bone and muscle growth
guard cells a pair of cells surrounding each pore, or stoma, in a leaf's lower epidermis
gullet see *esophagus*
gymnosperms see *conebearers*
habit learned, automatic behavior resulting from many voluntary repetitions of an act
habitat the kind of environment in which an organism lives
hallucinations seeing things that do not exist; often caused by the use of drugs
hallucinogens drugs that affect sensory perceptions
heart a saclike, muscular organ that is divided into four hollow chambers; pumps blood throughout the body
hemoglobin compound that carries oxygen in red blood cells; protein molecule combined with iron
herbivores animals that eat plants
hereditary diseases diseases that run in the family; inherited conditions
heredity passing on of traits, or characteristics, from parents to offspring
hip bones curved bones attached to the bottom of the spinal column
histamine the substance that causes allergic reactions like sneezing
hormones in plants, chemicals produced by cells that control body activities; in animals, compounds secreted by ductless glands that control body activities
horsetails types of vascular spore plants
host the organism on which a parasite feeds
humus decayed organic matter; part of soil
hybridization see *crossbreeding*
hybrids offspring of a cross of parents having contrasting traits; contain two unlike genes for a trait, and exhibit the dominant trait
hypothalamus a part of the brain that secretes hormones that control pituitary gland's activities
hypothesis an idea or possible answer to a problem that can be tested
identical twins two offspring that arise from one fertilized egg that splits after the two-celled embryo stage; always of the same sex
immune deficiency type of deficiency disease in which the immune system lacks or loses the ability to attack pathogens and destroy diseased cells
immune response reaction of the body's antibodies against antigens and pathogens
immune system the body's natural defenses that guard against diseases caused by pathogens
immunity inborn or acquired ability of body to withstand effects of disease-causing pathogens
impulses messages sent by nerve cells from one part of the body to another
inbreeding breeding method in which only closely related members of the same breed are mated
incomplete dominance blending inheritance; appearance of offspring is a mixture of the parents' contrasting traits
infectious disease disease caused by pathogens; usually contagious
ingestion part of nutrition process known as food-getting; taking in of food

inhalation taking in of air rich in oxygen; also called inspiration
inherited trait trait controlled by genes, passed from parent to offspring
inner ear located within hollow portion of skull bones; consists of cochlea, semicircular canals, and nerve cell endings from auditory nerve
inorganic compounds compounds composed of two or more elements, usually not including carbon
insects class of arthropods having three body sections, three pairs of legs, and two pairs of wings
insect-eaters plants having modified leaves that trap and digest insects
instinct inborn, automatic behavior consisting of a chain of reflex acts; aids survival
insulin hormone secreted by the islets of Langerhans in the pancreas; controls passage of glucose into the liver and other cells
interferon a protein produced by the body which interferes with the reproduction of viruses
internal fertilization type of fertilization in which sperm cells are deposited by the male inside the female, where they unite with and fertilize an egg
intestinal glands digestive glands located in the inner lining of the small intestine; secrete intestinal juice containing enzymes
invertebrates animals that lack an internal skeleton or backbone
involuntary movement movement that cannot be controlled by the will, such as pumping of heart
iris a circular, colored band located behind the cornea of the eye
islets (islands) of Langerhans groups of cells in the pancreas that secrete the hormone insulin
jawless fishes parasitic fishes having a skeleton of cartilage and lacking jaws, scales, and paired fins
joint area of skeleton where two bones meet
kidneys two bean-shaped excretory organs located along the back at waist level
kidney unit (nephron) structure composed of a capsule and attached tubule; removes dissolved wastes from the blood for excretion
kilocalories equal to 1,000 calories or one Calorie; units used to express energy provided by food
kingdom largest classification unit used to group similar living things
lacteals microscopic lymph vessels in the center of a villus that remove fatty acids
large intestine (colon) part of the food tube following the small intestine; stores solid wastes until they are eliminated from the body
leaf flat structure that grows out of a stem
lever a stiff rod that can be moved about some support or stationary point
lichen an organism composed of an alga and a fungus living in symbiosis
life activities (life processes) activities that enable an organism to stay alive
life cycle stages in the development of a living thing from fertilization and birth until reproduction and death
ligaments bands of tough connective tissue that bind one bone to another
limbs the arms and legs of a body
liver digestive gland that secretes bile; place where glycogen is made and stored
liverworts nonvascular spore plants
locomotion act of moving from one place to another
lungs two specialized organs that absorb oxygen from the air and release carbon dioxide and water vapor
lymph fluid that surrounds every body cell and fills all spaces between tissues
lymph nodes glandlike masses of spongy tissue through which lymph flows in the lymph vessels
lymphocytes special white blood cells that attack harmful foreign microbes
lymph vessels tubes that transport lymph through the body
malnutrition poor health caused by a lack of adequate nutrients
mammals class of warmblooded vertebrates that have hair and feed their young on milk
mammary glands special glands of mammals that produce milk for their young
mantle soft tissue that covers a mollusk's body; produces hard shell material
marrow soft blood-forming tissue located in the center of long, hollow bones
marsupials the pouched mammals; young complete their embryonic development in a pouch
matter anything that takes up space and has mass
mechanical digestion the grinding and softening of food; physical digestion
medulla oblongata part of the brain connected to the spinal cord that controls many automatic processes and reflexes
meiosis special kind of cell division that forms sex cells
melatonin hormone produced by the pineal gland; affects the body's natural cycles
meninges membranes that surround and protect the brain and spinal cord
metamorphosis a complete change of body form during development from juvenile to adult stages
metric system system of measurement that uses the meter for length, gram for mass, and liter for volume; based on multiples of ten
microbes (pathogens) microorganisms that infect other organisms
microorganisms tiny one-celled organisms
middle ear region between outer and inner ear; consists of eardrum, three little bones, and the Eustachian tube
mineral salts elements needed by the body for building tissues and aiding enzyme actions

mitosis complex process by which the original and copied sets of genes in a nucleus divide
mixture two or more substances mixed together that do not unite chemically
molecule smallest unit of a compound that can exist by itself and still have all the properties of that compound
mollusks phylum of aquatic invertebrates having a soft body enclosed by a mantle and shell
molting process in arthropods in which the hard exoskeleton is shed to allow growth
moneran kingdom of one-celled microorganisms; lack nuclei and membrane-bound organelles
mosses nonvascular spore plants
mucus slimy substance secreted by the mucous glands in the lining of internal organs
multicellular consisting of many cells
muscles tissues that move the bones of the skeleton to which they are attached
muscle tissue tissue made of cells that can shorten and cause body parts to move
mutant an individual bearing a new genetic trait
mutations changes in an inherited trait caused by a permanent change in a gene
myriapods arthropods having many legs; class of centipedes and millipedes
narcotics drugs that relieve pain and cause drowsiness
natural balance the tendency for the numbers of organisms in a natural community to remain the same
natural cycles repeated shifting of materials between organisms and the environment
natural immunity inherited ability to resist a disease
natural resources materials and living things in nature such as soil, water, trees, and minerals
natural selection process by which more fit individuals survive to reproduce
nematode roundworm having three cell layers and simple organ systems
nephron see *kidney unit*
nerve a bundle of nerve cell axons that lie side by side and are enclosed by connective tissue
nerve cell (neuron) specialized cell having more sensitive protoplasm than any other cell
nerve tissue tissue composed of nerve cells; found in brain, spinal cord, and nerves
nervous system system that controls the body's responses to internal and external stimuli
niche the way an organism lives within its habitat
nitrogen cycle circulation of nitrogen through the atmosphere, the soil, and living things
nitrogen fixation combining atmospheric nitrogen gas with other elements, forming nitrates
nonfectious disease condition usually caused by the presence or absence of some nonliving substance
nonrenewable resources resources that cannot be replaced after being used up or destroyed
nonvascular in plants, those lacking true roots, stems, leaves, and conducting tissue
nucleus round body near the middle of a cell; contains genes, which control all cell activities
nutrients compounds in food that can supply energy for cells or materials needed to carry out life activities
nutrition the life process that includes ingestion, digestion, absorption, circulation, and assimilation of nutrients
oils fats that are liquid at room temperature
omnivores animals that eat both plants and other animals
optic nerve nerve connecting the eye with the brain
order classification group below the class division and above the family subdivision
organ body part composed of several tissues working together
organelles tiny bodies within a cell's cytoplasm
organic compounds compounds that contain carbon atoms; made by living things
organism a living thing
organ system a group of organs that work together to perform a life activity
osmosis passage of water through a membrane
outer ear outer part of ear consisting of the fleshy projection on the head and a hollow ear canal
ovary in animals, the organ that produces egg cells; the female sex gland; in plants, the part of the flower in which ovules and egg cells are formed
oviducts tubes that lead eggs from an ovary to the outside of the female's body
ovules small round bodies located in the ovary of a flower; contain the egg cells
oxidation chemical combination of a substance, like digested food, with oxygen, resulting in a release of energy
ozone an irritating gas that is part of smog in the lower atmosphere; a natural barrier to ultraviolet rays in the upper atmosphere
pancreas digestive gland that produces pancreatic juice; contains islets of Langerhans, which secrete insulin
parasites organisms that live in or on another organism, from which they get their nutrients
parathormone hormone produced by the parathyroid glands
parathyroid glands produce parathormone; located in rear of thyroid glands
pathogens see *microbes*
peripheral nervous system numerous pairs of nerves located outside the brain and spinal cord
peristalsis automatic wavelike movements of muscles in the food tube that force food down
perspiration (sweat) liquid waste that is excreted by the skin
petals parts of a flower that are usually brightly colored
phagocytes white blood cells that engulf and destroy foreign bodies

pharynx throat; connects mouth and esophagus
phloem tissue in vascular plants through which materials are conducted downward
photosynthesis manufacture of simple carbohydrates in the presence of light, chorophyll, carbon dioxide, and water
phylum in classification, the largest division of a kingdom
pineal gland located in the brain between the left and right hemispheres; produces melatonin
pistil female reproductive organ in a plant's flower; consists of a stigma, style, and ovary
pituitary gland ductless gland on the underside of the brain; secretes several important hormones
placenta special organ in mammals that connects the embryo to its mother; exchanges food, oxygen, and wastes between the embryo and mother
placental mammals mammals whose offspring finish their embryonic development in the womb
plasma liquid part of the blood that carries dissolved substances and blood cells
platelets very tiny cell parts in blood that help in the process of blood clotting
platyhelminthe flatworm having three cell layers and some simple organ systems
pollen grains form the sperm cells of a flower
pollen tube extension that grows out of a pollen grain during fertilization in a flower
pollination transfer of a pollen grain from an anther to a stigma of a flower
pollution the addition of harmful substances into the air, land, or water
population all the organisms of the same species living in the same area
pores microscopic openings throughout the skin's epidermis
poriferans phylum of sponges; porebearers; they are composed of two main cell layers around a hollow cavity
pouched mammals see *marsupials*
predators flesh-eating animals that catch and eat other animals
prey animal that is caught and eaten by a predatory animal
primary consumers animals that eat plants; the herbivores
primates order of mammals that includes lemurs, monkeys, apes, and humans
producers plants and algae that absorb light energy and use it to make food; begin every food chain
prolactin hormone secreted by the pituitary gland that stimulates milk production
proteins organic compounds composed of amino acids; consisting of carbon, hydrogen, oxygen, nitrogen, and other elements
protist kingdom of simple organisms having nuclei and membrane-bound organelles
protoplasm all the living material in a cell
protozoans tiny one-celled animallike organisms
puberty sexual maturity; time of onset of male and female body characteristics
pulse rate rate at which the heart contracts and relaxes; about 70 times a minute
pupil dark, circular opening in the center of the eye's iris; lets in light
purebred offspring are just like the parents in possessing two of the same genes for a trait
rays the arms of a starfish body
receptors sense organs; contain receptor nerve cells
recessive character inherited trait that is overshadowed by a dominant trait and is kept from showing in a hybrid
recessive genes genes for a recessive trait; overshadowed by dominant genes
rectum area, at the end of colon, in which solid wastes are temporarily stored before elimination
recycle to reuse products made from depletable natural resources
red blood cells disc-shaped cells that have a hemoglobin molecule in their center; carry oxygen to the body's tissues
reduction division cell division in which each new cell receives one-half the species number of chromosomes; occurs during formation of sex cells
reflex unlearned, inborn, automatic, rapid response to a stimulus; aids survival
reflex arc nerve pathway over which an impulse passes when a reflex act occurs
regeneration regrowth of a body part that has been lost; form of asexual reproduction
renewable resources natural resources that can be replaced by natural processes eventually
reproduction process of giving rise to offspring
reptiles class of coldblooded vertebrates that have a body covering of dry scales; most lay eggs
respiration release of stored energy in nutrients as a result of oxidation reactions in living things
response reaction of a living thing to a stimulus
retina thin layer of nerve cells that coats most of the inner surface of an eyeball
ribs bones attached to the spine that form the rib cage
RNA nucleic acid sent out from the nucleus to instruct the cell to make proteins to carry out its life activities
rodents order of mammals having large front teeth for gnawing wood and other foods
rods retinal nerve cells that distinguish different shapes in dim light
root part of a vascular plant that anchors it in the ground; absorbs water and minerals; stores food
root hairs projections from a root's epidermal cells; absorb water and dissolved mineral salts
rotation of crops method of farming to conserve soil minerals by repeatedly alternating crops
roughage indigestible material in food, mainly vegetable fiber; helps body eliminate solid wastes
roundworms see *nematode*

saliva fluid secreted in mouth by salivary glands; moistens food and digests starch
salivary glands digestive glands that secrete saliva into the mouth
saprophytes see *decomposers*
scavengers animals that eat leftover parts of dead animals
scientific method an organized approach to problem-solving
secondary consumers animals that eat the primary consumers; carnivores that feed upon herbivores
secretin hormones secreted by the cells that line the small intestine, which cause the pancreas to release pancreatic juice
secretion production of useful substances by the body's glands
seed ripened ovule of a flower; composed of a seed coat, an embryo plant, and food tissue
segmented worms see *annelids*
segments the many sections of an annelid's body
selective absorption ability of a cell membrane to permit only certain substances to enter a cell
semicircular canals set of three bony canals in each inner ear; help the body maintain balance
sense organs body organs that are especially sensitive to stimuli
sepals leaflike, outermost parts located in a circle at the base of a flower
septum wall that separates the right side and left side of the heart
sex cells reproductive cells, one from each parent, that unite during fertilization and reproduction; have one-half the species chromosome number
sex chromosomes the *X* and *Y* chromosomes; the pair of chromosomes that determines the sex of an offspring
sex glands special glands that produce the sex cells and secrete the sex hormones; testes and ovaries
sex-linked traits inherited traits controlled by a gene carried on a sex chromosome; traits that occur more often in males than in females
sexual reproduction the production of offspring by two parents—a male and a female
simple reflex act rapid, inborn, beneficial response to a stimulus; see *reflex*
skeleton the bones of the body to which muscles for movement are attached; provides support and protection to body
skin the largest organ of the body; encloses and protects the body; functions in excretion of wastes
skull several curved, flat bones that enclose and protect the brain
small intestine part of the food tube connecting the stomach and the large intestine; digests nutrients and absorbs them through villi
smog hazy air pollution caused by human activities; contains ozone
social animals live in a colony with others of their same kind; members help one another survive
soil a mixture, composed of small rock particles, water, air, and organic matter, in which plants can grow
spawning reproductive process in which a female fish lays many eggs in the water and a male fish sprays them with sperm cells; type of external fertilization
speciation process by which new species evolve over time
species living things of the same, distinct kind
species number number of chromosomes characteristic of the body cells of a species
sperm the male reproductive cell
sperm ducts thin tubes that lead sperm cells from the testes to the outside of a male's body
spinal column backbone, composed of a series of 33 separate bones (vertebrae); part of internal supporting skeleton of all vertebrates
spinal cord tubelike mass of nerve tissue extending down from the medulla; protected by the vertebrae and meninges
spine backbone (see *spinal column*); sharp projections on echinoderm's body
spiracles openings on each side of the segments of an insect's abdomen that serve as breathing pores
spontaneous generation false idea about the production of living things from nonliving matter
spore cases protective sacs in which the reproductive spores of some plants and fungi form
spore formation asexual reproduction of some plants and fungi in which many small cells of equal size are formed
spores reproductive cells of some plants and fungi; have hard walls and can survive harsh conditions
stamens organs that produce the male sex cells of plants
starch large molecule composed of a chain of many glucose sugar molecules
stem part of a plant that supports the leaves and flowers, and connects them with roots
stem cutting piece of stem cut from a live plant to start a new plant by vegetative reproduction
stimulant a drug that affects the central nervous system, causing it to speed up body activities
stimuli changes in the environment of an organism that cause it to react
stoma one of the pores located in the epidermis of a leaf; lets gases in and out
stomach baglike digestive organ between the esophagus and the small intestine; part of the food tube
strip cropping farming method in which alternate strips of two different crops are planted to help save soil
supporting tissue plant tissue composed of long cells that have thick, hard cell walls
sweat see *perspiration*
sweat glands coiled structures in the skin's dermis; excrete perspiration through pores in epidermis

symbionts two different kinds of organisms that live together and aid each other's survival
symbiosis relationship between two different types of organisms living together
synapse region where the end brush of one nerve cell lies very close to the dendrites of another
tadpole immature frog that has no tail, no limbs, and breathes through gills
taste buds groups of special cells on the tongue's surface that detect different tastes
taxis response movement of a microscopic organism toward or away from a stimulus
tendons bands of tissue that bind a muscle to a bone
tentacles long, flexible arms that surround the mouths of coelenterates
terraces steplike, flat areas that are dug on steeply sloping farmland to help stop soil erosion
terrestrial community on land or land-dwelling
tertiary consumers animals (carnivores) that feed upon secondary consumers
testis the male sex gland that produces sex cells (sperm) and sex hormones
theory a scientific explanation of facts
thorax midregion of an insect's body that lies between the head and abdomen regions; the chest
throat (pharynx) part of the food tube that connects the mouth and esophagus
thymosins hormones that strengthen the immune system; produced by the thymus gland
thymus gland ductless gland located behind the breastbone; secretes thymosins
thyroid gland ductless gland located in the neck in front of the windpipe
thyroxin hormone secreted by thyroid gland; regulates body's metabolic rate
tissue group of similar specialized cells that act together
topsoil fertile layer of soil that supports plant life
toxins poisonous compounds produced by an organism
toxoid weakened toxin injected into a body to help produce immunity to a disease
transpiration evaporation of water from a leaf
trachea windpipe; tube through which air passes from the throat to the bronchi and lungs
triceps muscle on the upper side of the arm that works (in pairs) with the biceps muscle
tropism automatic response movement of a plant toward or away from a stimulus
tube feet tiny organs of locomotion in the echinoderms; also used to capture and feed upon prey
tuber food-storing underground stem that can grow into a complete plant by vegetative reproduction
ureter tube, one leading from each kidney, that connects to the urinary bladder
urethra tube leading urine from the urinary bladder to the outside of the body
urinary bladder sac that stores urine produced by the kidneys
urinary system body system for excretion of wastes (mostly water); maintains water balance in body
urine fluid waste composed of excess water, salts, and urea
uterus part of the oviduct of a mammal in which the embryo develops; the womb
vaccination injection of a vaccine or toxoid to cause the body to make its own antibody against the substance injected
vaccine weakened or killed microbes injected into a body to produce immunity to a disease
valves flaps of tissue in the heart that prevent the backward flow of blood
variable experimental factor that may be the solution to the problem being investigated
variation differences that exist among all organisms of the same species
vascular in plants, those having true roots, stems, leaves, and conducting tissue
vectors substances and living things that carry and transmit diseases
vegetative organ an organ of a plant that makes, transports, or stores food
vegetative propagation asexual reproduction of plants from their vegetative organs
vegetative reproduction see *vegetative propagation*
veins blood vessels that lead blood to the heart from capillaries in all parts of the body
ventricles the two lower chambers of the heart; receive blood from the atria and pump it out to all parts of the body
vertebrae the 33 bones of the spinal column
vertebrates animals that possess an internal skeleton and backbone
villi millions of microscopic projections in the lining of the small intestine; absorb end products of digestion
virus particle that can cause disease; on borderline between living and nonliving matter
vitamins nutrients needed in very small amounts to maintain healthy body; aid enzyme actions
voluntary act an action or behavior controlled by the will
voluntary movement response controlled by the will
warmblooded animals animals that maintain a warm body temperature regardless of the environment's temperature
water cycle circulation of water to and from bodies of water, the atmosphere, the soil, and living things
white blood cells special large blood cells that fight bacteria, viruses, and other infectious agents
windbreaks rows of trees and hedges planted to reduce the eroding force of strong winds
womb see *uterus*
xylem tissue in vascular plants through which materials are conducted upward
yolk stored food in an egg

Index

**Note: "t" and "f" denote table and figure references.*